The World within

14 95
P

The World within the World

JOHN D. BARROW

Professor, Astronomy Centre,
University of Sussex

Oxford   New York
OXFORD UNIVERSITY PRESS

Oxford University Press, Walton Street, Oxford OX2 6DP
Oxford New York Toronto
Delhi Bombay Calcutta Madras Karachi
Petaling Jaya Singapore Hong Kong Tokyo
Nairobi Dar es Salaam Cape Town
Melbourne Auckland
and associated companies in
Berlin Ibadan

Oxford is a trade mark of Oxford University Press

First published in 1988 by Oxford University Press
First issued as an Oxford University Press paperback
1990 (with corrections)
Reprinted 1990, 1991

British Library Cataloguing in Publication Data
Barrow, John D.
The world within the world.
1. Science—Philosophy—History
I. Title
500.2'01   Q175
ISBN 0–19–286108–5

Library of Congress Cataloging in Publication Data
Barrow, John D., 1952–
The world within the world/John D. Barrow.
Bibliography: p.
Includes index.
1. Physics—Popular works.   2. Science—Popular works.
3. Science—Philosophy—Popular works.   4. Cosmology—Popular works.—I. Title.
QC24.5.B37   1990   500.2—dc19   87–31539
ISBN 0–19–286108–5

Printed in Great Britain by
Richard Clay Ltd
Bungay, Suffolk

All these have never yet been seen—
But scientists who ought to know,
Assure us that they must be so . . .
Oh! let us never, never doubt
What nobody is sure about!

> *Hilaire Belloc*

Art is a lie that lets us
recognize the truth.

> *Pablo Picasso*

The art of writing is the art of applying
the seat of the pants to the seat of the chair.

> *Mary Heaton Vorse*

To Lois and Louise

Preface

In recent years there has been a vast increase in the publication rate of so-called 'popular' science books. Regardless of whether they do indeed turn out to be popular with anybody but their authors, their common aim has been to explain in simple terms the constant stream of new ideas and discoveries that have emerged in the fundamental sciences during the last decade. In this book the aim is not simply to pick again upon one of these esoteric fields at the frontiers of fundamental science and attempt to explain it in simple terms. Rather, it is to pick upon the traditional unspoken assumptions to which we owe all these abstract and pragmatic developments: that the Universe is ordered, that it is logical, that it is mathematical, that it is predictable, that it is governed by something outside ourselves which is the same everywhere and everywhen, but which has a deep resonance with the workings of our own minds: to explore something of the origin and possible meanings of the idea that there exist 'laws of Nature' and some of the unsuspected realms that such an idea has led us. This quest will lead us from pre-history to the evolutionary development of key notions in fundamental physical science and the attendant philosophy it has helped to shape. Time and again it will bring us back to old questions in a new guise: sharpened, widened, and deepened by the unvisualizable worlds of inner and outer space where we find Nature's logic laid bare by that complexity masquerading as simplicity that we have come to recognize as the hallmark of all natural phenomena. Are there really laws of Nature that exist out there independently of our way of thinking waiting to be discovered, or are they just the most convenient way of describing things that we have been able to write down? Are these natural laws the same everywhere? Are there places where they cease to hold? Can they change with time? Indeed, is it possible that there aren't really any laws of Nature at all? Why is the language of mathematics found to offer us such a ready translation of the Universe's workings? What are the implications of our own existence for our interpretation of the Universe's structure? Do we have the intellectual capability to understand the deepest principles behind the harmony and complexity of Nature?

Questions of this sort used to be primarily the preserve of philosophers, but in recent years physicists and cosmologists have found their theorizing taking them into realms where answers to such extraordinary conundrums as 'can the Universe be created out of nothing?' may have consequences for measurable quantities. Traditional dogmas as to what criteria must be met by a body of ideas for it to qualify as a 'science' now seem curiously inappropriate in the face of problems and studies far removed from the human enterprise. Some scientists believe that there have really been no new philosophical questions raised for hundreds of years except those provoked by new discoveries in science. While, for many others 'philosophical questions' has become a handy label to attach to a collection of vague or apparently unanswerable questions which only become worthy of serious consideration when they become scientific. This prejudice often goes hand in hand with the view that the goal of modern philosophy has been merely to show that more and more of the past 'problems' of philosophy are illusory semantic confusions. It is our hope that some of the unusual ways in which Nature operates may revitalize the consideration of the implications of scientific problems by philosophers and that they take far less seriously the strange views of those philosophers of science who, having rightly detected that there is a sociology of science, then wrongly conclude that there is thus no more to science than the transient fashionable whims of the society of scientists.

It has become fashionable to believe that the laws of Nature are both necessary and sufficient to explain what we see in the Universe and that hence some grand 'theory of everything' will provide us with the answers to all our cosmological questions. One of the messages of this work is that, although laws of Nature may be necessary for such grand explanations, they are by no means sufficient to explain what we see. Initial conditions, broken symmetries, organizing principles, selection effects, and human categories of thought all play an essential and irreducible role in augmenting any laws of Nature to determine a picture of the Universe in which we live.

The picture painted in the pages of this book is a personal and selective one. It is semi-popular. It is biased towards the mathematical and physical sciences. It makes no claims of completeness. Indeed, it is already far too long for the tastes of some. It has no gospel to preach, no angle to angle, merely a fragmentary story to tell. Its discussion of contemporary problems is a progress report that attempts to pick points of principle from the ideas of the moment, and by such modest standards it should be judged. For if a book is judged by an inappropriate yardstick then many confusions undoubtedly result—a fact of which one can most readily be persuaded by recalling an unfavourable review of *Lady Chatterley's Lover* that appeared in the American Huntin', Shootin' and Fishin' magazine *Field and Stream*: it judged that 'this pictorial account of the day-by-day life of an English gamekeeper is full of considerable interest to outdoor-minded readers, as it contains many passages on pheasant-raising, the apprehending of poachers, ways to control vermin, and other chores and duties of the professional gamekeeper. Unfortunately, one is obliged to wade through many pages of extraneous material in order to discover and savour those sidelights on the management of a midland shooting estate, and in this reviewer's opinion the book cannot take the place of J. R. Miller's *Practical Gamekeeping*.'

I would like to thank those individuals who have knowingly or unwittingly contributed in some way to the contents or the writing of this book; in particular I am grateful to past co-authors Joseph Silk and Frank Tipler, and thanks are also due to Suketu Bhavsar, Margaret Boden, Paul Davies, David Deutsch, George Ellis, John Maynard Smith, Sir William McCrea, Leon Mestel, Don Page, Arthur Peacocke, Martin Rees, Dennis Sciama, Robert C. Smith, Roger Tayler, Stephen Toulmin, John A. Wheeler, Sir Denys Wilkinson, and Stephen Wolfram. It is a pleasure to thank the staff of Oxford University Press for their efficiency and attention to detail which transformed lots of pieces of paper into a book in an impressively short period of time. I am also grateful to my wife for supplying the local external world in which this project could be continued and leave of it taken. Finally, some younger members of my family have long regarded the writing of this book as a major impediment to the construction of a model railway of totally unrealistic proportions. They are, according to the traditional formula, thanked for their innumerable critical comments on the manuscript, none of which I am afraid could be incorporated into the final version.

Brighton   J.D.B.
October 1987

Acknowledgements

The illustration on the cover has been reproduced with permission of Springer-Verlag from the book H.-O. Peitgen and P. H. Richter (1986). *The beauty of fractals*.

The extract on pp. 19 and 20 from Eric M. Rogers, *Physics for the inquiring mind: the methods, nature, and philosophy of physical science*. Copyright © 1960 by Princeton University Press and Oxford University Press (excerpt, p. 344) is reprinted by permission of Princeton University Press.

Figure 3.12 on p. 120 is reproduced from a diagram on p. 247 of A. Beiser (1967), *Concepts of modern physics* (revised edition) with permission of the publisher McGraw-Hill.

Figure 4.10 on p. 220 is a reproduction of Plate 189 from *The fractal geometry of nature* by Benoit B. Mandelbrot. Copyright © 1977, 1982, 1983. Reprinted with the permission of W. H. Freeman and Company.

Figure 5.8(a)–(c) is reprinted by permission of M. Henon from Figures 2–5 of his article ('A two-dimensional mapping with a strange attractor'. *Communications in Mathematical Physics* **50**, 69–77 (1976).

Figure 7.3 on p. 337 is reproduced from Figure 24 of Richard L. Gregory (1970). *The intelligent eye*, by permission of Weidenfeld and Nicolson, London.

Figure 7.2 on p. 333 is taken from J. L. Marroquin (1976), Human visual perception of structure. Master's thesis, MIT with the permission of the Massachusetts Institute of Technology.

Contents

4 Inner space and outer space

1

Prologue

O Nature, and O soul of Man! how far beyond all utterance are
your linked analogies! not the smallest atom stirs or lives in
matter, but has its cunning duplicate in mind.

Thomas Melville

Introduction

You're searching, Joe
For things that don't exist.
I mean beginnings
Ends and beginnings
Ends and beginnings—there are
no such things
There are only middles.

Robert Frost

Behind our ever-changing experiences of the world lies a changeless world of
order and certainty, impartial to our actions and desires; a world of which we are
a part, but upon which we can exert no discernible influence; a world that cares
not one whit whether its abstruse operations are intelligible to us, yet without
which there would be no Cosmos and no life.

For millennia Nature has slowly been fashioning first our minds, and then,
within them, an imitation of her workings that we have come to call the laws of
Nature. This imitation of Nature we have found effective from the microscopic
inner world of matter to the farthest reaches of the Universe. We have built it up,
step by step, through observation, and trial and error, from the mundane
everyday world around us to the extremities of inner and outer space.

In trying to make sense of our perceptions of the physical world we are faced
with deciphering the output of a cosmic computer. By scanning its patterns and
the responses it makes to our tinkerings with its keyboard, we try to piece
together the inner logic of its programming. It has always been suspected that the
impressive rationality of this software witnesses to the presence of an intelligent
author. But who is the author of the apparent rationality? Is it a Divine
Programmer, or just Man, the mundane keyboard operator, seeing the reflection
of his own mind as it struggles to organize the output?

What are we to make of this picture? Are there really laws of Nature that exist 'out there', independent of our way of thinking, waiting to be discovered; or are they just the most convenient way of describing things that we have seen? How did such a notion as the 'laws of Nature' arise? Are these laws the ultimate reality or merely pieces of administrative legislation enacted by ourselves to help us organize our knowledge of the world; are they just the signposts we erect behind us as we crash through the jungle of experience? Is it possible that there are no Laws of Nature at all? Maybe they and the Universe they appear to govern are entirely creations of our minds: an illusion that goes away when we don't think about it. And what would happen if there were no observers of the Universe?

We hope, like Einstein, that Nature is subtle but not malicious; but for all we know it may just be subtly malicious. We have parochial experience of a small portion of a possibly infinite Universe. The conditions that hold in our vicinity may not be typical of other sites in the Universe. Indeed, special conditions may be necessary for the evolution of 'observers'. Are the laws that govern in our locale the same as those elsewhere in the Universe? Are they fixed in time, or do they gradually change as the Universe expands and ages?

Questions of this sort used to be primarily the preserve of speculative philosophers, faced with the daunting task of drawing universal conclusions in the absence of detailed knowledge of all the minutiae to which they must apply. The positivist alternative appeared more depressing still: leaving us with a collection of the obvious and the tautological: a residue of knowledge hardly worth knowing. In recent years physicists and cosmologists have found their theorizing taking them into realms where answers to such metaphysical conundrums as 'can the Universe be created out of nothing?' have entered the mainstream of scientific discussion, and may even turn out to be observationally testable. The serious hunt for a 'theory of everything' is on. For the first time the candidates lie in the mainstream of physics, and not just on the desks of eccentric investigations by an Eddington or an Einstein. It is likely that the results of theoretical physicists' investigations over the next ten years will recapture the wide interdisciplinary interest in physics once created by the early quantum theory. Nevertheless, it is curious that this tendency of fundamental physics to move towards questions traditionally of interest to philosophers and theologians has developed in parallel with an increased lack of interest amongst physicists in the philosophical questions raised by these developments. To most scientists 'philosophical questions' has become a handy label to apply to any collection of vague or apparently unanswerable questions which only become worthy of serious consideration when they become scientific. Perhaps this is a legacy of scientists' perception of progress in philosophical inquiry as being tantamount to the art of showing that more and more of the past 'problems' of philosophy are illusory semantic confusions. It is our hope that some of the unusual ways in which Nature has been found to operate may revitalize the consideration of the wider implications of scientific problems.

The urge to predict and control

> The most direct, and in a sense the most important, problem
> which our conscious knowledge of Nature should enable us to
> solve is the anticipation of future events.　　Heinrich Hertz

One of the unifying ideals of the strange cults and counter-cultural philosophies
that have manifested the popular search for the transcendental in modern
societies is the desire to predict the future, whether it is to be found in the stars or
the tea-leaves, in a mystical vision or on the palm of your hand. No mass
circulation newspaper fails to provide its readers with their horoscopes. Those
attracted to such ways of thinking often appear curiously comforted to read that
'scientists are baffled', and grow more secure in a hope that established science
does not know all the answers after all.

But, paradoxically, science has progressed and been successful to the extent
that, largely unnoticed, it dominates the way in which most of us go about our
daily lives precisely because it is the only approach to Nature we have uncovered
that has been consistently successful in predicting the future. To an extent that the
ancients could never have dreamed of, we have uncovered a master-key to the
Universe: that by coding events into numbers we can predict the future. We can
predict the return of Halley's Comet, how aeroplanes will fly, when spacecraft will
get to the moon, when the Sun will wax and wane, the tides ebb and flow, and
what new species of elementary particle will be found in the detectors of
accelerators at CERN in Geneva. The fact that precise predictions like these can
be made so often, in such diverse circumstances, and in a manner so reliable that
we build our lives upon them in a thousand different ways, tells us something
profound about the way the Universe runs. It is these regularities in our
experience of Nature we have come to call the 'laws of Nature'. They have many
different forms, facets, and functions. Some are deep foundation stones of our
view of the Universe, whilst others we regard as stopgaps until we can discover
the real thing, or at least a superior edition of it. In the chapters of this book we
aim to introduce a few interesting aspects of these laws of Nature. We shall try
to stress the impact of new concepts and discoveries in physics which were
unknown when the conventional wisdom as to the character of natural laws was
conceived.

The ability to predict the future is a powerful weapon, and thousands of years
ago it was recognized that astrology paved the way for a powerful élite to rule in
ancient civilizations. The chosen few who claimed the ability to predict and
interpret the appearance of the heavens could exert control over the future of
individuals and entire nations. In more modern times we have seen the ability to
predict used in a less sinister way for economic advantage by foreseeing
favourable weather conditions for farming, or to warn of imminent natural
disasters. This emphasis upon predictive power necessarily engenders an interest

in seeing whether the predictions are right. For the astronomers of Imperial China a wrong prediction meant death. Today astronomy is not such a perilous occupation, although the disastrous and tragic consequences of bad scientific predictions have devastated the American space programme and the Soviet nuclear power programme in recent years, at great human cost.

An acid test of any scientific theory is its ability to stand up to the experimental test of its predictions. Scientists have come to regard predictability as an essential ingredient of any useful scientific law, because it provides a means of discovering if that law is false, and, if so, of correcting it. Thereby, we can force ideas to evolve subject to a form of selection which we impose. That form of selection is simple: conformity with what is. If a prediction is correct, the theory lives to fight another day. But success does not mean that the theory is correct, since its correctness in this one aspect cannot guarantee its correctness in other circumstances, although the greater the array of facts explained by one and the same theory, the higher is the probability that it will continue to be successful, and the more useful it becomes to us. Stephen Toulmin has pictured the evolution of a scientific law confronted with observation as resembling an up-and-coming tennis player: always looking for new opponents to challenge. But he particularly wants opponents that are just a little bit better than he is. Only in this way can he improve his game. 'The scientist, likewise, is on the lookout for events which are not yet quite intelligible, but which could probably be mastered as a result of some intellectual step which he has the power to take.' Attempting problems that are not ripe for solution—which are too hard, or which offer too little scope for observational check—is like playing the Wimbledon champion: exciting (maybe even raw material for a best-selling book) but ultimately unedifying.

Others, notably Karl Popper, place greater stress upon the failure of candidate laws of Nature to predict correctly. For in this case we learn something unequivocally: the current edition of the law is false. Most scientists would not be happy with the idea that a theory or law of Nature must be falsifiable, but they would be unanimous in regarding falsifiability as sufficient to qualify a statement as 'scientific' or meaningful. Here we must be careful. When we test the prediction of any theory, we can never subject one aspect of it to test in isolation. We are always testing the correctness of an interwoven collection of assumptions. If the expected result is not obtained, we have to choose which of the many entwined assumptions we have made is the wrong one. It may be an erroneous theory about the instruments being used to carry out the falsification test that is at fault, rather than the hypothesis supposedly under test. In practice, the art of good experimental design is to isolate the aspect under test by ensuring that all the other ingredients have been positively tested in a sequence of other experiments. You notice how this focus upon testing the predictions of laws of Nature ensures that science is really founded upon observations rather than upon 'facts', and so is a continually evolving structure. But although our experiments can test theories, and distinguish the successful from the unsuccessful, they cannot generate theories. Indeed it is usually the theories that suggest the experiments.

But a closer examination of our success in predicting natural events reveals something which appears at first sight somewhat paradoxical.

The paradox of predictability

> Given a written language, how large in terms of the total number
> of words must a book printed in that language be in order to
> contain complete information to manufacture the book?
>
> N. Rashevsky

Our ability to predict successfully appears to demonstrate that there is an aspect of Nature that lies outside our control: a part that is external to human influence. In order to see what this statement means, let us first explore how, in some circumstances, the prediction of future events is logically impossible no matter how prescient you may be.

However much psychology and physiology I learn, I could never come up with a prediction of your future behaviour that you had to regard as true, since you could always falsify my forecast after it had been revealed to you. It is not possible for my knowledge of human behaviour to constrain you unless you want it to. My prediction could none the less remain completely accurate if it was *not* shown to you.

A topical example of this situation is that media creation—the public opinion poll. A pre-election opinion poll that is published can never predict the result of a future election infallibly, because it is logically impossible for it to incorporate the effect of its own influence upon subsequent voting patterns. We could conspire collectively to demonstrate the inability of a particular poll to predict correctly by all voting for Screaming Lord Sutch after reading the poll's prediction that Sutch will attract no votes at all. In general, it is impossible *in principle* to take into account the effect of any forecast on the voting behaviour it is predicting. Notice that this does not mean that all such forecasts have to be wrong. Your forecast of my behaviour today might turn out to be completely accurate. I don't have to falsify it if I don't want to, but you can never be sure that I won't choose to do so. It could be, and forever remain, a completely accurate forecast of my future behaviour if you didn't show it to me. The key point is that predictive success cannot be guaranteed in these situations. I can always choose to falsify a prediction about my behaviour simply by disbelieving it. One can see that considerations of this sort require us to be careful when considering the question of whether free will exists. It is not simply a choice between free will or determinism, but free will or determinism relative to whom. It is amusing to note the corollary that successful horoscopes are impossible in principle unless you either don't read them, or deliberately choose to do the things that it is predicted that you will do!

If we were able to predict our own decisions by using some scientific formula, this would require us to have already made them. This is the logical dilemma.

These examples show us that election forecasting is not like weather forecasting. Election forecasting must have an unpredictable effect upon an election, but weather forecasting cannot affect the weather, which is governed by something *outside* our control. It is this difference that is responsible for the dramatic past success of the physical sciences and the equally dramatic failure of the economic and social sciences in applying the same methods of inquiry. Imagine how frustrating life would be for the chemist if molecules behaved like the subject of the economist's study!

This appealing feature of the studies undertaken by most physical scientists, that the things under study do not change their character when they are investigated, was regarded as a vital assumption to make about the nature of the world by Descartes and the Cartesians who followed him. This 'dualism' which separated the observer of Nature from the external world under study was not just a device to exclude religious and ethical questions from the sphere of scientific inquiry. It was motivated by a desire to exclude the magical antics of the alchemists from serious consideration. Alchemy required the experimenter to be in the correct frame of mind if favourable results were to be obtained. Oratory was required in the laboratory. There was, therefore, an unbreakable coupling between the observer and the observed. There could not be repeatable experiments. This is all somewhat reminiscent of the Uri Geller phenomena which, we are told, are only demonstrable to sympathetic audiences!

The legacy of the success of the dualist viewpoint in science until the twentieth century may be responsible for the unfortunate experiences of some physicists who have tried their hands at investigating paranormal phenomena. If deliberate fraud were involved in examples of such phenomena, then physical scientists are particularly ill-suited to detect it because it is their habit of mind to assume that the phenomena under study do not conspire against them. The psychology of the magician (or even the card-sharp) is more suitable for the investigation of such paranormal events than is the childlike faith in the regularity of Nature that characterizes the scientist.

The external world: a first approximation

> If the physical laws of this world are autonomous, we are not
> free; if we are free, then the physical laws are not autonomous.
>
> Karl Popper

The ability to predict natural phenomena successfully, be they falling apples or solar eclipses, identifies for us what we might loosely term *the external world*: the collection of things and events that we cannot influence by our own wills. This statement is not as anthropocentric as it might first sound, for it applies with equal force to forms of artificial intelligence like computers.

Yet, whilst there cannot exist binding laws on individual human action, or correct predictions of human behaviour, there can and do exist correct predictions of atomic behaviour, gravity, electrical currents, silicon chips, and so on. These events are predictable because they are not controllable by our own decisions. We cannot influence the intrinsic strength of gravity or the properties of electricity any more than we can change the weather by forecasting it. These external things appear to be legislated by things that we cannot wilfully influence in any way. By this argument we arrive at our paradoxical conclusion, that it is precisely because *we* cannot control some events that they are predictable, and the reason we cannot control them is that something else apparently does so already: that something we have come to call natural law.

We should also say that although predictability implies that something lies outside our control, inability to predict does not necessarily mean that events are under our control. We can fail to predict in practice for all sorts of other reasons, some of which we shall be examining in later chapters.

The argument we have just given is a statement of 'common sense' facts of life which were appreciated in an intuitive fashion by the Cartesians. It was an excellent working principle throughout the period when science progressed in leaps and bounds by formulating mechanical pictures of how the world worked. Unfortunately, we shall find in Chapter 3 that when Nature is probed to its extremities the barrier between the observer and the observed dissolves in a most peculiar way that has observable consequences. It transpires that we can know that the act of knowing changes things in an unknowable way.

Some of the most striking examples of these laws that prevent our control but allow our prediction of events are statistical laws; striking because there appears to exist no simple rule which acts upon each outcome of a repeated process to ensure that the long-term outcomes are predictable on the average. For example, if we repeatedly toss a fair coin, then, although we cannot predict with certainty whether each toss will land heads or tails, we can predict that as the number of tosses increases so the number of heads will tend to become equal to the number of tails, to a closer and closer approximation. It will be interesting to ask what it is that we cannot influence in this case which is responsible for reliable prediction. Statistical laws of this type play an important role in our everyday lives. They are what insurance companies use to predict the likelihood that we will suffer various misfortunes, and hence fix how much our annual premiums should be. Bookmakers too seem to live quite well by them. Indeed, it was gamblers who first created the subject of statistics.

It is interesting to recall a popular division of the external world from that of our minds developed by Popper. It is a modern form of dualism. Popper considers that there exist three 'Worlds'. World I contains real physical objects, both inorganic and organic: rocks, aardvarks, and you and me. It includes our brains and all the artefacts of human creativity: books, paintings, and sculptures. World II contains states of consciousness: the subjective experience we have of ideas, memories, dreams, thoughts, and perceptions. Finally, World III contains objective knowledge: all our literary records of scientific facts, speculations, philosophy, history, and the contents of computer memories, along with scientific arguments and mathematical theorems. Residents of World I cannot be represented in an objective way without moving into either World I or World II. Our scientific study consists of subjective experiences of the World II type about concrete entities in World I, and this leads to the formulation of the predictions, critical questions, and theories about the Universe which inhabit World III. The 'third' World objects are only made possible by the intervention of our minds. Our scientific theories about the Universe produce working representations of Nature's laws, usually in the form of mathematical equations, and these equations live in World III. We would like to know whether there really do exist laws of Nature in World I, or whether they originate solely in World II.

In the next chapter we shall attempt to trace the historical development of the idea that there exist laws of Nature, to see whether the notion has its roots in certain general human or specific cultural biases inhabiting World II. We shall also examine how this notion has played a significant role in the development of our religious and philosophical viewpoints.

Description or prescription?

> There is no evidence that Elizabeth had much taste for painting;
> but she loved pictures of herself.
>
> Horace Walpole

So far in this introduction we have been using the idea of laws of Nature in a vague manner. However, there exists a number of quite different ways of viewing laws of Nature, and it is important to be aware of these differences because different authors use the words 'natural law' or 'law of Nature' to mean quite different things. The simplest pair of alternatives are a frequent source of confusion.

To some the term 'laws of Nature' describes what does actually happen. They are an historical record of regularities. Such laws can of course never be broken, by definition. It is as if we recorded the flow of traffic outside our house and called this the 'law' of traffic flow. If one does view laws of Nature in this purely descriptive way, then it would be nonsensical to claim, for example, that miracles were impossible because they contravene the laws of Nature, for laws of this type can ordain nothing: they cannot be broken. On the other hand some reserve the honorific term 'laws of Nature' for something more special: our model of Nature. If by the laws of Nature we refer to various deductive relationships, mathematical formulae, symmetry principles, conservation rules, or rules that say 'if such as such is the case . . . then this will follow', then these man-made deductions or sweeping generalizations as to how Nature should behave can be broken. Such a violation would signal that we had previously arrived at the wrong rule to codify events, perhaps through faulty logic, erroneous experimental data, or, more likely, the occurrence of a phenomenon that had not been encountered before.

We should also beware of trying to force reality into compartments into which it was not designed to go. It may be convenient for the purposes of exposition to draw a line between the two alternatives of laws as descriptions or explanations, but identical statements can sometimes play either role depending upon context. 'I am writing this book' can be a description if it is the answer to the question 'what are you doing?', but an explanation in response to the question 'why don't you do the washing-up?'.

This ambiguity between the descriptive role of science, as epitomized by experiment, and the explanatory function of theory is probably a healthy sign. It illustrates the balance that exists between theory and experiment in our science. Were either explanation or description obviously dominant, then it would signal that one facet had become overdeveloped with respect to the other. In some subject areas, like cosmology or elementary particle physics, theory does indeed race ahead of experiment because of the vast scale and cost of building the crucial experiments. In others, notably zoology and biochemistry, observational and experimental data proliferate to an overwhelming degree. It is very difficult to formulate a wide-ranging law, because it must accommodate such an enormous number of facts just to get off the ground. Hence, physical sciences gravitate towards prescriptive laws, whilst life sciences use descriptive laws. Sometimes we must offset this problem by picking what we consider to be the primary factors requiring explanation, and ignoring others in the hope that a picture that successfully explains the primary factors can be steadily improved to accommodate the minor disagreements. Ernst Mach's warning about being too ambitious is worth heeding: 'If all the individual facts—all the individual phenomena, knowledge of which we desire—were immediately accessible to us, a science would never have arisen'.

Even when a vast array of facts are made known to us we must exercise skill, judgement, experience, and intuition in selecting the salient ones. Our experience shows that only a small subset of the physical Universe needs to be studied in order to elucidate its underlying themes and patterns of behaviour. At root this is what it means for there to exist laws of Nature, and it is why they are invaluable to us. They may allow an understanding of the whole Universe to be built up from the study of small selected parts of it.

The different views of science

> If the purpose of scientific methodology is to prescribe or expound a system of enquiry or even a code of practice for scientific behaviour, then scientists seem able to get on very well without it.
>
> Peter Medawar

Philosophers of science have argued long and hard over the meaning of the concept of a 'law of Nature'. These arguments have had no discernable influence upon the practice of science; most scientists being sympathetic to the prejudice that 'the philosophy of science is about as useful to scientists as ornithology is to birds' (notwithstanding the observation that some species of bird owe their continued existence to the interest of ornithologists). The first point to clarify before we delve into these disputes is one of definition. The difference in terminology concerning laws of Nature we have just mentioned is also reflected in other aspects of science. The term 'quark' may sometimes be used by a physicist to mean the thing in itself, and sometimes to refer to our theoretical model of it. The possibility of such nuances of meaning has spawned several completely different views of scientific laws. None of these interpretations is refutable, and so whichever one any particular experimental physicist adopts should not in any way influence the results he or she obtains, although it does affect the type of investigations a theorist will embark upon, the subjects they regard as worthy of study, and the type of questions regarded as addressable by science. It will significantly colour the picture they give to lay persons in popular expositions of their work, and the extra-scientific implications they draw from it. Despite the fact that none of the alternative views of science and laws of Nature can be definitely refuted, we shall see that there does exist persuasive circumstantial evidence in favour of some views and against others. It will be useful to bear these perspectives in mind when we examine some of the new scientific developments in later chapters, because these views of science were conceived in an age when science was a smaller and simpler subject. It will be interesting to see if the subject can still be squeezed into these viewpoints.

Empiricism. The empiricist maintains that our knowledge of the world consists only of assemblages of individual facts upon which the observer exerts no subjective influence. All meaningful concepts can be reduced to sense data, and our scientific theories are simply convenient resumés of our observations. The observer makes no personal contribution when acting in the capacity of investigator.

This approach is a close cousin of the pre-war philosophical movement called *positivism*, which argues that in our investigation of the world we only encounter particular instances, never universals. We create the universals by giving a common name to many particular instances that share certain common features, but the features that they share in common have no existence apart from residing in our collection of particulars. Laws, according to the empiricist, are only exhibits of these common groupings. They are convenient descriptions, which do not guarantee anything will happen in the future. The 'criminal element' is not a universal entity, but a collection of persons brought together by reason of their delinquent disposition. This approach to the world is sometimes referred to as 'nominalist' to stress that it regards laws of Nature as purely descriptive: nothing more than names for particular sequences of events.

Empiricism implies that we can never really 'know' anything about things that cannot be observed. Any knowledge we claim to have obtained about the world by pure thought simply reflects our ways of using words. It tells us nothing about the real world. Another offshoot from this theme is the argument that there is no point in arguing for the truth of a statement unless you know what it means. This led to the school of thought that stressed the importance of *verification*. Indeed, the meaning of a statement was defined as its method of verification. Two means of carrying out such a process of verificiation were envisaged: the observation of experience, and the logical links between words.

Operationalism. This is another breed of positivism. The operationalist argues that science is nothing more than prescriptions for effective investigation of the world in the laboratory. The only meaningful concepts are those which can be defined by a sequence of practically realizable steps, termed 'operations'. Other concepts, like 'intelligence' for example, are meaningless. According to this doctrine, scientific laws are not used to describe, but to manipulate the world, and theories are only machines for generating predictions about it.

The operationalist places a little more emphasis upon the role of theory relative to observation than does the empiricist. Good theories are ones that allow us to manipulate the world most accurately and advantageously. The most useful ones have the greatest survival value, but they cannot be credited with anything like 'truth' or 'falsity'. The operationalist would also stress that all we ever do with these theories, even when we have formulated them successfully, is to use them to find a newer, bigger, and better theory. So, in practice, they are primarily tools.

Instrumentalism. Operationalism eventually evolved into an offshoot called instrumentalism. The empiricist resembles the philosophical positivist, and the operationalist has much in common with the type of linguistic philosophy that positivism mutated into, with its stress upon determining how words are used in order to ascertain their meaning. All that matters about a scientific statement is how it is used, not what it means or describes. This creed is termed *instrumentalism*, to stress the notion that scientific theories and laws of Nature are to be regarded merely as instruments for learning about the world around us: they are not to be taken literally, nor endowed with any truth in themselves. What matters is not whether they are true or false, but whether they are *useful*. Theories are lights in the dark; signposts in the higgledy-piggledy of experience.

Unlike the empiricist, the instrumentalist does not maintain that the only valid concepts are those reducible to sense data. Theory and the human theorizer are allowed to play a role. The activity of the scientist is to be seen to be not only the recording and sorting of observed facts, but the invention and improvement of models which can organize the plethora of data.

Idealism. The idealist maintains that, since all the knowledge we claim to possess is filtered through our minds, we can never be sure that there exists any direct correlation between reality and our ideas about it. It is, in effect, an elaborate dream: an *idea*. Hence we must draw a distinction between Nature as it really is and our perception of it. The entities in our theories do not really exist in the Universe; they are entirely imposed upon the chaos of sense data by our minds. An extreme version of this approach is *solipsism*, which maintains that the external world experienced by each observer is entirely a product of his own imagination.

Realism. The realist believes that the external world exists, and that the observations and laws of Nature found by scientists are directly related to reality. The entities which correct theories describe exist independently of observers, and consequently all theories are either true or false irrespective of whether they are known to exist. Where the operationalist and the empiricist places stress on observability, the realist focuses upon intelligibility and analysability.

The realist assumes that universal laws of Nature exist 'out there' apart from our perception of them, and the thing that is observed makes the main contribution to knowledge, not the observer or the process of observation itself.

Pros and cons

> There is nothing so absurd that it has not been said by philosophers.
>
> Cicero

Although we have stated that it is not possible either to rule out or logically prove any of these rival dogmas (how could it be, for are they not at root all diatribes on the nature of proof?), nevertheless, it is certainly possible to provide persuasive arguments against some and for others.

Let us begin with the empiricists. To the working scientist the empiricist view that all a theory can ever aspire to be is a description of data does not ring true. Only occasionally do theories arise in this way. The elementary-particle physicist, attempting to find the key which will unlock for him the secret of what elementary particles exist, and how they interact with one another, will formulate mathematical theories which he has picked out because of their compelling symmetry properties. These properties will require that particular particles must exist and interact with each other in special ways in order that the symmetry be respected. Our particle physicist will attempt to produce an economic mathematical description of what we already know, but the predictions of his new theory will derive from symmetry principles far removed from current experimental capabilities. No, scientists neither use nor think of theories as mere catalogues of data. No particle physicist would confuse a *theory* of elementary particles with the Particle Data Group directory of particle properties. Nor would he want to restrict a fundamental concept like that of the 'photon', for example, just to its measurable aspects, because it offers such a wealth of explanations of other observable phenomena in situations where it is not itself measurable.

There are many examples which support this conclusion: general relativity was conceived by Einstein as a pure theory in response to 'thought experiments', and not as a description of sense data, for virtually no relevant data were known to him at the time. In fact, when contrary observational data were presented to Einstein on two occasions he concluded that it was the observations that must be incorrect (and he was right). Many of our most fruitful concepts in the physical sciences, quarks for example, are not observable. The scientist who just collects and organizes facts is like a stamp collector whose aim is simply to amass stamps indiscriminately.

Along with new experimental data, our most important discoveries are of new ways of solving problems, unforeseen links between different subjects, powerful analogies, and new, unsuspected types of problem: collections of new possibilities as well as collections of new facts. And if we follow the empiricist into believing that only the particular facts have any real claim to exist, then what do we do with the most pertinent member of this collection of particular facts: the fact of their various inter-relationships and common properties?

Empiricism sounds like harmless sophistry at first, but upon closer examination is found to possess all manner of embarrassing consequences. If followed rigorously, many useful physical concepts would be outlawed on the grounds of unobservability. Any general law of Nature would also appear to be forbidden at the outset, because we could in practice only have confirmed its validity in a small number of cases. It could never be verified. This seems to lead to the demise of science in fairly short order. The first line of retreat was to allow a statement to be regarded as meaningful if it could give rise to verifiable consequences when employed in conjunction with other statements, but this ends up by not excluding anything at all.

The operationalist viewpoint was encouraged by the realization that some everyday notions which scientists had taken for granted as obvious were in fact meaningless. For example, the special theory of relativity formulated by Einstein in 1905 revealed that the familiar idea of simultaneity in time possessed no absolute meaning. When one moves away from our blinkered experience of bodies moving at speeds far less than that of light (186,000 miles per second) then an observation that two events are simultaneous is not shared by other observers in motion relative to it. The lesson learnt from this is that just because a concept sounds meaningful this does not mean that it is. Simultaneity turns out to be a relative concept in the way that size is. If you look at two objects from one angle one may appear bigger than the other, whereas from another angle they may appear to be identical. In order to interpret correctly a statement from a witness to the effect that one object was seen to be bigger than the other, we require more information. We need to know the sequence of measurements carried out by the witness to ascertain the sizes.

Faced with a concept, the operationalist wants to know 'Can one measure it? Is there any experimental procedure which can establish it constructively step by step?' The operationalist requires the answer 'yes' to both these questions. But this is dangerous ground, for almost all quantitative science is written in the language of mathematics, and this language involves all manner of concepts that are not measurable or physically constructable in the sense desired by the operationalist. Limits, irrational numbers, complex numbers, infinitesimals, infinities: all are vital concepts in the mathematical description and explanation of Nature, but they are not directly measurable ones. Still worse, we must seriously question whether we can measure operationally even the simplest conceptual quantities like a 'length' without doing so in terms of some theory about that quantity. Our fundamental standards of mass, length, and time are all in fact defined, not in terms of comparisons with macroscopic man-made artefacts, but relative to sub-atomic standards whose existence was established by theory. And if we reduce everything to a jigsaw puzzle of experimental operations that are necessary to specify it uniquely, we are denying that we can really discover anything new at all. How can we ever develop our understanding of a particular entity to a new and deeper level? More disastrous still would seem to be the total fragmentation of science that would result from the operationalist doctrine. Every time that we measure the same quantity by a different method we must regard it as a different quantity. One of the ways in which science tries to escape from the bias of individual experimental procedures is to measure quantities by different methods, and compare the answers. This would be forbidden to the operationalist. Since every experiment differs in some way, however small, every step-by-step experimental construct must be regarded as different. We must dispense with the uniformity of Nature.

Operationalism also appears to shoot itself in the foot in that it ignores the fact that we need to know what counts as an allowed 'operation'. In order to arrive at this we need to introduce all manner of logical, mathematical, and physical concepts. Moreover, we must decide what counts as a possible (as opposed to an impossible) operation. The operationalist needs to be a clairvoyant. For what is impossible today may be possible tomorrow. Operationalism requires a conspiracy between our level of theoretical speculation at any one time, and our experimental capability. It tells us that we cannot use certain concepts until we have built slightly more sensitive pieces of equipment that allow more accurate measurements to be made.

The operationalist and the instrumentalist do not maintain that our laws and theories of Nature are merely discovered. They may be invented as well. The observer is admitted to play a larger role in the game than the empiricist will allow him. And hence this view of science accords a little better with what scientists actually do. But it faces a real dilemma none the less. Ultimately, the instrumentalist holds that a theory tells us nothing about the world, but surely the reason why some theories are found to be more 'useful' and reliable instruments for conducting us towards new knowledge about the world than others is because they correspond better with certain unalterable facts, rather than with our unprincipled fancies. When does a scientist ever designate a theory as 'useless' except when he is saying that it is either false or vacuous? The only valid criterion of 'usefulness' is correspondence with the facts as they are encountered. Astrology might have all sorts of useful psychological functions, but it is certainly not on the grounds of its usefulness or lack of it that astronomers reject it. It is because they maintain it to be false.

In some areas of human activity it is often very important that two mutually exclusive views be allowed to coexist in the interests of stability. For example, you might be in favour of your country maintaining an adequate nuclear arsenal for defence purposes, and yet also be in favour of the existence of strong anti-nuclear pressure groups. The existence of the latter can place an important constraint upon the former. Yet in science we do not have such equilibria. While there do indeed exist mutually exclusive theories attempting to account for, say, the existence of galaxies, they only coexist because there does not yet exist evidence able to falsify one of them at the expense of the other.

The idealist's claims are the most radical. 'Your entire description of the world is mind-imposed,' he claims, 'including your wrong-headed idea that it isn't.' But there are embarrassing facts which weigh against this possibility. Scientists working independently in different cultures come up with the same laws of physics. They may differ in the formalism which is used to represent them, but they do represent the same thing as surely as do the symbols '8' and 'VIII'. This convergence of our scientific findings is strongly suggestive of the existence of an underlying concrete reality independent of ourselves. Sometimes scientists arrive at theories they do not understand, which is a curious turn of events if those theories are entirely products of our minds; and sometimes they are led to develop laws that are simply too ingenious for us to have thought up without Nature's prompting. And why do the products of our minds agree with some sets of data but not with others? There are an infinite number of wrong answers to any question, but only one correct one: what is it that enables our minds to produce answers that are so often judged correct? This would seem to require a remarkable set of coincidences if idealism is true. And why is it that we are stumped by certain observations and can provide no explanation for them at all?

Although the idealist view has been fashionable among non-scientists in modern times, it is at root a primitive view. Human thought took millenia to mature to a state which did not regard Man as the centre of the Universe. The idealist seeks to take a giant step backwards, and reinstate the human observer in his pre-Copernican position as a fulcrum of reality.

The hyper-idealist view called 'solipsism' cannot (by definition presumably) be discussed. No significant philosopher has defended it (although this might perhaps be taken as a consistency argument for it, or at least for the rationality of its supporters, since no believer in solipsism could have any reason whatsoever for defending it!). Even Bishop Berkeley, who is often cited as an originator of a strongly idealistic world-view, was far from being a solipsist. He maintained that the world existed perpetually in the mind of God.

We shall in what follows subscribe to the 'common-sense' realist view until we run into definite evidence against it. Almost every working scientist is a realist— at least during working hours. Although, if he is honest, he has probably never given the matter much thought, because he has found his investigations and researches to be almost entirely independent of his views, yet if questioned at the weekend he might not wish to defend the realist position too strongly. It appears that science is best done by believing that realism is true, even if in fact it isn't. Although this last statement sounds weird it resembles Immanuel Kant's view of Nature. He believed that, although we cannot prove that Nature is purposefully organized, we must co-ordinate our observational data as though it were so organized. Such a systematization is only possible if we believe in the existence of a 'Principle of Natural Purposefulness', and so investigate Nature as though it were ordered by an intelligence other than our own. Whether or not they subscribe to such an idea explicitly, this is what scientists actually do; otherwise they would not become scientists. This is in itself a logical difficulty for rival philosophical viewpoints. If all scientists work believing realism to be true then, by the criterion of the idealist, realism *is* true; and the fact that science is evidently most effectively done by adopting the realist perspective makes it the most useful view to adopt, and hence by the instrumentalist's criterion also the correct one.

Yet one can object to an over-confident subscription to the realist view. It ignores the limitations of our biased and often totally erroneous human understanding. We cannot forsee how subtle and elusive that underlying reality we are seeking may ultimately be. It might be, quite literally, inconceivable. The limit of an infinite sequence of numbers often possesses properties not shared by any member of the sequence of terms that converges to it. Many would want to qualify their subscription to realism so as to distinguish our attitude towards things themselves from our attitude towards scientific theories about them. So, for example, one might well be a realist about the existence of some type of elementary particle called the muon, but adopt a pragmatic anti-realist view of our theoretical description of it because of a belief that we will never possess an absolutely correct description of it. Theologians are very adept at adopting this schizophrenic discrimination between entities themselves and our models of them. They will be realist about the concept of God, but anti-realist about any attempt to pin the Deity down by any axiomatic definition, choosing instead to list only the things that He is not.

There is another line of thought which might lead one to find scientists believing that realism was true regardless of whether it is. Our minds and percepts are the result of a process of natural selection which has led to the evolution of a way of interpreting the world that helps us to survive in it. Realism is clearly the type of belief that is not only the simplest and most straightforward (the most *British* perhaps?), but also the one most likely to survive. Imagine the following scenario: a primitive tribe lived by its wits in a dense jungle which it shared with creatures of a more hostile persuasion—lions, tigers, cobras, bears, and suchlike. Now within the tribe there suddenly arose, as a result of a series of philosophy lectures recently delivered by an itinerant lecturer from the British Council (who was subsequently eaten), a division of opinion as to the true nature of their perceptions of the world. A split occurred. One half of the tribe, of the idealist philosophy, moved to the other side of the forest, leaving the remainder with their old-fashioned realism. Gradually the idealist population became depleted. They had come to the conclusion that bears, lions, and tigers (who had not received a visit from the philosopher, and none of whom were idealists) were simply creations of their own minds. They held learned seminars on this topic after dark around the camp-fire. They beat their swords into ploughshares and their spears into pruning hooks, and they sent unarmed scouts out to check the self-consistency of their ideas. None ever returned. The realists, meanwhile, thrived. Under no illusions as to the reality and realism of tigers, they prospered and multiplied. Their descendents propagated the realist assumptions about the world, and it goes unquestioned in their villages to this day. In the primitive world the wages of being unrealistic is death.

The worrying angle to this little parable is that it appears that a belief in realism would tend to be selected for in the evolutionary process even if it were false. On the other hand, maybe the parable convinces us of the reality of lions and tigers. The other anti-realist philosophers can be incorporated into this parable in various ways. With regard to linguistic philosophers we could begin by recalling William James's remark that 'the word "dog" does not bite'. The rest is left as an exercise for the reader.

These arguments provoke one last speculation. We can appreciate how a certain robust realism has a very high survival value compared to idealism in a conscious life-form facing a hostile environment. If everybody is an idealist, then they either starve to death or are displaced by the invasion of one realist predator. But as life evolves to higher levels of intellectual sophistication it becomes possible for living beings to start thinking about thinking itself, rather than merely thinking about things. It is at this stage that idealism, along with other abstractions not selected for by natural evolution, can emerge and survive as abstract ideas. However, an advanced civilization drawn inexorably away from realism by such speculations would undoubtedly be handicapped in any conflict with a somewhat less philosophically sophisticated society. Maybe intellectual evolution leads inevitably to an idealist form that is overcome by a realist 'predator'?

The operationalist could attack the realist about his attitude to theories by pointing out that his theories usually contain concepts which even the realist would not claim really exist; *lines* of magnetic force, '*bags*' of quarks, and so forth. They have been introduced because they are useful props in visualizing what is going on. Whereas the realist would regard these objections as evidence that he should not claim to be a realist about theories, but only that there was something real to theorize about, the operationalist would ask how to decide where to draw the line between the props we have already identified in theories, and the remaining components of them which you confidently expect to be 'really' there. The realist would also want to draw attention to the way in which the same concept of, say, 'electrons' or 'atoms' turns out to be useful in so many completely different circumstances. This ubiquity in the face of diverse external factors which ought to influence any subjective concept of an electron or atom, suggests that there is a bedrock of reality associated with the concepts being employed that transcends mere usefulness.

A more subtle difficulty for the realist is that he must imagine that universal concepts exist apart from the particular cases of them which we record. This seems to imply that there exists some arena in which these disembodied universals exist abstractly as a collection of blueprints. Of this dilemma we shall have more to say when we discuss the role of mathematics in Nature.

We shall adopt the realist view about entities, but not about theories, as a working hypothesis, and continue with it until it runs into problems—as we shall find it does. That is, we shall regard laws of Nature and scientific theories as our own inventions. They may be only an approximation to reality, but it is a reality independent of our minds. The aim of our theorizing is to come up with models of reality which are realistic. We hope that they are always converging towards a closer and closer approximation to reality, and we conduct continual comparisons between what is seen in the world and what is predicted by our models, to confirm that this convergence is truly occurring. None the less, we do not deny that some elements of our picture of Nature are idealistic, just as some parts are undoubtedly just plain wrong. Einstein regarded the most effective scientist as a polyglot who, to the philosopher, appears as a pure opportunist because

he appears as a *realist*, insofar as he seeks to describe the world independent of the act of perception; as an *idealist* insofar as he looks upon the concepts and theories as the free inventions of the human spirit (not logically derivable from that which is empirically given); as *positivist* insofar as he considers his concepts and theories justified only to the extent to which they furnish a logical representation of relations among sense experiences.

One can see the real problem with these debates concerning the status of laws of Nature. Each position is internally self-consistent, and there is no way of discovering which of the alternatives is correct. We might, of course, favour one or the other for entirely different reasons. The idealist stance invites attack with Occam's razor because it seems to be claiming that we should treat every observation as though it were seen in a distorting mirror. Indeed it could, but since then we have absolutely no way of knowing what the original reality is, why bother about it? Just study the appearances. An interesting example of this conundrum is provided by the debate that arose in the nineteenth century as to how belief in a 'young' Earth, just several thousand years old, could be reconciled with the growing evidence for fossil finds of very great antiquity. One suggestion was that when the world was recently created it appeared ready-made with fossils of great apparent age. (Somewhat earlier this dilemma had emerged in scholarly debates about religious art when it had to be decided whether Adam should be portrayed with or without a navel.) In fact this type of thinking is still prevalent in some quarters. Many American fundamentalists support such a pre-aged view of the creation of the Earth or of the entire Universe. There is no way of disproving it of course, but whether it is true or false is irrelevant to scientific practice, because scientists would still study the universe *as if* it really was fifteen billion years old, even if it was created only ten years ago with all the appearance of being fifteen billion years old. We can say nothing about a concept as elusive as 'ultimate reality'. We study appearances. They are *our* ultimate reality.

Labels

> What's in a name? that which we call a rose
> By any other name would smell as sweet.
> Shakespeare

By way of light relief we want now to indicate, with the help of a dialogue composed by Eric Rogers, the way in which some of our superficially different ideas about the same phenomenon can amount to nothing more than a difference of words used to label them.

You and Faustus have developed a difference of opinion regarding the nature of friction. Whereas you maintain that a rolling ball is stopped by friction, Faustus is trying to persuade you that it is really stopped by a hoard of cunning little demons:

YOU: I don't believe in demons.

FAUSTUS: I do.

Y: Anyway, I don't see how demons can make friction.

F: They just stand in front of things and push to stop them from moving.

Y: I can't see any demons even on the roughest table.

F: They are too small, almost transparent.

Y: But there is more friction on rough surfaces.

F: More demons.

Y: Oil helps.

F: Oil drowns demons.

Y: If I polish the table, there is less friction and the ball rolls farther.

F: You are wiping the demons off; there are fewer to push.

Y: A heavier ball experiences more friction.

F: More demons push it; and it crushes their bones more.

Y: If I put a rough brick on the table I can push against friction with more and more force, up to a limit, and the block stays still, with friction just balancing my push.

F: Of course, the demons push just hard enough to stop you moving the brick; but there is a limit to their strength beyond which they collapse.

Y: But when I push hard enough and get the brick moving there is friction that drags the brick as it moves along.

F: Yes, once they have collapsed the demons are crushed by the brick. It is their cracking bones that oppose the sliding.

Y: I cannot feel them.

F: Rub your finger along the table.

Y: Friction follows definite laws. For example, experiment shows that a brick sliding along the table is dragged by friction with a force independent of velocity.

F: Of course, same number of demons to crush, however fast you run over them.

Y: If I slide a brick along the table again and again, the friction is the same each time. Demons would be crushed in the first trip.

F: Yes, but they multiply incredibly fast.

Y: There are other laws of friction: for example, the drag is proportional to the pressure holding the surfaces together.

F: The demons live in the pores of the surface: more pressure makes more of them rush out to push and be crushed. Demons act in just the right way to push and drag with the forces you find in your experiments . . .

. . . and so on . . .

Faustus's idea is certainly a useful instrument for developing a systematic set of laws. Where you put forward a 'law' of friction he proposes a rule of demonic sociology. Within the context of the above dialogue the two proposals are actually indistinguishable. One seems 'scientific' and correct whilst the other (demons, hopefully) appears crazy because of other connotations that the ideas have outside the context of this dialogue. We must be careful not to regard differences of terminology as signifying something more. 'Laws of Nature' is a concept that can very easily suffer from this abuse. A theist will often want to ascribe the laws of Nature to the activity of God, or even equate them with God. The atheist might ascribe the workings of the Universe to 'logic', or 'self-consistency', but within the limited scientific context each is just an equivalent label, like the demons and the atomic theory of friction.

Accidental, legal, and statistical laws

> Being built on concepts, hypotheses, and experiments, laws are no more accurate or trustworthy than the wording of the definitions and the accuracy and extent of the supporting experiments.
>
> Gerald Holton

The word 'law' is used in so many different ways in the societies in which we live that it is important to point out some similarities and differences between these vernacular usages and the idea of a law of Nature. We have already remarked on the dual meaning of natural laws, which can be taken as describing what happens naturally—as *effects*—or alternatively can be taken as a set of rules which determine what can or will happen—as *causes*. The first of these alternatives views the statements that we elevate to the status of natural laws as nothing more than a mental shorthand for particular sequences of events that we often observe to occur in conjunction, and so as an essential product of our way of thinking and observing. But this cannot be the whole story, because it does not exclude accidental generalizations from its compass. For example, I may observe that all the children in my street own bicycles, or that there is not a single diamond on the Earth's surface exceeding one hundred tons in mass. These may be accurate descriptions of our experience, but we would not want to give them the status of laws of Nature any more than an astronomer would regard Bode's law of planetary distances as having the same status as Newton's law of gravitation. The supporter of the descriptive approach might counter by saying that Newton's law just happens to be a very good law which provides a description of a far wider and more interesting class of events than my laws of bicycle ownership and diamond size, or Bode's law of planetary distances, which are simply examples of bad laws. Yet the extreme generality and power which we associate with laws like Newton's (even though we have learnt that they do not hold exactly, let us for the moment assume that they are consistent with all known observations) moves us to believe that they incorporate or describe some necessary rather than accidental generality. We see that those statements, like Newton's which we elevate above my other examples have the conditional form 'if . . . then'. They are rules for what effect will result from a particular cause, or they may be relationships that must hold between different quantities under particular conditions, or functional relationships telling us how one quantity will vary if another changes its value, or, alternatively, how quantities will change their values in time and space under particular influences. The common feature of these prescriptions is the fact that they can be cast into the linguistic form 'if . . . then': *if* the force acting upon an object is doubled in strength, *then* the resulting acceleration of the object will double; *if* a gas exists in an insulated enclosure with a particular pressure and volume, *then* Boyle's law gives its temperature; *if* two planets orbit in space, *then* the gravitational attraction between them varies inversely with the square of the distance between them.

At first it appears that our accidental descriptive law of bicycle ownership ('all the children in my street own bicycles') sounds much like other laws of Nature that say *for any A, if it is a B then it must be a C.* If any child lives in my street it must be a bicycle owner. This sounds just the same as Newton's law; if you take any object and double the force on it then you will double its acceleration. But pseudo-laws, like my 'law' of bicycle ownership, do not support the consequences of a contrary fact. They do not lead to a true deduction of the form: *if A were a B then A would be a C*. For example, it need not be true that 'if that child (who lives in another street) lived in my street then she would own a bike'. But if that object (which has not had the force on it doubled) were to have double the force applied to it, it would double its acceleration. Purely descriptive statements of accidental regularities do not support these contrary-to-fact conditional statements.

Another deficiency of the purely descriptive view of natural laws is that it implies that all scientists wish to do is to make an exact record of Nature. This seems as impoverished an idea as the notion that artists are interested only in making exact photographic reproductions of their subjects—a mere copy of Nature which would be either right or wrong. It leads nowhere. We select a particular set of observed features that are thought to be connected in some way, but ignore others that we feel to be inessential to the phenomenon at hand. We never attempt to describe everything that is occurring in a given experiment. In practice, scientific laws are simultaneously descriptive, explanatory, and predictive. It makes no sense to try and pin them down further, for on different occasions we employ them in each of these capacities.

It is often said that legal laws tell us how we should behave whilst natural laws tell us how we do behave. There are apparently no penalties for breaking natural laws, and indeed no possibility of violating them. But legal laws can be restated to reveal a closer connection between the two. The speed limit could be connected to the statement that if drivers do not limit their speed on the motorway then the probability of their survival is lowered. In this sense, legal constraints on our behaviour are abstracted evolutionary laws in statistical form. If these laws of behaviour are violated sufficiently often certain consequences follow with enhanced probability.

This incorporation of probability into our account of things leads us to mention another possible reaction to laws of Nature: are they simply statistical statements? Indeed, we shall see later on that chance is the common factor behind the observed stability and consistency of many macroscopic phenomena. To a considerable extent, our knowledge of the world is statistical. No set of mechanical rules or laws about the behaviour of matter can be verified by us in any way except statistically. Our measurements cannot be one hundred-per-cent accurate; there must inevitably exist uncertainties, and these can only be reduced by repeating the measurement under similar conditions many times. Our laws of planetary motions are descriptions of an average of data taken over a long period of time. One might be tempted to conclude that if we end up formulating statistical laws of Nature, then this is just a manifestation of our ignorance of the exact underlying laws, but equally it might be argued that our discovery of exact laws may be an illusion created by only looking at the world in a very coarse-grained way that smears out a vast sea of microscopic effects that are truly statistical in character.

An exact law can, of course, never be verified precisely, but only to some particular level of precision. On the other hand, it can definitely be falsified by observation. A statistical law does not share this property. It can never be falsified, because the sequence of observations could always begin to behave differently in the future. The statistical law does not predict anything about individual events in the future. For example, it is a statistical law that the number of times a fair coin falls 'heads' will equal the number of times it falls 'tails' over an infinite number of tosses, but this equal likelihood tells us nothing that enables us to predict the result of the next toss of the coin.

One of the things we shall see in the next chapter is that there used to exist a closer relation between the ideas of legal and natural laws in the minds of leading scientists. In the West, it was an article of faith for many that natural laws were the decrees of a Divine Lawgiver. It has even been suggested that such a view played a key role in the successful development of science in the Western cultures, and did so because they were influenced by the Judaeo-Christian tradition which fostered faith in the underlying rationality and orderliness of Nature during periods of history when human ideas were inbred by all manner of magical and occult notions. If one believes in nature-gods, it is difficult to generate much enthusiasm to study the regularities of Nature's workings. Yet the Western tradition, which regarded the laws of Nature as a manifestation of God's will and His maintenance of the world order, ran into some confusion over this idea. Appeal to the regularities of Nature, the 'laws which never shall be broken for their guidance hath He made' (in the words of the famous Foundling hymn), as the exemplar of God's existence and personality became a cottage industry after Newton's great discoveries. Rather than point to particular events of an ambiguous and subjective nature, natural theologians picked upon the laws governing the *regularities* of the mechanical watch-world to argue for the existence of a divine intelligence behind it. But paradoxically these advocates coexisted with other theologians whose arguments for the existence of the Deity were grounded upon the evidence for miraculous events—that is, violations of the laws and regularities of Nature. Today such arguments are out of fashion with both scientists and theologians. They have gone the way of the medievals' attachment to curious ontological arguments for the existence of God. Their inability to persuade is due to the fact that they were wielded in argument by theologians whose own faith in no way rested upon them. Their personal faith in the existence of God rested somewhere else entirely. A refutation of their particular scientific or philosophical 'proof' of the existence of God would not have shaken their own faith one iota. By the same token they should not have expected their 'proofs' to change other minds.

Intelligibility

> In every age the common interpretation of the world of things is controlled by some scheme of unchallenged and unsuspected presupposition: and the mind of any individual, however little he may think himself to be in sympathy with his contemporaries, is not an insulated compartment, but more like a pool in one continuous medium—the circumambient atmosphere of his place and time.
>
> A. N. Whitehead

One of the things that makes science such a fascinating subject is that the natural world is fashioned in an 'interesting' way. By this we mean that its structure is not so complex that our study of it is a doomed and hopeless task, nor is it so simple that it is rapidly completed and found trivial, like a game of 'noughts and crosses'. It is both a challenging and a challengeable problem. This combination holds a special attraction to the human mind, as is witnessed by the fantastic commercial success of puzzles like Rubik's Cube. When we embark upon the solution of other puzzles we invariably do so under the assumption that certain ground rules are being adhered to. We expect all the pieces to be present in a jigsaw puzzle. There is assumed to be a word at the heart of a game of twenty questions. The Rubik cube is assumed to be solvable from the starting position. These assumptions are so habitual that they are unspoken.

The practice of science also rests upon a number of presuppositions about the nature of reality. We usually take them for granted. This is not only because of unconscious familiarity, Michael Polanyi notices that

> the metaphysical presuppositions of science . . . are never explicitly defended or even considered by themselves by the inquiring scientist. They arise as aspects of the *given* activity of enquiry, as its structurally implicit presuppositions, not as consciously held philosophical axioms preceding it. They are transcendental preconditions of methodological thinking, not explicit objects of such thinking; we think *with* them and not *of* them.

They enable us to proceed most effectively from simple experience of the world to knowledge about the world. The most obvious are the following:

1. There exists an external world which is external to our minds, and which is the unique source of all our sensations.

2. This external world is ultimately rational. 'A' and 'not A' cannot be true simultaneously.

3. The world can be analysed locally without destroying its essential structure.

4. The elementary entities do not possess what we call free will.

5. The separation of events from our perception of them is a harmless simplification.

6. Nature possesses regularities, and these are predictable in some sense.

7. Space and time exist.

8. The world can be described by mathematics.

9. These presuppositions hold in an identical fashion everywhere and everywhen.

The first assumption saves us from solipsism, and admits the existence of an object of study that does not respond to the whims of our free will. The second underwrites the scientific enterprise as a worthwhile activity, and supposes that the laws governing the external world share something in common with those that govern the functioning of our minds. It is interesting to recall that Kurt Gödel showed that is possible to prove that mathematical systems based upon arithmetic contain statements, which can be expressed in their symbolic language, which cannot be proved either true or false within the language. Not all statements in the language of arithmetic are true, and hence it can be consistent. This follows because any inconsistent logical system would allow any statement to be proved within it.

The third axiom of faith is more hopeful. It asks us to believe that we can study the world locally in small pieces, or in small regions of space or time, and infer the global structure by putting these small-scale features together. If the world were of an intrinsically holistic character this would be false. It could, for example, resemble a Rubik cube or juggling skittles: a problem that can only be meaningfully considered as a whole.

Number four is really a twofold assumption; first that some elementary entities exist, and second that these obey some rules which they cannot influence, and so are predictable whether or not that prediction is in causal contact with those entities.

Assumption five has given philosophers sleepless nights for centuries. It maintains that the act of observing the world does not radically alter its intrinsic character. It is the belief of the realist. As we indicated above, whether or not this is actually true, we have to assume it to be true in order to begin to do science.

The sixth assumption underlies our scientific method. It does not prejudge whether these predictions will be exact or statistical, or even limited in some essential way. It also allows our tentative predictions to be falsified. Assumption seven grants that space and time are conditions of our experience of the world, and so cannot be derived from that experience. We have to start somewhere.

Our eighth assumption is a curious one which we shall discuss at length in a later chapter; it is a product of our, possibly rather limited, experience of the world. Mathematics is the most sophisticated language we know of which possesses a built-in logic and a way of deducing its own limitations. It is a recipe for writing down analytic truths, or tautologies, and science aims to show these to be equivalent to various natural events which on the face of it appear to be non-analytic, or non-tautological truths. All our precise current knowledge about Nature is at root mathematical, but we cannot be sure whether this bears witness to the intrinsic character of the world or to the fact that mathematical properties are the only ones which we have been able to find systematically.

Finally, item nine is our subscription to a belief that the ultimate structure of the world which we discover does not depend upon where we are in the universe of space, or the time at which we investigate it. The events we see, the quantities we measure, and the conditions we experience may, of course, change with time and place of observation, but the basic laws of Nature should not. This rather speculative assumption is essential for any progress in extraterrestrial sciences. We need to assume that the laws we find to govern Nature locally are also true globally in order to repeat comparable experiments at different times and different places. Only in this way can the scientist, like the Prince of Denmark, 'be bounded in a nutshell', but count himself 'a king of infinite space'.

Not all of these nine statements are totally independent, and the list makes no claim for apodictic completeness or ultimate correctness. It serves to convey to the reader the frame of mind that the average working scientist has. Although most of the presuppositions seem obvious to us, in the sense that we have no alternative but to adopt them as guiding principles if we wish to make any progress at all, we shall see in the next chapter how a reluctance to accept some of them led to the premature extinction of many a culture's early interest in science. Something more than a desire to investigate the world may be necessary in a culture in order to endow a subgroup of its members with even the most basic assumptions about the rationality of the world. Curiosity alone is not sufficient.

In the chapters to follow we shall find our confidence in the nine axioms of faith steadily eroded. Yet unless investigators commence with the presumption that these axioms are true they can never proceed reliably to the conclusion that they are false. This procession is essentially the path from classical to modern physics. But, before we can follow it forward, we must trace it back to its likely source.

2

Time past

It has been said that though God cannot alter the past, historians can; it is perhaps because they can be useful to him in this respect that He tolerates their existence.

Samuel Butler

Primitive beginnings

The waking have one world in common, but the sleeping turn aside each into a world of his own.

Heraclitus

Primitive Man came to terms with the realization that his well-being was wedded to his ability to conquer the changing elements of his environment, to foresee and exploit them for his advantage in a harsh and competitive world. He came not to view this world as external. Rather, it was alive with the passions and personality that he saw within himself and his fellows. Likewise, his personal feelings and experiences took on a cosmic significance. Everything was inhabited by a personal spirit. The world was only a manifestation of his own mind. There were no inanimate objects. Dreams were as real as rainstorms. All phenomena possessed a multitude of different intrinsic meanings, of the sort that we would now associate only with symbolic representations of them in art, sculpture, or poetic diction. Primitive Man did not seem to view himself as distinct from Nature in the abstract. Often it appears that things are being personified with human characteristics but that assumes them to differ from him in a way that he would not have recognized.

We experience modern interest in the occult merely as distant spectators not as true participants. We cannot appreciate what is must be like to live immersed in a world that is viewed as entirely magical.

At first, like the beasts, primitive Man could learn only the hard way, by the consequences of experience. Gradually, as he noticed daily and seasonal regularities, he was able to recognize those appearances which seem always to follow from others. He discovered that events do not all arise in an unconnected and arbitrary manner, but exhibit a degree of predictability that could be exploited for advantage. He learnt to simulate the results of his possible future actions by thinking about them. If the result of some behaviour was foreseen to be adverse he could avoid it. He no longer needed to learn solely from the consequences of his mistakes. He had passed from experiencing the world to thinking about it.

From the discoveries of anthropologists we have learnt which presuppositions were shared by large numbers of primitive cultures. There was a nascent belief in the uniformity of Nature, and in the succession of cause and effect. These words have a familiar modern ring, but their emergence was through the practice of magic rather than scientific experiment. At first such a consequence appears strange; surely, if magic is a focal belief then one must believe that anything can happen? But this is not necessarily true. Dancing the rain dance brings rain, sacrificing to the fertility god brings an abundant harvest, and fêting the god of war brings the defeat of your enemies. Through such simple formulae there is displayed a real belief, however misplaced, in the idea of an orderly and predictable world in which certain acts invariably create particular effects. In such mythopoeic cultures there was complete determinism. Nothing happened without a cause and a reason; everything had an interpretation. The concept of chance could not arise.

Our habit of thinking of cause and effect in Nature as the consequence of some set of impersonal laws is foreign to primitive cultures. Because there was seen to exist no separation of the personal and the subjective from the things of Nature, all events were ascribed to wilful acts by something or somebody; for example, as the outcomes of conflicts between opposing forces of good and evil. Where we are content to view a single isolated event in the external world as possessing a cause and an effect, the primitive mind saw everything interwoven into a tapestry which was expressed by a story. Rather than analyse events into the particular, they sought to fit them satisfyingly into the whole. This exercise is the forerunner of what we call abstract thinking: the manipulation of signs and symbols in a self-consistent logical fashion according to certain rules. One further difference in emphasis for primitive Man in his approach to cause and effect in Nature was the focus upon the idiosyncracies and peculiarities of Nature, where we would look for the regularities. To the modern meteorologist the pertinent question is 'why does it rain?', but to the primitive mind it was 'why is it raining *here* and *now*?'. This is what matters when events are only experienced, and not studied. Primitive Man was a participant in Nature, not an observer. It was the Greeks who would introduce the idea of 'theory', a concept derived from the practice of thoughtfully watching, rather than competing, at the Games. It is the derivative 'theatre' that retains the true meaning in modern English. Primitive Man was an actor in a cosmic drama, and the whole world was his stage.

Magical ideas are not easy to overthrow. How do you falsify the idea that some ceremonial invocation of the rain-god's services causes it to rain? If the rains do not come then either the ceremony was wrongly performed or it was overridden by some more powerful magic. Why? Because your witch-doctor tells you so.

Magical views of this sort divide a community into two groups: the inner ring of those initiated into the arts of interpretation and the occult techniques, and everybody else. All the second group can do is wonder how the others do it. It is very difficult for anyone to say that 'the Emperor has no clothes', and even harder to have any reason to want to.

These early tribal cultures also appear to have shared another type of presupposition which we do not: the notion that things which have been in tactile contact with each other on one occasion will thereafter retain some vital relationship. This is a natural view to hold if one believes also in a completely animated world of warring spirits. If you meet somebody, do you not retain some impression of the meeting? This impression influences your future thoughts and behaviour. As a consequence, is it not natural to believe that objects, or even people, can be controlled by manipulating things that either resemble them or have previously been in physical contact with them? These artefacts were not viewed as symbols of the person from whom they were taken: they were the person. If the first of these ideas had its origin in the observation of kinship and herding instincts amongst humans and animals, then perhaps its residue remains in our debates as to the relative importance of inherited and acquired human abilities. The belief in the possibility of exercising control over others by building effigies of them, which usually contain a piece of hair or some other part of their body, still exists in certain quarters of the Caribbean where voodoo practices survive (this idea is the original meaning of our term 'jinx'; to put the jinx on someone was to bring them misfortune by this means). It hints at the idea of simulation which modern scientists use so successfully. Often one builds a 'model' of some part of Nature, whether it be in the form of mathematical squiggles on a piece of paper or an electronic recreation of its behaviour in a computer, and this simulation enables us to derive understanding and control of the real thing. It is most significant that the primitive belief reveals the idea that a part of something contains the essence of the whole.

The last fairly universal presupposition about the world that we should recognize was a belief that naming things endowed the namer with authority or control over them. This was clearly an early Jewish belief, shared with surrounding cultures, as witnessed by the significance attached to Adam's naming of the animals in the book of Genesis as a sign of his subsequent 'dominion' over them, the renaming of Jacob, and the Jews' veto on speaking the name of God. The Egyptians sought to obliterate their enemies by inscribing the names of their rulers on pots that were ceremonially smashed. This action, they believed, would inflict real pain upon their enemies. We even possess records of a conspiracy by members of his family to kill the Pharaoh Rameses by making and breaking images of him.

Perhaps this association of the name with the named arose out of the practice of parents of naming their children at birth, and then raising them to follow in their ways, so that the physical and temperamental features shared by parents and children then created a belief that the naming process after birth created some special link between them. On the other hand, it is equally plausible that the parents' naming of their children was a consequence of more primitive beliefs in the power of the namer over the named. We shall never know. (It is tempting to wonder if it is the linguistic philosophers and empiricists who have inherited this peculiar occult belief!) Even today there persist strange cultural differences in the use of a person's name. One of the most unsettling things for any new English visitor to the United States is the way in which Christian-name terms are immediately adopted by complete strangers—even by computers.

These speculations are of interest simply because magic was one of the forerunners of modern experimental science. Even by the eighteenth century so great a scientist as Newton still possessed strange magical notions which seem totally alien to us when laid alongside the Newtonian mathematics and physics which we share. But Newton saw his work, whether it be in mathematics or optics, alchemy or biblical criticism, as part of a single enterprise, and it is this strange eclecticism which provoked Keynes to write that

> Newton was not the first of the age of reason. He was the last of the magicians, the last of the Babylonians and Sumerians, the last great mind which looked out on the visible and intellectual world with the same eyes as those who began to build our intellectual inheritance rather less than 10,000 years ago.

What is regarded as magic changes with time. To a citizen of the Middle Ages would not the technological achievements of modern science be indistinguishable from his idea of magic? The 'Connecticut Yankee in King Arthur's Court' was a rival to Merlin, not to the King's engineers and philosophers. Yet it would be wrong to argue that magic was simply the forerunner of modern experimental science. One of the aims of this chapter is to illustrate the extent to which the wider system of beliefs within a culture determines whether or not it is fertile for the growth of science. The practice of magic was indeed a form of experimental technology, but it was born of a belief that is alien to modern science—that the natural course of Nature had somehow to be altered or perverted in order that power over it or over other persons could be obtained. The ancient practice of magic implicitly recognized a natural way of the world governed by what we might call 'laws', because it sought always to reverse it. It had to call up the evil spirit to oppose the good. This we see reflected in the fact that so many occult practices emphasize the inversion of the natural order—recall the 'black mass' in which the Catholic mass was chanted backwards.

For these unnatural practices to be efficacious they had to be performed with the right mental attitude. Modern science does not seek phenomena that respond to the state of mind of the experimenter, nor does it recognize the possibility of somehow breaking the natural order of things. Such an idea makes no sense within the framework of modern presuppositions, for the natural order is nothing more nor less than the set of all events that can or do happen. The scientific view aims to explain as much as possible by a single type of logical principle. It seeks to avoid the magician's division of the world into 'ordinary' everyday phenomena governed by one set of laws, separate from an extraordinary, occult world governed by a different form of law and logic: one that it was believed could be influenced and conjured up by the intensity of the human will.

Social and religious precursors

> Thou shalt not make unto thee any graven image or any likeness
> of anything that is in heaven above, or that is in the earth
> beneath, or that is in the water under the sea.
>
> Exodus 20: 4–5

The development of civilized societies possessing a strong form of central government occurred at many times and in diverse places, and had an indirect influence upon the presuppositions which underpinned the development of science. A picture of the Universe as an ordered state, subject to rules and regulations introduced for human well-being, owes something to this course of events. Practical necessity, dictated by the collective needs of large societies whether for agriculture, navigation, or the provision of military defence, motivated many scientific investigations. But it is not so clear what conjured the spirit of more abstract enquiry. One could even speculate that an outwardly prosaic culture might treat philosophical excursions into abstract questions concerning the origin and nature of the Universe, divorced from the world of everyday experience, as a stimulating exercise for the imagination, rather as we might regard reading a novel or a visit to the theatre.

In the West the scientific enterprise seems to have evolved most successfully in an environment in which there existed a strong belief in the role of law and order in the widest sense. Two possible situations immediately suggest themselves: states with a strong civil legal system and central government, and those cultures in which there existed a strong monotheistic religious belief. The former allows an analogy to develop between the ordered working of Nature under the jurisdiction of natural laws, and the ordered running of society according to civil law. The latter fosters a belief that Nature is governed by the decrees of an omnipotent and Divine Law-giver. A culture displaying both of these attributes is an especially advantageous environment for the emergence of a firm belief in laws of Nature, and in the rationality and ordered character of the world: a firm belief that there exists something worth investigating.

Despite the persuasiveness of this idea it does not seem to be the key. It is more significant to recognize that there was a gradual metamorphosis away from the early belief in laws as arising from the intrinsic character of the thing that they controlled, toward the view of the early monotheistic cultures that laws were imposed upon Nature by an external Law-giver. In the Greek setting we shall find this metamorphosis going hand-in-hand with the rejection of teleological modes of explanation in favour of those based upon causal ones. This turn towards the notion of imposed law is pregnant with ideas that transcend the scope of the idea that laws of Nature are merely the observed habitual successions of events. It emphasizes a common factor behind Nature, and establishes Nature's universality. More important, it establishes the invariance of some element of Nature in the face of the flux of events. If the laws governing the motion of stones arise from intrinsic properties of each stone, then different stones may move differently, and the same stone may alter its dynamical behaviour as it weathers and changes its appearance. There is no constant factor in Nature, and there could be no unchangeable laws. We should, therefore, examine how the idea of imposed laws was nurtured by the great monotheistic traditions.

In general, for the ancients the idea of a law of Nature was different to what it is conceived to be today. For the modern scientist the most useful sort of law is a set of rules, usually enshrined in mathematical equations, which specify how something changes—either with the passage of time, or as it moves from place to place, or both. Yet, for the Greeks it meant something that does not change: an immutable, static, and perfect harmony. Often, as we shall see, we can still show that there is a close link between any set of equations which describe change in time and space, and some unchanging harmonious symmetry, but the symmetries involved often turn out to be extremely abstract.

The biblical view is important, because it exerted a powerful influence during much of the period when science grew into its modern form. The majority of leading British scientists until the beginning of the twentieth century were devout Christians, and we shall see that this had a particular influence upon their scientific work.

The Old Testament pictured God fashioning the world out of a formless void, ordering it in special ways, and instituting certain definite prohibitions. This both inspires and reflects a distinctive attitude towards the natural world. Jewish religious ritual and belief was quite different to practices elsewhere in the ancient world, and to the earlier magical view of Nature we have just outlined. For the early Hebrews God was not *in* the objects of Nature themselves, hence the outlawing of idols and the construction of 'graven images'. The nearest the Jews come to anything entheistic is the association of mountains and 'high places' with a closeness to God. For them there was an external world distinct from the ordering force behind Nature, and both were separate from their subjective experience. The prohibitions against the casting of spells and divination prevented the belief that the performance of Nature could be dictated by human willpower, and reinforced the separation of the subjective human mind from the objective world of Nature.

The striking commandment against the construction of 'graven images' given at the start of this section has many consequences. Ever since then, 'idol' has had a pejorative sense. The instruction forbids not only the worship of objects as gods, but also the *representation* of the one true God. By this stipulation about their religious practice there is ingrained a perpetual contentment with an *abstract* view of things, and a distrust of having useful or symbolic representations of things. There is no scope for any idealistic view to grow in contemplations of religion or Nature.

Much of Old Testament history tells of the struggle of the Jews to maintain such differences between themselves and their neighbour states. A large part of the prophetic writings are warnings of the evil consequences of such idolatrous practices.

The Hebrew view of Nature was aesthetic and celebrative rather than coldly analytical or manipulative. Nature was seen primarily as a sign and symbol of its Creator, rather than as a puzzle to be solved or a source of power to be harnessed. To the pious Jew Nature was a signpost pointing up to God, fashioned to inspire joy, worship, and awe. It was not a representation *of* God. Some features of its make-up were evidences of his goodness and beneficence, while others displayed his displeasure. Nature was not a thing apart to be dissected like a specimen under a microscope; still less something to be altered or exploited. Rather, it was something to be within and altogether enjoyed. Dominion over other creatures had been legitimized in the Garden of Eden, but not dominion over Nature itself. Moreover, although the natural world was acknowledged to be the handiwork of the Creator, it was not held that a knowledge of the works of creation led necessarily to a deeper understanding of God. Deep understanding, or 'wisdom', was to be found in the moral world, and came 'not through the earthquake, wind or fire', but through the 'still small voice' of the Spirit of God. There was no religious reason for the Jews to take any special interest in the natural world.

This relatively prosaic view of nature contrasts sharply with that found in most pagan histories and mythologies in that the natural world is seen, not as a temperamental deity to be worshipped or feared, but as an artefact of the Creator to be respected, admired, and husbanded by His appointed stewards.

Science could not develop directly within such a world-view, but nor could a host of mystical notions which corrupted the growth of rational enquiry into the laws of Nature elsewhere. Nature gods are anathema to the Jews, and this taboo has the important consequence that the study of Nature is legitimate. It is secular. Solar astronomy runs into awkward methodological problems if society worships a Sun-god. Animal husbandry is difficult to reconcile with sacred cows. For all the religious failings attributed to them in the Old Testament, the practice of astrology was never one that ensnared the Jews, in contrast with almost all other ancient cultures. The first chapter of the book of Genesis establishes very rapidly that the Sun and the Moon are secondary creations, 'and He made the stars also'. This passage may well have been written at a time when the Jews were under the domination of a Babylonian empire whose views concerning the origin of the world were steeped in astrological notions about the power of the Sun and Moon. In the Gilgamesh creation-story, which is often claimed to have influenced the account in Genesis, the Babylonian Sun-god Marduk is described as controlling the stars by his commands:

> Then Marduk created places for the great gods.
> He set up their likenesses in the constellations.
> He fixed the year and defined its divisions;
> Setting up three constellations for each of the twelve months.
> When he had defined the days of the year by the constellations,
> He set up the station of the zodiac band as a measure of them all,
> That none might be too long or too short.

This account is also interesting because it too contains the concept of a law-giving deity. The writer requires a reason for the regularities observed in the heavens.

The Old Testament contains many explicit references to Yahweh in the role of Divine Lawgiver exercising power over natural phenomena: separating the land from the sea, ordering the night and day, the seasonal variations, moving the winds and the tides. These attributions show that there was an awareness of long-standing regularities in Nature which were regarded as remarkable, rational, and worthy of divine explanation. Yahweh was regarded as the necessary and sufficient explanation for all natural phenomena, requiring no assistance from other complementary principles or the innate tendencies within things. The recognition of miraculous events indicates that the distinction between what was 'natural' and what was not was fully appreciated. So was enshrined a continuing faith in the rationality, intelligibility, and constancy of Nature *imposed from without*, from which the concept of natural laws could grow.

Why did the Jews not have any scientific interests during these early times? The fact that they never developed any seafaring tradition may be a significant factor. Such a neglect avoids the need to indulge in any systematic study of the heavens for the purposes of navigation. A long period of nomadic existence and continual skirmishing with neighbouring countries must also tend to stunt technical progress. There is motivation for neither architecture nor industry. Life remains tied to agriculture and tradition. Work is necessary and sufficient to produce the daily bread; life is too hard for there to exist the luxury of enquiry for its own sake. Of those without the worry of poverty only Solomon is recorded as taking a detailed interest in the natural world, but his motivation appears to have been more lyrical than 'scientific'.

It is clear that astrological inclinations are two-edged. On the one hand they led some ancient cultures to study the heavens and make detailed records of what they saw there, but on the other there always stood the interpreters, who resisted any inclination to view things in a new way, since that would amount to introducing a new religion. Over long periods of time such a belief-system tends to become complicated and contrived, every new observation being fitted into the existing plan no matter how ill the fit, for the real role of the belief-system is to maintain social divisions between those who know how to interpret the heavens and those who do not. Not until there exists a universal and precise mathematical language is it possible for absolutely anyone to take part in science.

Chinese science

> There is no confidence that the code of Nature's laws could ever be unveiled and read, because there was no assurance that a divine being, even more rational than ourselves, had ever formulated such a code capable of being read.
>
> Joseph Needham

To argue that the development of a strong faith in laws of Nature and the possibility of scientific investigation of the world is facilitated by monotheistic religious beliefs divorced from Nature gods and astrology is one thing, but could it turn out that the absence of such ideas would prevent the development of natural laws?

It is not possible to answer this question with any great certainty. Yet, there is one interesting and well-documented example. It appears that the idea of a single supreme Deity was foreign to the early Chinese, and as a consequence the fate of natural science in that culture was a curious stillbirth. For the Chinese there existed no concept of a divine being who acted to legislate what went on in the natural world, whose decrees formed inviolate 'laws' of Nature, and who underwrote the scientific enterprise. Despite sophisticated technological developments in rocketry, printing, and the widespread use of magnetic compasses in their sailing ships, these inventions provoked no urge to explore natural regularities or the geography of the globe. Their introduction failed to foment periods of revolutionary scientific change and enlarged intellectual horizons, as they did in Western society during the same period.

A central idea of Chinese thought from earliest times until the present appears to be that of the spontaneous development of order in the world. This notion could have had its roots in observations of the natural world, for example of floral patterns, or the purposefully organized collective behaviour of insect colonies. In these examples there arises a mysterious harmony between many separate parts without external human interference. Alternatively, we might find its roots in the gradual appearance of social order within small peasant groups who found themselves 'naturally' evolving a stable and organized way of life within their communities, without the imposition of rules by some external central government. Rules arose by negotiation and compromise rather than dictatorial decree.

There were two dominant systems of thought and conduct indigenous to the Chinese. Confucianism arose in the sixth century BC partly in reaction to the destructive social chaos that existed, sustained by continual civil warring. The followers of its founder, Khung Fu Tzu ('Confucius' is the Latin form), laid great stress upon the importance of correct and just social behaviour, together with the intuitive etiquette and custom that leads to the right-ordering of society. In complete contrast, their later opponents, the Taoists, were interested primarily in the order displayed by Nature, and stressed above all the unity of the natural world, and its independence of all the human standards of behaviour so valued by the Confucians.

Confucianism eventually attained official status, and developed into a paternalistic form of liberalism. It was responsible for engendering a view of the world that did not seek rules for the behaviour of Nature in logical analysis or through systematic observational studies, but looked instead to the analogy with the harmonious social customs that evolve out of collective human activity. These cannot be predicted. They contain spontaneous and intuitive aspects that could only be arrived at by constantly ensuring that one was co-operating with everything and everybody in the proper fashion. They are simply too complicated to be predictable in practice.

This Confucian liberal ideal, that society should be ordered spontaneously by good and harmonious customs, was called *li*. It was a foil to the contemporary legalist philosophy that society should be governed by the positive legal dictates, or *fa*, of some supreme ruler or judge, rather than by these less precise natural constraints that everyone intuitively shared, what we would today term 'natural justice'—that lowest common denominator of feeling about what is right shared by most members of a stable and homogeneous society.

For the Confucians, the principal role of the civil ruler was to provide an impeccable example for others to follow. An order arrived at spontaneously by mutual human interaction and common consent was considered greatly superior to one imposed from outside by some external diktat. There was not that great respect for positive statute law that has traditionally existed in Western cultures. If an edict of *fa* were found to run contrary to what society regarded as *li*, then the former was regarded as being undesirable*.

Li operated across the entire spectrum of life; it was the reason for the motions of the Moon and the stars, for the successful exercise of self-control in human

---

*As an aside we might suggest that the most enjoyable way to attain a feel for the practice of law in ancient China is to read the famous 'Judge Dee' mysteries written in the 1950s by Robert Van Gulik. These trace the career of a Confucian judge in sixth century China from his first appointment as a local magistrate to his experiences as Lord Chief Justice and Regent for the Emperor when Peking is stricken by plague. Van Gulik was a Dutch diplomat and an accomplished oriental scholar who based these mystery stories upon real cases that arose in ancient China—suitably embroidered by the story-teller's art, retaining a traditional Chinese literary style (although mercifully he does depart from their tradition of revealing 'who-dun-it' at the beginning of mystery stories rather than at the end), and immersing the reader in a complete recreation of everyday life in ancient China. These stories are used as background reading in many degree courses in oriental studies as an entertaining and accurate introduction to the ambience of everyday life in early China, and some of the best have recently been republished in special editions for that purpose by the University of Chicago Press.

dealings, and for the social divisions of rich and poor. It was 'the greatest of all principles'. In all these realms, what we in the West might have called 'laws' of Nature were supposed to emerge gradually but quite spontaneously, and to persist as manifestations of stable and orderly behaviour arranged for the mutual benefit of everyone and everything.

The Taoist tradition opposed this search for the order of Nature in social behaviour. It arose through the influence of nomadic magicians and their disciples. In reaction to other philosophies of life the Taoists retreated and lived as recluses outside the milieu of society. They believed that only by being at one with Nature could the order within it be understood. Man and his idiosyncratic customs were inessential in this quest to commune with Nature. The order they were seeking, the Tao (the 'Way'), did not admit of a precise definition; it was beyond our understanding, and arose as an equilibrium of the mysterious interplay of opposites. Again, the emphasis was upon the spontaneous ordering that arose through the holistic interplay of all of the constituents of Nature. This holistic view also denies a clear-cut notion of an external world of physical reality, because we are ingrained into the harmony of Nature in a way that is essential to the order of the whole. To begin a study of Nature creates a problem of self-reference. Laws are latent within things, not imposed from without.

The magical background of the Taoist tradition was a motivation for 'experimental' practices, whereas the Confucian scholar was an aristocrat who looked down on manual work and anything akin to the experimental investigation of the world. Yet, despite their technology, the Taoist philosophers never framed any statements that we might call laws of Nature. They had no confidence in the ability of reason to unravel the Universe, or in the existence of a created order to be uncovered by investigation, and ultimately no faith in the intelligibility of Nature. Their liberal view that the Tao could not exert any coercive influence over Nature, and their theology of pantheistic naturalism, ran counter to the entire concept of a God or gods controlling the Universe.

The unusual results of these philosphical traditions have been highlighted particularly by Joseph Needham's extensive investigations of the history of science and technology in ancient China. The Chinese world-view appears to have stifled the development of science within this great culture. The evolution of harmonious customs in both the social and the natural realms was seen as arising from the intrinsic character and dictates of each member working for the good of the whole, rather than by the decrees of any external Supreme Being. There was no evolution of the idea of an analogy between civil legislation and ordained constraints upon the allowed workings of Nature, and no tradition of belief in any sort of supreme legislator or omnipotent Creator. Everything had the ability to bring itself into being, and so there was no psychological desire to introduce a personal Creator.

Thus, it has been suggested that the emphasis upon *li* rather than the more Western tradition of *fa* suppressed the notion that Nature followed definite laws laid down by God, and gave the early Chinese no reason to believe in an underlying rationality in Nature that might be uncovered and understood by detailed observation and codification. The Universe was believed to be far too complex for such an enterprise to be even considered. This holistic view would also deny the idea that by studying Nature on a small scale, piece by piece, one could arrive at an understanding of the whole. Nature was all of one piece, every part playing its part in a huge self-consistent pattern, yet each followed only the influence of its own intrinsic compass—a society of ants rather than one of men. The idea of order never carried with it any implication that there must inevitably exist constraining laws and an ordaining Law-giver. Despite their early technological superiority, the scientific attitude faded and died unfulfilled in ancient China because the notion of 'laws of Nature' never arose.

The Greeks

> The real importance of the Greeks for the progress of the world is that they discovered the almost incredible secret that the speculative Reason was itself subject to orderly methods.
>
> A. N. Whitehead

It is often claimed that modern civilization owes everything to the ancient Greeks: that all our mathematics, science, and philosophy has sprung from the efforts of a few gifted individuals who lived and worked in a tiny region of the Mediterranean some two thousand years ago. But the more closely one looks into this notion the less substance one can find in it. Despite the presence of philosophers and mathematicians of imagination and genius, the Greeks failed to produce any significant corpus of scientific and technological knowledge. The discovery of laws of Nature came near their grasp, but remained tantalizingly beyond it. Like most generalizations this last statement does have its exceptions, but they are surprisingly few and far between, and when in the post-Aristotelian period, one does find the pre-eminent genius of an Archimedes or a Ptolemy, it does not lead to a revolution in human thought as did that of a Newton. The great advances are made in astronomy, geometry, and in logic and philosophy. It is challenging to ask why the intellectual atmosphere of Greek culture, with its stable government and highly developed systems of civil law, did not lead to the uncovering of the great code of Nature's laws.

At first sight it appears that many of the trends in Greek thought were taking the perfect course for the ushering-in of a scientific revolution. The shackles of mythopoeic thinking were steadily eroded, and replaced by an emphasis upon rational thought in which there was an appreciation of the relationship between cause and effect. Nature was no longer believed to act in an arbitrary or irrational manner, and there was a faith in the intelligibility and rationality of the world itself. It was a subject worth studying. A further distinctive feature of Greek culture was the fact that the enterprise of inquiry into Nature was for the first time not exclusively in the hands of a minority of priestly types, who alone could dispense its interpretation to ordinary people. Science became secular: a lay activity that could be pursued by anyone fortunate enough to be born a freeman rather than a slave.

The great emphasis on the search for 'absolute truth' in the activities of numerous Greek philosophical schools witnesses for the first time to a systematic desire to seek out and comprehend the truth for its own sake, divorced from any military or practical stimuli, or from the acquisitive desire to amass a vast panoply of facts. Surely this is the seed from which the tree of knowledge grows?

The reasons for the failure of the Greeks to develop a fruitful notion of laws of Nature and commence their systematic discovery are manifold, and their mutual interrelationships complicated. None the less, we can isolate a number of dominant trends in Greek thought and practice that appear collectively to have erected a barrier through which their thinking about the natural world could not pass.

Let us begin with some of the general trends that were present during that period from about 600 BC until Aristotle's death in 322 BC, during which all the significant philosophical developments took place. Later, we shall pick upon the individual biases of particular, influential schools of thought. The most striking general feature to be found is the extent to which what we would now call 'science' is a subculture within philosophy. Indeed, there existed no distinction between what we would now term 'science' and 'philosophy': there was only a general inquiry into the nature of things, a natural philosophy. Before Aristotle's influence there did not exist any organized division of this inquiry into different subject areas. The principal distinction that did exist was between the array of natural things, which displayed change and complexity, and the static and absolute truths which were of a mathematical type, epitomized by geometry. Groups of thinkers were linked by their connection with a particular school of thought, their thoughts about natural phenomena conditioned by the dominant in-house philosophical line. It is as if departments of physics in universities today were run by leading philosophers of varying persuasions. At the University of Brummage the idealists might hold sway, and attract colleagues and visitors who toed this line, while at the University of Slippery Rock everything would be done in strict accord with the dictates of the operationalists. Science would become natural philosophy in the worst sense (this is not to deny that there are parochial influences in modern scientific work, and 'centres' where a particular school will research the consequences of a particular theory, or work in a particular style set by their leader; it is just that these modern schools of science are not shaped by philosophical views concerning the entire spectrum of human enquiry).

This philosophical umbrella over investigations of the natural world was, as we mentioned earlier, initially vital if problems and phenomena were to be discussed in a rational manner, but the goals of the early philosophers were, in retrospect, far too ambitious. The Greek philosophical schools were interested in ascertaining the whole and ultimate truth about the world. Their focus was always upon the general rather than the particular. This ambition has one very unfortunate effect: having begun with theories about the nature of everything, it is very difficult then to proceed to detailed theories about particular things. A theory that can tell us something about everything will, by its very generality, be too vague and diluted to tell us very much about anything in particular. Science only began to make dramatic progress during the Renaissance when it limited its objectives, and started to address the particular as the necessary prerequisite to any understanding of the general.

Whereas today scientists may disagree about explanations for the different things that we see in the Universe, they are agreed upon the subject areas for study. But for the Greeks the term 'nature' had many different meanings, and their philosophers preferred to argue about the meanings of the idea rather than observe what happened in the world.

There was another unfortunate emphasis in the quest for the ultimate nature of things that dominated the work of the early materialist philosophers. Although they were the only Greek philosophers to believe that things could be understood from the evidence of our senses, rather than depending on some higher form of reality and truth that is not accessible to the senses, the things they applied this logic to were totally beyond the reach of observation: questions like the origin of all things, the origin of life, and so on. Their materialistic reductionism led them to take the view that, if all things were made of air or water, then the properties of anything from a mind to a mallet could be entirely explained by knowing the properties of air or water. Later, both the Platonic and Aristotelian viewpoints were a strong reaction to these notions; they held that ultimate reality was something totally immaterial and accessible only to pure thought.

The predominance of the philosophic attitude led to another superficially positive feature becoming a snare. The emphasis, in particular that of Aristotle, upon formal logic as the paradigm to be emulated by all reasoning about Nature creates sterility. Formal logic is admirable in mathematical argumentation where one is interested in finding unusual connections between types, or numbers, or the relationships between the lengths of the sides of triangles. It allows a network of deep interconnections to be set up between things that one already knows. It shuffles and relates things that are already defined; but logic alone cannot reveal to us the existence of new types of entity. And, of course, it has little use for experiment or observation. A reverence for geometry fosters a belief that the most important properties of the world are static, and prevents one focusing upon the dynamic aspects of its structure. It reinforces the Greek attitude that what is important is what things really 'are', rather than how they behave or how they are related to other things. A scientist who uses only logical reasoning from 'self-evident' principles will soon find himself in a cul-de-sac—like a blind landscape artist. For our knowledge of the workings of the natural world to progress we must be opportunistic. We shall see in a later chapter that science owes a remarkable and mysterious debt to mathematics, but the Greeks were to some extent impeded by their very reverence for its inexorable logic.

Greek mathematics was unashamedly pure mathematics. It introduced the use of 'axioms' to represent truths that were regarded as self-evident, unworthy of demonstration. They realized that analysis had to have a beginning. The avoidance of applied mathematical thinking was coupled with a sociological bias that in retrospect appears as a major impediment to the development of science within their culture.

The Greeks had a low view of manual labour. They were aristocrats indeed; and slavery was a key element in their economy. In some communities we know that there existed more than one slave for every two free men. These inequalities within their society created a great division between the intellectual and the practical world. The result was a society devoid of technological progress. Some have argued that the presence of slavery within a society removes the stimulus to look for labour-saving devices, and is sufficient to explain the absence of any systematic technological progress in that direction by the Greeks. Although this is an argument that could be true about some societies, it is not very persuasive. Did not the Egyptians and other great empires of the ancient world use slave labour, but advantageously harness it for their technological projects? It is more likely that the existence of a slave labour force will serve to encourage vast technological projects like the construction of the Egyptian Pyramids. For the Greeks, the effect of slave labour was indirect and psychological. Manual work was what slaves did. It was mundane. To make devices and experiment was beneath the dignity of the aristocrat; it was an activity to be pursued by those who could not think. The results of such a high-minded prejudice are unfortunate. To seek the truth for no other reason than to know it is admirable, but if it leads to the view that a search for the truth for some other reason is inferior, then it becomes a snare. The result of this mentality was that all concern with Nature was dominated by theory. There was no experiment. The study of Nature was very much a spectator sport. It was literally the activity of the 'theoretician' who sat in contemplation of the activities of the competitors in the ancient Olympics. He was the one sent to bring back the revelation from the Oracle at Delphi. He did not participate, nor did he dig out the facts for himself.

The subscription to a slave labour force does have one direct practical consequence for the development of applied science. Although it could, if exploited correctly, accelerate the progress of vast architectural or engineering enterprises, it hinders the development of a large pool of independent skilled craftsmen within a society. From the workshops of such individuals, and by their motivation to manipulate Nature to build bigger and better artefacts and tools, did the practice of applied science receive its stimulus during the Renaissance.

There appears to have been more to this early Greek aversion to experiment than a distorted sense of values. Although Greek thinkers did not appeal to nature gods to provide explanations for the phenomena they witnessed around them, they retained inhibitions ingrained into their culture by previous mythological beliefs. Nature was not entirely secular, and the idea of manipulating it in order to learn things would have seemed peculiar. It was a living organism, not a mechanism. Understand the world by all means, but do not try to alter it or use it for mundane purposes. It is probably no accident that most Greek scientific progress was made in astronomy, where no manipulation of Nature is possible. Moreover, the Greek gods—an impetuous and moody lot—did not encourage the idea that the workings of Nature were dictated by a universal and divine decree. There were many gods, each with his own area of jurisdiction, and they inhabited a world in which decisions were made by argument, negotiation, and compromise rather than omnipotent fiat.

The mythological basis of their earliest ideas about the world perhaps had another important consequence for their scientific method which ensured that there was little emphasis upon experimentation. Many of the physical theories and cosmologies of the Greeks read like rational revisions of the early myths. They are invariably stories and explanations of what had happened in the dim and distant *past*. They were not hypotheses or predictions about the future. All that was required of them was the absence of internal contradictions. They were exercises in deductive ingenuity; the question of observational consequences in the future never arose.

The founding of schools and academies for the promotion of discussion and inquiry had many obvious positive benefits. For one thing it provided thinkers with critical colleagues—something that a Copernicus or Galileo was to lack. The informal discussion of conjectures and refutations with fellow scientists of all opinions has become the basis of modern research work. Despite external appearances, the principal function of large scientific conferences today is not the timetable of official lectures around which the meeting is framed, nor even the texts of those lectures in the published proceedings, but the bevy of informal conversations and arguments that abound in the hectic conference atmosphere. Most take place in hotel lobbies, on trains or coaches *en route* to the lecture hall, over lunch or dinner breaks, or in small informal groups meeting over coffee.

Yet, we have seen that the collective behaviour of the Greek schools also had a detrimental effect because of the overriding role played by the allegiance of different schools to particular philosophical systems. Another bad side-effect, best illustrated by the practices of early pre-Socratic groups like the Pythagoreans, was a curious desire for secrecy. They were secret organizations in the way that the Freemasons are today.

In around 531 BC Pythagoras of Samos had founded a monastic brotherhood in which the philosophical beliefs of the school dictated their way of life on an everyday basis. Today we would call such a thing a cultic religion. Indeed, it possessed many rules and regulations about daily life that are not dissimilar to the Jewish ceremonial laws found in the Pentateuch. This order lasted for about two hundred years, and was extant during the careers of Plato and Aristotle. Whereas the first Greek philosophers in Miletus had been guided purely by curiosity in their musings about air, earth, and water, the Pythagoreans had a mystical imperative. They were seeking out the divine harmony of the world in order that they might become one with it. The Pythagoreans were the first of the Greeks to refer to the world as a 'Cosmos', so emphasizing the harmonious order it represented to them. They believed in the reincarnation and transmigration of souls, and that the elevation of their social status in the next life could be guaranteed by rigorous adherence to their code of conduct in the present one. Eventually the cycle of birth and death might be broken by their sublimation into eternal union with the gods. The contemplative mood was the highest of ideals in all aspects of everyday life. For example, Pythagoras would have regarded the spectators as a more important component of the Games than the competitors. For the spectators come neither for material gain nor to seek power and status. Of course, Pythagoras' spectating was not just idle viewing, but the active thoughtful appreciation of the art-lover looking for beauty and geometrical harmony in a sculpture or a piece of architecture. It involved the mind as well as the eyes.

The Pythagoreans' great discoveries in mathematics and musical harmony are well known, but the mystical motivation for their investigations had some peculiar consequences. The discovery of irrational numbers in geometry shook the foundations of their beliefs, and so was for some time kept a closely guarded secret. Their religion of numerological beliefs dictated that observations of the world had to fit in with their ideas of harmony: observation could not serve as the basic guide to the way the world was. For example, Aristotle later wrote of them that:

> They held, for instance, that ten is a perfect number and embraces all the powers of number. On this view they asserted that there must be ten heavenly bodies; and as only nine were visible they invented the 'counter earth' to make a tenth.

Superficially, the Pythagorean belief that everything was 'made of numbers' appears not dissimilar to the modern notion that the Universe is best described by mathematical quantities. But there is a real difference between admitting, as we do, that it is possible to describe the workings of Nature by mathematical laws, and believing, as the Pythagoreans did, that there was something divine about the numbers themselves. We have found the fruitful idea to be that there are numerical relations *between* things, but we do not endow the numbers themselves with any magical or sacred status as did the Pythagoreans. What now seems so strange to us is that Pythagoras would not have understood how there could be any distinction between his mathematical ideas and his mystical beliefs about immortality and oneness with the Cosmos. Such is the unbreachable gap between the past and the present. Yet this mystical trend always survived amongst some group of natural philosophers right up until the time of Galileo. The pursuit of numerology was a rival, and ultimately unsuccessful, route towards the mathematization of Nature. Whereas science eventually grew by identifying causes with effects in a mathematical way, the hermeticistic* successors of the Pythagoreans sought the inner meaning of Nature through mathematical symbols; but they regarded Nature as an encrypted message which they could divine by insight, rather than a system dictated by externally imposed laws which they could determine to increasing precision by experiment. Their image of Nature as a book to be decoded and read would ultimately be displaced by the image of the world as a machine, in which every event had a cause and an effect rather than a meaning.

In the end the positive trends in the Pythagorean approach, the use of mathematics to describe the world, and the search for causes of events, were dissipated by their mystical motivations. But the cause of the school's demise was not internal self-inconsistencies; rather, it was the character of the dominant Platonic school of thought that succeeded it, and the way in which it developed earlier Pythagorean thinking about the visible universe.

Plato

> Beauty is a sequence of hypotheses which ugliness cuts short when it bars the way that we could already see opening into the unknown.
>
> Marcel Proust

Plato was born into a noble Athenian family in about 428 BC. His contributions to philosophy and literature were immense, but we wish to focus upon just one key aspect of them that impinges upon our quest for laws of Nature. Although Plato inherited from the Pythagoreans a respect for the role of arithmetic and geometry in human inquiry, his principal motivations were not what we would call scientific, nor was he trying simply to solve problems raised by earlier generations of natural philosophers. His main interests were in moral, social, and political questions.

Plato introduced a distinction between the visible world experienced by our senses, and a theoretical world of pure 'Forms' or 'Ideas' which were the perfect blueprints of which all that we saw and experienced in the world were imperfect copies. The real truth about the Universe, Plato maintained, was only to be found in this world of the perfect Forms, not in the natural phenomena we observe with

---

*The so-called 'hermetic tradition' derives from fifteen anonymous treatises written in the first three centuries and traditionally ascribed to the Egyptian god Hermes Trismegistus. They are a mixture of mystical and religious writings which include elements of Greek, Jewish, and Persian philosophy and had a significant influence upon some renaissance thinkers.

our senses, nor even in the mathematical quantities that we have discovered. Matter only existed with a secondary status: to realize the Forms. These Forms were held to be eternal. They would, in some way, have to remain in existence even if the physical universe and observers within it were to disappear. This notion, that ultimate reality was only to be found in things unseen, was implicit in the earlier Pythagorean thinkers as well as in the fragmentary writings of Heraclitus, Parmenides, and Anaxagoras. It introduced a dualism between the real world and an apparent one. Of the former real, changeless world of perfect Forms only a rational knowledge was possible, but there could be empirical knowledge of the constantly changing but imperfect world of our experience. Much of Plato's thinking attempted to relate these two strands. The realist strand was followed by Aristotle, and developed into practical science. The rationalist and Idea(l)ist theme that derived from Plato would one day merge again with Aristotelian empiricism to create the world-view of the medievals, but subsequently they would diverge again.

Plato retained the Pythagoreans' strange ideas about reincarnation, for he held that our senses first perceived these Forms in a past existence in which the soul was unencumbered by the imperfections of the body. This precognition enables us to recognize and recollect the versions of these perfect Forms we see in this present incarnation.

Some have argued from this evidence that the Platonic school was an unmitigated disaster for science. It effected a complete divorce between theory and observation. No workman could invent a new device, because he had to wait for the gods to originate the perfect Form on which it would be based. Technical inventors of the past were told that they had merely copied the blueprints of the gods! New invention had to await a sort of divine initiative as though the gods had to declassify ideas before humans could have them. The idea that some type of natural philosophy should be pursued with the aim of material gain or the saving of labour, rather than as part of the pure search for ultimate truth, would have been unacceptable to Plato's school. Not only was the idea of technology and experiment anathema because of the abiding prejudices against manual work, but it was now being claimed that there was no primary evidence concerning the structure and workings of Nature to be found in sensory perceptions of it. Modern scientific practice (although not necessarily all philosophical interpretations of science, as we saw in Chapter 1) is predicated upon just the opposite: the primacy of sensory perceptions, whether they be made by human eyes or by Geiger counters. Thus it appears that Plato's ideas do more than just not stress observation of the natural world: they positively discourage it as a misleading and imperfect guide to that true nature of things which is to be found only by pure thought. But before endorsing such a negative view too enthusiastically, it is interesting to look a little more closely at Platonic Forms to see if we may have missed something rather subtle in Plato's own conception of them. Look at the following dialogue of Plato's between Socrates and Glaucon from *The Republic* (528 E ff. Cornford's trans.), which clearly states his views about the relative merits of theoretical and experimental knowledge. The discussion is initially about astronomy:

And now, Socrates, I will praise astronomy on your own principles. . . . Anyone can see that this subject forces the mind to look upwards, away from the world of ours to higher things.

Anyone except me, perhaps, I replied. I do not agree.

Why not?

As it is now handled by those who are trying to lead us up to philosophy, I think it simply turns the mind's eye downwards.

What do you mean?

You put a too generous construction on the study of 'higher things'. Apparently you would think a man who threw his head back to contemplate the decorations on a ceiling was using his reason, not his eyes, to gain knowledge. Perhaps you are right and my notion is foolish; but I cannot think of any study as making the mind look upwards, except one which has to do with unseen reality. No one, I should say, can ever gain knowledge of any sensible object by gaping upwards any more than by shutting his eyes and searching for it on the ground, because there is no knowledge of sensible things. His mind will be looking downwards, though he may pursue his studies lying on his back or floating on the sea.

I deserve to be rebuked, he answered. But how did you mean the study of astronomy to be reformed, so as to serve our purposes?

In this way. These intricate traceries in the sky are, no doubt, the loveliest and most perfect of material things, but still part of the visible world, and therefore they fall far short of the true realities—the real relative velocities, in the world of pure number and all geometrical figures, of the movements which carry round the bodies involved in them. These, you will agree, can be conceived by reason and thought, not seen by the eye.

Exactly.

Accordingly, we must use the embroidered heaven as a model to illustrate our study of these realities, just as one might use diagrams exquisitely drawn by some consummate artist like Daedalus. An expert in geometry, meeting with such designs, would admire their finished workmanship, but he would think it absurd to study them in all earnest with the expectation of finding in their proportions the exact ratio of any one number to another.

Of course it would be absurd.

The genuine astronomer, then, will look at the motions of the stars with the same feelings. He will admit that the sky with all that it contains has been framed by its artificer with the highest perfection of which such works are capable. But when it comes to the proportions of day to night, of day to night to month, of month to year, and of the periods of others stars to Sun and Moon and to one another, he will think it absurd to believe that these visible material things go on forever without change or the slightest deviation, and to spend all his pains on trying to find exact truth in them.

Now you say so, I agree.

If we mean, then, to turn the soul's native intelligence to its proper use by a genuine study of astronomy, we shall proceed, as we do in geometry, by means of problems, and leave the starry heavens alone.

That will make the astronomer's labour many times greater than it is now . . .

Only we must constantly hold by our own principle, not to let our pupils take up any study in an imperfect form, stopping short of that higher region to which all studies should attain, as we said just now in speaking of astronomy. As you will know, the students of harmony make the same sort of mistake as the astronomers: they waste their time in measuring audible concords and sounds one against another . . . I am thinking of those Pythagoreans whom we were going to consult about harmony. They are just like the astronomers—intent upon the numerical properties embodied in these audible consonances: they do not rise to the level of formulating problems and inquiring which numbers are inherently consonant and which are not, and for what reasons.

Here we find expression of the conception that there exist three types of knowledge about the world. At the one extreme there is the most mundane knowledge; that is, the amassing of bare facts, catalogues of astronomical observations. At the other is the contemplation of the perfect Ideal Forms behind the world. In between there is something else, the description and representation of the bare facts of astronomical observation using arithmetic and geometry. This can act as an aid to transport us from the undigested listings of observations to the Ideal Forms which they imperfectly reflect. This middle course of using mathematics to arrive at working laws was an 'instrumental' one. It was useful, but by no means unique.

In the opening exchanges Socrates takes the expected Platonic position, insisting that the mere observation of extraterrestrial phenomena is not the way to a deeper understanding of reality. Such a depth is not to be found in observations of concrete things by means of our senses, but only through theoretical contemplation. The Ideal Forms which he is urging the astronomer to pursue could represent what we would now call the laws of Nature, or they might be envisaged as what the modern physicist would term an 'ideal gas', a 'frictionless surface', or a 'perfectly inelastic collision'. That is, they are idealizations, known to differ from what is observed, but the differences are essentially harmless; the 'ideal gas' captures the main features of real gases, and allows simple theoretical rules to be deduced.

Plato's idealism regarding perfect Forms is linked to a type of instrumentalism. He suggests that astronomers should devise geometrical models which 'save the appearances', that is which describe what is seen, but make no claim to be representations of the underlying reality. Later, this programme was carried out by Ptolemy (AD 65–165), who adopted the anti-realist Platonic view that his descriptions of the heavenly motions were useful devices for summarizing what was seen. Various different descriptions could achieve this, but the simplest was naturally chosen as the most useful in applications.

The immediate scientific consequence of the doctrine of Ideal Forms, or Platonic Idealism as it became known, was to reinforce the neglect of applied science, and turn attention towards the mystical search for what things really 'are' rather than how they behave. Even if it did possess the vision of laws of Nature as a greater truth than the individual instances of them, it placed these laws outside of the world of observation and measurement.

To conclude our brief discussion of Plato's influence upon the search for the blueprints of Nature, we should probably agree that the immediate consequences of his introduction of the doctrine of Forms were disastrous for the development of scientific inquiry. Observed phenomena were regarded as misleading. The ultimate reality was unobservable and immaterial. Undoubtedly there were benefits in such a dramatic development in abstract thinking, and the difference between reality and appearance would eventually be important, but this could not come about until there had been a period of detailed information-gathering. Plato was to be succeeded by a philosopher with just such a practical bent.

Aristotle

> We are all inclined to direct our inquiry not by the matter itself, but by the views of our opponents; and, even when interrogating oneself, one pushes the inquiry only to the point at which one can no longer find objections.
>
> Aristotle

Aristotle was born in 384 BC in the city of Stagira, an Ionian colony on the northern Aegean coastline. His father was physician to the court of King Amyntas II of Macedonia, and although both he and his wife died while Aristotle was still a boy, this childhood contact with practical scientific matters seems to have left an indelible imprint upon him. At 17 he entered the Academy as a student of Plato, and began his studies by composing dialogues in the Platonic mould, rather as modern research students might begin their apprenticeship in research by working out some practice problem or a simple generalization of a problem solved first by their research supervisor. After Plato's death he left Athens, probably because he disliked the growing trend towards making philosophical thinking about Nature into a branch of geometry. He did not believe that it was adequate for science merely to 'save the appearances' by giving *some* helpful geometrical account of what was seen. He was a realist who wanted to see theories of Nature taking their place within a wider philosophical scheme of things.

During this self-imposed exile his later interests in flora and fauna began to germinate, but his most famous experience was as tutor to the young son of Philip of Macedonia—a teenager who was one day to become known as Alexander the Great. When Alexander took power on his father's death in 335 BC, Aristotle returned to Athens, and founded a new school of study that became known as the

Lyceum. Within this ancient university he set about the detailed study of everything there was to study. The Lyceum continued in Athens and Alexandria for around 600 years, while its prototype, Plato's Academy, remained in Athens for more than 900 years. The names of both institutions have coloured education ever since, and the world still abounds with 'academies' and '*lycées*'.

Modern scientists have acquired an extremely negative attitude towards Aristotle and his work. Much of this has arisen because he has been encountered, not as Aristotle writing in his own words, but through the dark glass of 'Aristotelianism', that labyrinth of thought developed and propagated by the medieval Scholastics. It is this dogmatic Aristotelianism that was to draw such a contemptuous reaction from the liberated scientists of the Renaissance. Yet, despite all their problems with the institutionalized views of their day, Rennaisance scientists like Galileo had the greatest admiration for Aristotle himself as both a philosopher and a scientist. Even in the nineteenth century Charles Darwin regarded him as still the greatest of all naturalists. Undeniably, he has turned out to be the most influential thinker who has ever lived.

Aristotle's influence upon the idea of laws of Nature is slightly ambiguous. On the one hand he effected a vital change of emphasis towards empiricism after Plato's idealist influence, whilst on the other he introduced a barren way of thinking about the causes of natural phenomena that was to be an impediment to the right understanding of Nature for nearly two thousand years. His writings were to become the supreme and final authority in Western thinking until the Renaissance, and were invested with a dogmatism that he himself, judging by his continually evolving ideas, would have surely found repellent.

Whereas Plato had turned his gaze away from mundane, everyday phenomena, and directed the attentions of his followers towards the imaginary world of perfect Forms, Aristotle was a man of this world. An ardent realist, he was interested above all in the detailed observation of the natural world, and organized the work of his school accordingly. Botany, biology, geology, astronomy: all these subjects were studied systematically and in great detail. Animals were dissected, and specimens collected and catalogued. Aristotle was engaged in some of these activities personally, whilst others he placed in the hands of his most able colleagues.

To establish these practical studies as important directions of enquiry, and to persuade his ablest colleagues to engage in such work was quite an achievement. He had inherited individuals steeped in the Platonic prejudice against looking at material things at all, let alone the study of bugs and beetles. Young students dreaming of putting forward new and grandiose speculations about the ultimate nature of the unseen perfection behind the Universe, now found themselves consigned to digging up worms and studying the internal organs of dead dogs. Indeed, how many professors of philosophy would find such a redirection of their activities appealing even today?

This complete change of emphasis away from abstract, unfounded, theoretical speculation about unknowable things towards the down-to-earth investigation of observed things was a remarkable and vital achievement. Aristotle realized that there was a need for speculations about Nature to be judged by some criterion more rigorous than our ability to defend their logical consistency in philosophical debate with our opponents.

So, given Aristotle's passion for observing natural phenomena, his disdain for the fantastic philosophy of the Platonists, and his objections to the mystic geometrization of Nature by the Pythagoreans, why was there not a scientific revolution?

Aristotle's work emphasized the collection and classification of data, but he was not interested in relating facts together in the way that a modern scientist would seek to do. Moreover, there was no attempt to manipulate Nature artificially by carrying out experiments to test particular hypotheses. Aristotle was not looking for laws of Nature that codified changes or recognized regularities in the chains of cause and effect. This blind spot was not an accidental omission, but an inevitable consequence of his wider views about the nature of things, and the laws that governed their behaviour.

Whereas Plato had divided things into the appearances and the other-worldly Forms, Aristotle regarded only the observable things of this world as real and worthy of study. But he distinguished between two aspects of the things we see and touch. These he called 'Form' and 'Substance'. He believed, like Plato, that there were timeless, unchanging Forms, but they were intrinsic properties of the objects in this world, not abstract templates lodged in some great design-shop in the sky. 'Substance' was matter, the stuff that we touch and manipulate, but 'Form' was the specific realization it had been given, and our minds can elicit the Form within any observable object by abstract reasoning. Form was the thing into which the Substance was fashioned. The difference is clear to us: the Form cannot exist without the Substance that is necessary to express it, and any Substance must have some Form—although that Form could change with the passage of time.

Suppose that we construct a shed in our garden. The Substance may be wood (the word Aristotle uses for 'substance' actually means 'timber', the usual building material of that time), and will remain so, but the shape and design that distinguishes the end-result from a pile of logs is its Form. This Form may change with time as wind, rain, and rot alter the appearance of the shed.

This idea of Form has in some ways influenced our own language, for when we speak of 'informing' someone we mean that we wish to transfer some abstract idea into their mind, and we call a collection of such ideas 'information'. This information can be stored on different substances today—in the neuro-circuits of our human memories, on paper, on magnetic tape, or on a blackboard. Unlike Plato's abstract Ideal Forms, Aristotle's Forms were *in* things. This way of thinking led to his distinctive ideas about the causes of natural phenomena. Roughly speaking, Aristotle viewed causes and effects as inherent properties of the things themselves, rather than as relationships between events as would modern scientists. So, for instance, he would view a lunar eclipse as an attribute of the moon rather than as a consequence of its motion. From this basis he proceeded to distinguish *four* distinct 'Causes' of things.

The rival Atomist school of thought claimed that a thing was completely explained when one knew what it was made of. Against this Aristotle argued that such information about its constitution specifies only its Substance, and so only provides us with what he termed its 'Material Cause'—the wood of which my shed is made. In order to understand a thing completely we need to know three further Causes. A 'Formal Cause' must be identified. This specifies the Form or design it possesses. Next, we seek its 'Efficient Cause': to determine the agent which directly produces the object by embodying the Form in some material substance—the builder of my shed. This 'Efficient Cause' is closest to what modern scientists would call a 'cause'. Finally, there existed what to Aristotle was the most important cause, the 'Final Cause'—the purpose for which the object exists. From the idea of Final Causes there arose a teleological view that events evolved necessarily towards some goal or purpose, as if they were magnetically attracted by it. This idea was to create a world-view that dominated the West for more than a thousand years.

Broadly speaking, the Greeks viewed the Universe as a living organism rather than as a mechanism like a watch. This has much to do with Aristotle's identification of Final Causes, for living beings do indulge in purposeful behaviour, and if one believes the World to work by analogy with a living organism then one will arrive at a belief in natural events having purpose. Regularities in the mechanical and inorganic world do possess the clear feature that past causes determine future effects: we kick a stone, and it then moves. But in the realm of living things the relationship is superficially quite the opposite. The future seems to determine the present: we do things with some future purpose in view, squirrels gather nuts in order to prepare for their winter seclusion. This teleological aspect of the natural world, and Aristotle's particular interest in animal behaviour led to the incorporation of Final Causes into his fourfold scheme of causes in a leading way.

What you conclude from this view of the world will depend upon what you come to think about the 'end' to which events are being purposefully directed. Aristotle characterized this 'end', not just by what happens last, but as a perfect state possessing unique harmony. It is such a state that things naturally try to attain, and all motion takes place in accord with this tendency.

This view was to prove a major stumbling-block for scientific progress. The idea that the Universe should be viewed as a living organism rather than as a machine (as it tends to be in modern times) is a strange one, but hardly an unexpected one amongst thinkers preoccupied with studying living organisms for the first time within a culture that had no experience of technology and machines. It illustrates what has always been the major source of erroneous scientific reasoning: not so much incorrect data or wrong theories, but the mistaken belief that an entire body of ideas that apply successfully to one phenomenon can be taken over and used to describe another quite different one—the false analogy.

The introduction of Final Causes set an authoritative seal of approval on a way of thinking that had existed sporadically before Aristotle, but which was to dominate later Western thinking. If Nature does work towards some purpose, then what better purpose than the good of Mankind? From these beginnings there developed the tradition of 'Design Arguments': explanations that things were as they were because they were constructed by the Deity either for the benefit of Man or to attain the most perfect harmony. The laws of Nature existed to make the Universe a fit place for Man to live in. They had their particular forms because they were goal-directed. Subsequently, those who sought to use Aristotle's philosophical scheme for theological purposes usually added an additional 'First Cause' to the four Aristotelian 'Causes'. This would be the Creator or (less grandiosely) the initiator of the garden shed in our earlier example. One can see how appealing the symmetry of First and Final Causes might be to a theology that speaks of God as Alpha and Omega. The mystical ideas of Teilhard de Chardin, which aroused a surprising amount of interest in recent times, described a teleological cosmological view in which evolution proceeded towards a Final Cause which he termed the Omega Point.

The introduction of the idea of Final Causes was disastrous for the immediate development of science in the modern sense. A healthier outcome in the short term would have required attention to be focused upon trends and regularities in the character of Efficient Causes. Although the early Atomists did confine their interest solely to Efficient Causes, their views were not influential until they were revived long afterwards. The future synthesis of Aristotle's philosophy with the Thomist Roman Catholic tradition was to stress the barren aspect of Final Causation as being a sufficient explanation for what was seen. This doctrine became a celebrated argument for the existence of God, and grew extremely anthropocentric in its emphasis, regarding the well-being of Man as the goal of all natural phenomena, and basically *the* law of Nature.

Yet Aristotle was not so naïve in his original formulation of the idea. He was neither anthropocentric nor animistic in his idea of Final Causes. Although a stone tends to fall towards the Earth, it does not *desire* to do this. A Final Cause has been introduced because Aristotle believed that if only the other three causes were invoked to describe motions, then this left natural processes crucially undetermined: anything could still happen. This ambiguity is avoided by linking every Efficient Cause with some Final Cause. In his astronomical studies of celestial motions the Final Cause of the motion really amounts to what we would call a Law of Nature, in the sense that if the state of motion at one particular time is known, together with the Final Cause, then this determines how it will move in the future. The actual future motion is completely specified. At first, it appears that Efficient Causes have become laws of motion while Final Causes act as boundary conditions to make the motion unique (although final rather than the now conventional initial boundary conditions), but it is the Final Causes that have in modern times become our Laws of Nature, while Efficient Causes have turned into the natural forces which those laws describe.

With the benefit of hindsight it is also clear to us that, although Aristotle's system of various causes does clarify different abstract aspects of things in general, it is something of an obfuscation when we focus upon things in particular. There is no end to the list of indirect 'causes' that we could associate with the building of my shed. The lumberjacks who hewed the trees, the natural environment that allowed the trees to grow, the person who first conceived the idea of a wooden shelter: the list is endless. There are too many possible causes if one is interested in 'why' things happen rather than simply 'how'. It was by a considerable restriction of objectives that the scientific method eventually became both fruitful and unambiguous.

Aristotle's laws of motion

> A given weight moves a given distance in a given time; a weight
> which is heavier moves the same distance in less time, the time
> being inversely proportional to the weights.   Aristotle

Aristotle formulated laws of motion in words rather than mathematical symbols (that did not become common practice until after Galileo's time); but it is useful to express them symbolically. He claimed that the force (F) exerted on a body is proportional to the resistance (R) opposing its motion, and the speed (V) with which it moves is also proportional to that impressed force, F, so long as the force causing motion exceeds R. If it does not then there is no motion. Hence, algebraically,

$$V = kF/R \quad \text{if } F > R$$
$$V = 0 \qquad \text{if } F \le R \tag{2.1}$$

where k is the constant of proportionality. In addition, for bodies falling under gravity, the speed attained is proportional to the body's weight (W), but inversely proportional to the air resistance, so

$$V = kW/R. \tag{2.2}$$

When cast in this form Aristotle's 'laws of motion' appear to be scientific laws in the modern sense, but this is a misleading view. They were really mere descriptions of how moving bodies fitted in with the doctrine of Final Causation. The freely falling bodies fell to earth because that was the natural thing for them to do. Such reasoning of course explains nothing. It merely translates mysteries into other forms of words. The laws of motion described above were never used in any way to learn more about how bodies move, nor were they subject to experimental check. Had they been checked against what really happened discrepancies would have emerged. These laws are not in any way strange. They are what one might expect clever sixth-formers to arrive at if they had never encountered any other ideas about laws of motion than common sense and everyday experience. Some years ago an American physicist published the results of a survey he had carried out among American college students who had received no formal education in physics. His questionnaire sought to determine their intuitive beliefs, based on everyday experience, about the motion of bodies subject to forces. The consensus revealed a remarkable accord with the views of Aristotle, views that we would rightly regard as the common-sense approach. Modern science is not, of course, founded upon the common-sense view, primarily because it is above all seeking to refute, criticize, or improve the current view. The 'common-sense' view of something is a description that crystallizes from what is already known, and implies a certain unwillingness to admit any change in it, the implication being that any deviation from it would be uncommonly senseless. Superficially, it seems eminently reasonable to suppose that if you place a ball on the table, and push it gently, then at first it does not move—there is a resistance to motion—but that suddenly, when the force you are exerting becomes greater than that critical resistance, it will begin to move. The harder you subsequently push it, the faster it should move.

Nevertheless, a closer look at the law given by formulae (2.1) displays an awkward hiatus. If F starts with a value larger than R, and is steadily reduced, then V is predicted to fall continuously at first; but according to (2.1), when F reaches R it must jump instantaneously from the value k to zero. There exists a discontinuity (see Figure 2.1). This fact was regarded as problematic by some medieval commentators, and in the fourteenth century Thomas Bradwardine of Oxford suggested it be revised to a smooth form that we would now express as

$$V = \log(F/R). \tag{2.3}$$

This has the advantage of a continuous limit with $V \to 0$ as $F \to R$ (the mathematical notion of a logarithm was not formally introduced until the seventeenth century—Bradwardine used geometrical arguments to arrive at a result equivalent to (2.3)); see Figure 2.1.

A sixth-century commentator, Johannus Philoponus of Alexandria, had earlier suggested a revision of (2.1) which also overcomes continuity problems by replacing it with

$$V = F - R \tag{2.4}$$

so again $V \to 0$ as $F \to R$.

A further awkward prediction of Aristotle's law (2.1), and also of Bradwardine's revision (2.3) but not of Philoponus', is that the velocity V of a body should

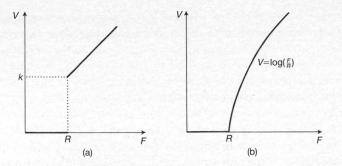

Figure 2.1. (a) Graph of Aristotle's law of motion giving the velocity attained under the action of an external force F and subject to a resistive force R. There is an unrealistic discontinuity as F approaches the value R from opposite sides. (b) An attempt to eliminate the discontinuity with the logarithmic law of motion proposed by Thomas Bradwardine in the fourteenth century.

become infinitely large as the resistance is reduced to zero. Aristotle was aware of this unphysical consequence, but used it as a proof that there could not exist any motion in a vacuum (in which one clearly has $R=0$), and so deduced that the vacuum did not exist. By this argument he opposed rival atomist philosophical schemes in which motion took place through the vacuum of the 'void'.

Following the Aristotelian period, and the continuation of the traditions of the Academy and the Lyceum, there were many great scientists in the Greek world. Men like Archimedes, Aristarchus, and Ptolemy must be numbered amongst the greatest minds of all time. Some, like Archimedes, did work of a technological and experimental nature, yet they regarded the exercise of pure thought and the mathematical aspects of reasoning as superior to their practical work. As a result, they did not influence the search for laws of Nature in the next historical period. It was the teleological Aristotelian perspective that was to dominate thinking about the natural world for more than a millenium.

The Aristotelian legacy

> It would be very singular that all Nature, all the planets, should
> obey eternal laws, and that there should be a little animal, five
> feet high, who, in contempt of these laws, could act as he pleased,
> solely according to his caprice. Voltaire

If Aristotle had been merely a giant amongst many giants then his scientific reputation today would undoubtedly be the greater. His massive contributions to knowledge and logic should have been seized upon by his successors, and used as starting-points for future research. From this stimulus could have flowed a gradual extension of knowledge and thinking about causes. Had it done so we would undoubtedly herald Aristotle as the innovator who started the ball rolling.

Unfortunately, no momentum of this sort was subsequently generated. Although two of Aristotle's successors as directors of the Lyceum were critical of the unfettered use of Final Causes to explain why things happen in the way that is observed, neither Theophrastus nor Strato's views were widely accepted. Thus, Aristotle's teleological ideas were able to fossilize into a relic that was used as the template against which new ideas were tested.

The template came to be venerated as perfect. For a thousand years after his death Aristotle's written opinion on any matter was taken as perfectly conclusive. His logical writings were compiled by medieval scholars into a canon known as the 'Organum' or 'Instrument', because of their perceived character as the key to all knowledge. In 1620 when Francis Bacon, the soon-to-be-impeached Lord Chancellor of James I, wrote an influential tract on the experimental scientific method, he entitled it the *Novum Organum* (the 'New Instrument') to emphasize its alleged superiority over Aristotelian dogma.

Both the followers of Christ and Mohammed had converted Aristotle to their respective faiths with equal ease. Unfortunately, the aspect of Aristotle's thinking that appears to have attracted the theologians was the doctrine of Final Causes. It was this misguided emphasis, together with the picture of the Universe as a living thing rather than a mechanism, that dominated institutional thought during the Middle Ages. There gradually arose the convention that Final Causes were looked for and interpreted as evidences for the existence of a God or gods. The earliest well-developed work of this kind is to be found in Cicero's dialogues entitled *The Nature of the Gods*. It is also seen in a slightly different form throughout the writings of Galen, whose medical work was to acquire the same authority as Aristotle's work in other areas. The fact that human organs are either well-suited for the purpose to which they are put or exhibit exquisite engineering is taken as evidence that they were perfectly designed by the Deity for the role that they perform. It is then but a small step to argue that the existence of these beautiful examples of 'design' that we invariably witness in Nature—the intricacy of the human eye, the way in which animals seem tailor-made for their habitats, and so forth—is actually evidence for a Designer who is not only unique, but is that God in whom one believes for other reasons. Laws of Nature are anthropocentric, and observations are interpreted in their light rather than used to test their correctness.

The acquisition of such a way of thinking hampers all useful enquiry into the laws by which Nature operates, because it becomes so easy to produce 'explanations' which superficially appear very satisfying, but which upon closer analysis add nothing whatsoever to our understanding of the world. If the Sun shines because its purpose is to supply heat and warmth to Mankind, then no further inquiry into its nature seems necessary. This way of thinking is rather attractive to the philosophical mind that wishes to take on board Aristotle's general view of the nature of things, but does not want to be burdened with the mundane business of observing the natural world in detail. Even if one is naturally inclined to observe rather than theorize about the world in the Platonic manner, the bias towards interpreting the world as a manifestation of anthropocentric purpose colours every observation one makes. One searches for evidences of design, rather than documenting what is observed in a dispassionate and all-encompassing manner. Examples that appear contrary to one's prejudices about the purpose of natural things are conveniently ignored. No longer is the study of Nature a secular search for the truth; Nature possesses an interpretation again. The study of Nature ceases to be an investigation into the way the world works; it becomes a search for evidence to support your belief in why it works. 'How?' simply serves to confirm your belief in 'Why?'

Right up until medieval times the idea continued that material objects contained certain natural sympathies or antipathies that made them strive towards their natural places in Nature. Whereas today we stress the relationship *between* things as the expedient way to code how they behave, the medievals searched for the innate sympathies within things in order to elucidate the reason *why* they behave as they do.

By endowing every observational fact with a meaning and an interpretation this teleological outlook discouraged the pursuit of observation by a large body of ordinary investigators like those who today are responsible for the steady progress of the mainstream of scientific knowledge. Genius can be relied upon to evade the shackles of outmoded ideas and naïve prejudices, but it is not genius that carries through the great groundswell of a scientific progress; it simply initiates it.

Alongside the teleological but realist views of Aristotle there travelled the instrumental Platonic view that our mathematical laws of Nature were only useful descriptions. The instrumental view appears to have been dominant until the Renaissance. Ptolemy had regarded his astronomical picture as a useful fiction which enabled him to predict planetary positions, and the same view was drafted into the scholastic synthesis of Aristotle and Judaeo-Christianity by scholars like Maimonides and Thomas Aquinas who sought to reconcile their belief in Aristotelian physics and Ptolemaic astronomy. The latter incorporated 'imperfect' non-circular motions, and so had to be judged either as wrong or only an imperfect representation of the true astronomical system. The Copernican revolution was a revolution in part because it suggested that the heliocentric theory was not just a representation of the appearances, but a description of how things *really* are. We recall that Andreas Osiander's anonymous preface to Copernicus' work (added without his knowledge, and not seen by him when, on his deathbed, he received the first printed copy of *De Revolutionibus* on 24 May 1543) tried to offset the considerable pre-publication publicity by contradicting Copernicus' realist intentions, claiming that 'it is not necessary that these hypotheses should be true, or even probable; but it is enough if they provide a calculus which fits the observations'. Although contrary to Copernicus' intention, Osiander's evaluation is of some substance, if only because Copernicus incorporated a very large number of epicycles in his theory in order to 'save appearances' in the Ptolemaic manner. However, the Copernicans subsequently insisted upon the reality of the description of Nature. They needed to defend this view because they had for the first time used the exact quantitative laws of terrestrial physics to explain the motions of the heavenly bodies.

Laws and rules of nature

> One of the most curious and exasperating features of this whole magnificent movement is that none of its great representatives appears to have known with satisfying clarity just what he was doing or how he was doing it.
>
> E. Burtt

The explicit metaphor 'law of Nature' took a long time to emerge in scientific work. It was not an expression used regularly by the Greeks, and appears first in its modern guise in the optical studies of Roger Bacon (1210–92) when he speaks of laws of reflection and refraction, and of things 'not following the laws of Nature'. He also uses the term 'law' (*lex*) to describe regularities in Nature in much the same way that we do. But he does not take this idea from the religious notion of a single Divine Lawgiver, as one might expect. Rather, he sees many different rules for particular classes of phenomena. He does not advocate the more general idea of a single set of unifying laws of Nature with one source. Thus he uses the terms *lex* and *regula* when speaking of Nature. Whereas *lex* described legislation laid down by authority, *regula* (rule) was a guideline or a standard to judge things against, and this meaning has given rise to our expression 'as a rule', meaning 'usually'. Eventually the two terms would diverge in meaning, with *lex* coming to signify something inherent in things causing them to behave in particular ways, while *regula* became regularity, the property of events or sequences of instances of natural phenomena. For Bacon, his laws of light signified regularities of Nature, and not Divine decrees.

The more theological view that laws of Nature were Divine stipulations* arises

---

*It is interesting to recall the medieval view, well represented by Aquinas, which viewed the innate Aristotelian tendencies as aspects of the natural world which were providentially employed by God. However, in this cooperative enterprise their basic character was inviolate. According to this view, God's relationship with Nature is that of a partner rather than that of a sovereign as it becomes in the mechanical view when laws of Nature are imposed upon Nature from outside. The latter view took prior claim over the former following the condemnations of certain Aristotelian views by the Bishop of Paris, Étienne Tempier, in 1277. The condemnations focused upon those Aristotelian conceptions which appeared to limit the free choice of God in the ordering of the Universe—for example the prohibition on rectilinear motion of the Universe or the creation of a void—and opened the door to the mechanistic doctrine of freely imposed laws of Nature. Much later, this view was found conducive to the Reformation theologians, like Luther and Calvin, who laid great stress upon the sovereignty of God and the inflexible predestination of events ordained by God. The Reformers believed that no innate tendency of matter could determine the motion of things: only God could dictate such behaviour; and the laws of Nature were the expression of that dictatorial control.

more prominently among astronomers, who are not manipulating Nature to extract information but gazing at the heavens. Tycho Brahe (1546–1601) even claimed that 'the wondrous and perpetual laws of the celestial motions, so diverse and yet so harmonious, prove the existence of God'. A similar merger of the evidences of regularity in Nature with the biblical notion of a celestial Lawgiver underwriting the uniformity of Nature was adopted by Descartes and Kepler. But gradually these investigators found that they had little need of a view as to the origin of the regularities in Nature; given that they believed such regularities to be present, they made progress by observing and coding their observations in mathematical summaries without requiring a philosophy of science. After glimpsing the regularity behind the world, Kepler and Copernicus were motivated primarily by the desire to uncover more of it, not to interpret it or use it to support an extra-scientific philosophy. The meaning and the method of science were beginning to go their separate ways.

Kepler (1571–1630) was one of the first scientists to exploit a faith in the underlying simplicity and harmony of Nature to guide his thinking towards laws of Nature. An ardent realist, he believed that God had fashioned the Universe using certain archetypes which were also present in the mind of Man because he was made in God's image. This meant that it was reasonable for him to hope to understand Nature.

The idea that the laws of Nature that he and others uncovered were just useful temporary descriptions of no ultimate truth-value Kepler found unacceptable. He argued that there could not exist different, but equivalent, representations of the laws of Nature in this sense. If they were pursued far enough, and compared with enough observations, all except one would be revealed to differ with observation upon at least one crucial point. Thus one and only one of the so-called equivalent representations was the correct one. Although he had formulated his laws a posteriori from the facts, he believed that they could have been deduced a priori.

Kepler was also responsible for expressing laws of Nature as mathematical equations. So successful were these expressions that this subsequently became the *modus operandi* of science: the true laws of Nature were now invariably mathematical laws. The medieval view had begun by regarding physical events as symbolic, but it ended when events were replaced by mathematical symbols. Such symbols were universal. They allowed science to grow in different places and be able to communicate its findings unambiguously. Things were no longer considered explained if their teleological purpose could be uncovered: they were explained if they could be reduced to a single set of mathematical patterns. In time this would become the implicit definition of scientific explanation.

Kepler's formulation of mathematical laws founded upon observational data was not wholly satisfactory. His first law of planetary motion stated that planets move in elliptical orbits, but a law of this sort does not allow one to predict where a planet will be at some future time on the basis of the knowledge of its current position. Such a development required the wedding of Kepler's astronomical picture with a more powerful mathematical technique. The first steps in this direction were taken at Pisa by Kepler's brilliant correspondent—Galileo Galilei (1564–1642).

Galileo was not content to deduce laws of Nature from observation, but set about manipulating Nature in order to make the laws of Nature transparently evident. In order to capture nuances of the fall of bodies under gravity he rolled them down inclined planes so that the effects were slowed down and amenable to study. He was able to frame and test conjectures with great skill, and gradually built up a successful mathematical description of motions by a systematic experimental interrogation of Nature. He showed that Aristotle's laws of motion did not agree with experiment. This, Aristotle could have discovered; but what Galileo was able to do that Aristotle could never have brought himself to do was to conceive 'thought experiments' in which bodies fell with no resistance (in a vacuum!). By this idealization of a vacuum he isolated the essential features of the phenomenon of motion from the inessential. This habit of framing imaginary mental experiments shows the extent to which the new scientific method dictated the way in which scientific questions could now be decided. The result of these steps was, for the first time, a unified physics which was applied to all natural phenomena. It was the first science to 'look' and feel modern. Its underlying methodology was quite different from that of previous natural philosophy; Herbert Dingle judged that its signature

> . . . is that self-control—the voluntary restriction to the task of extending knowledge outwards from the observed to the unobserved instead of imposing imagined universal principles inwards on the world of observation—that is the essential hallmark of the man of science, distinguishing him most fundamentally from the scientific philosopher.

And, having achieved this and so much else, Galileo did something else distinctly modern: he set about writing 'popular' accounts of his discoveries for the general public.

Newton, the Newtonians, and Newtonianism

> Very few people read Newton, because it is necessary to be learned to understand him. But everybody talks about him.
>
> Voltaire

The most fascinating development in the advancement of our notion of laws of Nature came with the work of Isaac Newton (1642–1727). His contribution to what is known about Nature was the greatest that has ever been made by a single individual, but it is not for this reason alone that he is of interest to our story. Unlike other great scientists living shortly before him, Newton's work had a dramatic influence upon an entire culture. His genius was not met with the censure and persecution that greeted the outspoken and tempestuous Galileo; it was troubled only by constant personal arguments over priority, first with Hooke and then with Leibniz. It created a whole philosophy of Nature, and started the popularization of science for the general public in the English language. Surrounded by the favour of Church, Queen, and Government, Newton's pre-eminence became the centre around which the Royal Society re-established its fading scientific prominence and intellectual respectability.

Newton arrived at a propitious time. The Copernicans had set science upon a course that no longer saw Man as the focus of things. Galileo had developed the mathematical method to the extent that there now existed well-posed problems. Communications and nascent scientific societies fostered the exchange of ideas and information. The social respectability of science attracted the patronage of wealthy and influential figures. The rise of active experiment, rather than passive observation, had attracted craftsmen of great skill to the scientific enterprise. Instruments were designed and built for the sole purpose of observing the world in wider and finer detail. What the microscope and the telescope revealed fired curiosity beyond measure.

Whereas a practical genius of the Renaissance like Leonardo da Vinci displayed a vast diversity of interests, and catalogued and drew examples of everything he saw, Newton saw many things with a deep unifying understanding. He invariably recognized the essential common factor behind superficially different phenomena, and as a result there was henceforth to be a strong emphasis on the mathematical regularities of Nature rather than its eccentricities. This single-mindedness enabled Newton to isolate a collection of profound laws of Nature that survive today as an excellent approximation to the behaviour of bodies moving at speeds much less than that of light. So successful was the Newtonian theory of the world that realism appeared with a new force. It was widely believed that Newton had found the ultimate laws of the Creator.

Newton saw clearly that the Aristotelian heritage which had been bolstered by the nit-picking of the Scholastics was sterile. It was an argument about the innate properties of things which sought the reasons for these properties in the peculiar intrinsic strivings of the things themselves. Newton was interested in finding general rules which determined *how* things happened;* he was not interested in

*Here, there is a contrast with the traditional Aristotelian view that matter was steered by innate tendencies rather than by externally imposed laws. In 1693 Newton wrote explictly of this to Richard Bentley 'That Gravity should be innate, inherent and essential to Matter . . . is to me so great an Absurdity, that I believe no Man who has in philosophical Matters a competent Faculty of thinking, can ever fall into it.' Newton saw these external laws as imposed directly and completely by God just as society imposed laws upon its citizens. This gave rise to conflict with Leibniz who reiterated the Aristotelian view that matter possessed innate tendencies and led to Samuel Clarke's claim in correspondence with Leibniz that in the Newtonian picture there are 'no powers of nature independent of God'. We are so familiar with the concept of imposed laws (whether or not we appeal to God as the legislator) that it is easy to overlook the type of mental block there might have been to this idea. At a time when vitalist notions separated the living from the non-living world and the mind of man set him above everything in Nature, one would have to accept that God's laws could be 'understood' and responded to by inanimate objects by analogy with the human response to moral and social laws.

the insoluble problem of *why* they happened, because he believed that it was possible to say 'how' without any reference to the issue of 'why'. In the introduction to his *Principia* he writes that in the past philosophers were 'employed in giving names to things, and not into searching them out'. He goes on to establish a scientific method that is intended to rectify this imbalance. Of this method he claims: 'these Principles I consider not as occult Qualities, supposed to result from the specific Forms of Things, but as general Laws of Nature, by which the Things themselves are formed.'

Newton's method was not totally revolutionary, and we can be confident that he spelt out its basic axioms long after he had evolved it by intuitive use in solving problems. Nor is it enough to possess a correct philosophy of science in order to make scientific discoveries. As a prelude to considering Newton's laws of motion let us take a look at some of those devised by René Descartes (1596–1650), who died a few years after Newton's birth. These laws Descartes referred to as 'rules of Nature' in 1644, but after 1647 adopted the terminology 'laws of Nature'. They 'look' mathematical. They are derived from observation. They deny any system of final causes. None the less, they are incorrect. However good the scientific method of the scientist, it is still necessary for the world to be observed *correctly*.

The terminology 'laws of motion' seems to have become common following the papers of Descartes, Huygens, Wallis, and Wren on the behaviour of colliding objects. The law suggested by Descartes is interesting as an example of an incorrect law of motion that was to be superseded by Newton's. Descartes' law of motion has two parts, and predicts the results of a collision between two masses:

1. *If two bodies have equal mass and velocity before they collide then both will be reflected by the collision, and will retain the same speeds they possessed beforehand.*
2. *If two bodies have unequal masses, then upon collision the lighter body will be reflected and its new velocity become equal to that of the heavier one. The velocity of the heavier body remains unchanged.*

Descartes had derived both these laws on the basis of apparent symmetries and a notion that something must be conserved in the collision process. Unfortunately, Descartes' proposals possess the same defect that marred Aristotle's proposals: the problem of discontinuity. This was first pointed out by Leibniz.

Let us carry out one of the 'thought experiments' that Galileo made so fashionable. Imagine a sequence of different collisions involving separate pairs of masses, each approximating closer and closer to one involving one particular pair of masses. Then the effects of these collisions should also approach closer and closer to the effect of that particular example. We call such a property '*continuity*'. If $E(A)$ represents the effect of some cause, A, then continuity is the requirement that as the cause A gets closer and closer to some other cause B then $E(A)$ gets closer and closer to $E(B)$. It is clear that Descartes's laws of motion do not have this property. If $E(M)$ is the velocity of the first body, of mass M, after collision, and v its velocity beforehand, and we denote the mass of the second body by m, then Descartes' second law says that

$$E(M) = v \text{ whenever } M > m \tag{2.5}$$

and

$$E(M) = -v \text{ whenever } m \leq M \tag{2.6}$$

Now observe that if, in our sequence of thought experiments, we first reduce the magnitude of M so that it approaches closer and closer in value to m, but then increase it from values initially below m we shall arrive at a contradiction because (2.5) and (2.6) require that v gets closer and closer in value to $-v$, and this is only possible if v is equal to zero! Careful observation of collisions between balls of almost identical mass would have revealed the incorrectness of the Cartesian 'law' of collisions.

Newton observed the world both more carefully and more widely than did Descartes, and also had the advantage of Descartes's own wide insights before him. The laws of motion that Newton laid down were published in 1687, although they were worked out long before that. Unlike the present-day scientist, Newton was in no hurry to publish or otherwise announce his discoveries. Indeed, it may only be a consequence of prodding by his friends that much of his work became known at all. One certainly suspects that, faced with the prospect of the public and official opposition ranged against a Copernicus or a Galileo, Newton, 'with his prism and his silent face', might have taken many of his discoveries to the grave in secret. Fortunately he did not.

Newton's three momentous laws of motion were stated by him as follows:

1. *Every body continues in its state of rest, or uniform motion in a straight line, unless it is compelled to change that state by forces impressed upon it.*

2. *The change of motion is proportional to the motive force impressed; and is made in the direction of the straight line in which that force is impressed.*

3. *To every action there is always opposed an equal reaction; or, the mutual actions of two bodies upon each other are always equal, and oppositely directed.*

The most interesting of these statements is the first law. We normally encounter it at school in more familiar language (Newton wrote originally in Latin), as the statement that 'bodies acted upon by no forces remain at rest or move with constant velocity'. In an earlier draft Newton expressed it first as

if a quantity once move it will never rest unlesse hindered by some externall cause,

and then as

By its innate force alone a body will allways proceed uniformly in a straight line provided nothing hinders it, before its penultimate rendering as

> By reason of its innate force every body preserves in its state of rest or of moving uniformly in a straight line unless in so far as it is obliged to change its state by forces impressed upon it.

In fact, this law owed much to Descartes, who was the first to discard the ancient notion that motion was some type of process and recognize the sense in which the states of rest and steady motion were similar. In 1644 his famous *Principia Philosophiae* contained the predecessor of Newton's first law:

> If [a body] is at rest we do not believe it is ever set in motion, unless it is impelled thereto by some [external] cause. Nor that there is any more reason if it is moved, why we should think that it would ever of its own accord, and unimpeded by anything else, interrupt this motion.

Newton's deductions differed from Descartes's in the way he had arrived at them—by a variety of experiments and observations. In their support he cites observations of projectiles, tops, and planets. They also went farther because Newton recognized that whenever motion became non-uniform a force was acting. He also brings in the element of universality by his telling opening phrase, 'Every body'. We should not miss the intended contrast with Descartes's *Principia Philosophiae* when Newton calls his great work *Philosophiae naturalis principia mathematica*.

Newton's first law states that when there are no impressed forces there will be no accelerations. If you see a body moving with a changing speed or along a path that is not a straight line, then a net force is acting upon it. You recall how different was Aristotle's law of motion (2.1); Aristotle claimed that force led to *unaccelerated motion* on Earth, and he also maintained (but only for philosophical reasons) that the natural celestial motion was circular. Circular, rather than straight-line motion was the natural state of the Aristotelian celestial world. Rest was the natural state of motion of objects in the terrestrial world. Newton would have told Aristotle that force was the 'efficient cause' of acceleration.

What is so interesting about Newton's remarkable statement is that neither Descartes, nor Newton, nor anyone else had ever seen a body that is acted upon by *no* forces. Everything feels the force of gravitational attraction exerted by the other bodies in the Universe, and in any particular situation a body will usually feel all sorts of other inevitable forces as well. There is no known way of insulating bodies from all forces. We cannot just turn off the forces of Nature. Indeed, some of these forces hold solid bodies together. This means that Newton has done something far more sophisticated than his predecessors; he has not simply written down an empirical description of what is seen in Nature, because his first law describes a situation that has never been seen, nor ever will be. It is not a straightforward realist's discovery of what the world is like, because there are no objects in the world meeting the stipulations of the law; although Newton did believe he was describing an underlying reality, not just 'saving the appearances'. It is not an operational statement either: it does not tell us how we should measure forces or velocities. It seems closest to the spirit of the Platonic idealist. Newton is envisaging an ideal situation which he has conceived by his observation of large numbers of non-ideal situations. Modern physicists would call this a 'model'. He has approximated circumstances to those where there are opposing forces acting upon a body but they almost cancel each other out, and leave no net force acting upon the body, to a very high degree of approximation. His first law is a creation of the mind in that spectacular sense. It is an abstraction that captures the essential elements of the real. It involves an intuition as to what the various forces acting all are, and to endow them with equal status.

Later scientists were to follow this course on many occasions. The art of formulating good laws of Nature was to recognize which aspects of a situation were inessential. No statement of a law of Nature will be able to take into account all the observational factors involved in a particular natural phenomenon. It is too complicated. Rather, the mark of the potent scientist is to see through the inessentials to the dominant feature. But the law that results must be such that it would remain true if all the inessential neglected secondary features of the world were incorporated into our description. One is not ignoring these features because they are awkward counter-examples, but because they introduce no new points of principle. If their inclusion were to render the derived law false then they would no longer be inessential.

As Newtonian physics grew there appeared the notions of frictionless surfaces, ideal gases, perfect conductors of electricity, inelastic collisions, perfect insulators, perfect spheres, and so on. None of these entities exists in the real world, but a law is most usefully formulated in terms of what the behaviour of such 'ideal' objects would be if it was the limiting case of a continuous sequence of situations approximating more and more closely to that of the ideal one. Thus, we make use of the principle of continuity employed so tellingly against Descartes's law of motion to arrive at more general aspects of physical laws. The ideal object is regarded as the limit of a sequence of observable ones, rather than as an other-worldly blue-print. This approach maintains the divorce of physics and philosophy created by Descartes. Newton was never involved in debates about Platonic idealism. Laws of this kind allow one to assess the behaviour of real things by the extent of their deviation from the ideal behaviour. As the situation we wish to describe approximates more and more closely to the ideal one postulated by our law, we expect the behaviour of the real situation to approach increasingly closer to the behaviour stated in the law. In practice it is not possible for our measurements and observations to be perfectly accurate and so there is no way in which we can ever confirm whether a law is *almost* rather than *precisely* true.

Newton's second law of motion also evolved into its present form over many years. First, as

A body must move that way which is pressed,

then to

The alteration of motion is ever proportional to the force by which it is altered,

and penultimately

The change in the state of movement or rest [of a body] is proportional to the impressed force and takes place along the straight line in which the force is impressed.

Newton's third law (action equals reaction) is also an interesting departure from ancient prejudices. Forces are recognized as arising from the interaction of bodies. But there is no subjective decision as to what is 'really' acting on what: the door is knocking on my hand just as much as I am knocking on the door. No medieval or Aristotelian could have arrived at such a democratic view. The 'purpose' of the knocker, and the difference between the inanimate and the animate participant would have been pre-eminent. Newton does not deny that these factors exist, but he realizes that they are irrelevancies. The removal of intention and purpose from consideration in natural science was part of the ongoing Copernican revolution, with its removal of things human and things divine from the scientific method. We know of a preliminary formulation of the third law by Newton as follows:

As much as any body acts on another so much does it experience in reaction. Whatever presses or pulls another thing by this equally is pressed or pulled.

Newton was interested primarily in how things happened in Nature, not in philosophical arguments as to why they happened or for what ultimate purpose they were intended. Nevertheless, he did not regard these latter issues as meaningless or irrelevant; indeed, he spent the greater part of his time pursuing strange metaphysical and theological interests*. Newton regarded these matters as questions whose answers could not be obtained by the careful observation and experimental interrogation of Nature. Judging by his alchemy and religious writings, which considerably outnumber those concerned with scientific matters, he clearly believed that some of these questions were answerable by other methods.

Newton considerably narrowed the objectives of natural philosophers, but he did so to great advantage, for he was able for the first time to confine their attention to the realm of *soluble* problems. By so narrowing his objectives he was able to make the remarkable claim that the uncertain speculations so beloved of

---

*It is amusing to consider why one never finds Newton ridiculed for calculating and publishing a date of 3988 BC for the creation of the world, in the light of the derision that Bishop James Ussher (a contemporary of Newton's) still seems to attract with a date of 4004 BC!

his predecessors, 'hypotheses' as he termed them, were unnecessary to his enterprise. There was a foolproof way of deriving laws of Nature, or scientific theories, from experience alone: experimental investigation. This scientific method is the one we still employ today. It seems so obvious a way to proceed that it is hard for us to see how anyone could ever have thought differently—but they did. The Aristotelian tradition had encouraged observation but not experiment. When Galileo's opponents were confronted with the evidence of observations made with his telescope, they maintained at first that they were unreliable because of the distortion of reality created by lenses and other observing instruments. The idea of manipulating events to extract information about the world was a foreign idea to the ancients. It grew up in Europe through the influence of craftsmen and engineers as much as through that of natural philosophers. Newton's methods allowed him to investigate the behaviour of phenomena by carrying out controlled experiments in order to test his ideas systematically. Many of Newton's contemporaries were still sceptical of determining the nature of things from the data of practical experience rather than from some all-embracing philosophical principle. Their doubts were not entirely without foundation. They believed, as classical ideas about the purpose of things would tend to encourage one to believe, that there was no unique interpretation of the observations one made. This encouraged a manifold of speculation with no motivation to determine a unique and correct explanation. In fact, Newton was not opposed to interpretations or speculations as such, but only to those who would use such hypotheses as an excuse for not carrying out real experimentation when it was possible. His distaste for 'hypotheses' is the natural reaction of a man in possession of a far superior instrument for winnowing truth from error. From now on, general philosophical statements could no longer assume the status of laws of Nature. Only those statements that had passed the test of experimental check against the manifest facts of experience would receive this accolade. Newton completed the divide between the meaning and the method of science.

The distinction between Newtonian science and 'hypotheses' was clearly delineated by his second great innovation: the use of mathematics. This was the method by which new possibilities would now be generated for test by experiment. Newtonian physics was mathematical physics. In all his investigations, Newton sought to express laws of Nature in mathematical language, so that they would be completely unambiguous. When the necessary mathematics did not really exist he developed it himself. In this way some of the most important mathematical tools ever invented were introduced into the physicists' armoury. The most powerful was the calculus, by which continuous changes could be described by mathematics. Given the starting configuration, the future could be predicted or the past reconstructed. This step completed a gradual change of emphasis in scientists' thinking about the ultimate nature of the world they were studying.

Amongst Newton's many achievements there is one which, in retrospect, stands out as being of the greatest importance for the future course of thinking about the laws of Nature: the law of gravitation. In this development he made good use of important earlier discoveries about celestial motions made by Kepler and Huygens. He was aware of a law of this type in the mid-1660s. Of this creative period he wrote nearly half a century later that, 'I deduced that the forces which keep the Planets in their Orbs must [be] reciprocally as the squares of their distances from the centers about which they revolve'. However, the first written version of Newton's law of gravitation that has been found is that contained in an unpublished and untitled manuscript now given the title *On Circular Motion*, with a probable date of 1665. It was this manuscript that Newton cited in his correspondence with Halley as evidence of his priority over Hooke with respect to the discovery of the inverse-square law. There the 23 year-old Newton deduced the inverse-square law from Kepler's third 'law' of planetary motion and the definition of the centripetal force required to sustain circular motion, and concluded that,

> Since in the primary planets the cubes of their distances from the Sun are reciprocally as the squares of the number of revolutions in a given time (Kepler's third law): the endeavours of receding from the Sun will be reciprocally as the squares of distances from the Sun.

Of course, this demonstrated the proposition only for the special case of *circular* motion. Later, in Book I of the *Principia* Newton would generalize this to the pertinent case of motion along orbits described by conics. Moreover, the inverse-square dependence on distance is not the whole story; the full law of gravitation contains a direct proportional dependence upon the masses of the attracting bodies. The final product is very simply stated: given two bodies with masses M and m whose centres are separated in space by a distance r then the attractive force of gravity between them, F, is proportional to the product of their masses and inversely proportional to the square of the separation, r, of their centres in space, and acts along the line joining those centres,

$$F \propto Mm/r^2. \qquad (2.7)$$

Newton's description, in Book III of the *Principia*, is that gravity acts

> according to the quantity of the solid matter which they [the sun and the planets] contain, and propagates its virtue on all sides to immense distances, decreasing always as the inverse square of the distances.

The important point of principle that was to flow from this generalization about Nature was the constant of proportionality implicit in the formula (2.7). Despite the fact that it does not occur in the *Principia*, and only makes its explicit appearance into the literature at the hand of Laplace in the eighteenth century, it has ever after been known as Newton's gravitational constant, and later denoted by the symbol G by Laplace, thus

$$F = GMm/r^2 \tag{2.8}$$

By introducing this constant element in a mathematical formula, Newton took a dramatic step forward. He claimed that all bodies, be they celestial or terrestrial, were subject to this selfsame law of Nature. They all felt the same intrinsic strength of the gravitational force, no matter what their separations or masses.

By his unification of all the effects of gravity Newton was responsible for identifying 'G' as the first *constant of Nature* in a law of Nature. It still holds such a status today. Even though Newton's theory of gravitational forces was subsumed within Einstein's theory of general relativity in 1915, this new theory still retains Newton's 'constant' G in a similar role.

What do we mean when we say that a quantity is a *constant of Nature?* That it is independent of the identity of the bodies whose masses are M and m, of all their physical properties, other physical conditions—the time and place of measurement, the temperature of the laboratory, and so on. The identification of such quantities was to set science upon a fruitful course that has culminated in a perspective in which a large part of our picture of the structure of the Universe is contained in the numerical values of a small number of fundamental constants of Nature, which still includes Newton's G. It is the set of particular values which these quantities are observed to take that gives our Universe its particular physical qualities. As yet we do not know why these constants of Nature possess the particular numerical values that they do. Of this we shall have more to say in later chapters.

Newton's introduction of a universal constant of gravitation in (2.8) highlights another important point. The proportional dependence upon the masses and the separations is a relationship that can be determined by theoretical reasoning from more elementary principles, but the value of the constant of proportionality can be obtained only by observation. This division of laws of Nature into functional dependencies between quantities (here it is between force, mass, and distance) and universal proportionality constants is a continuing feature of science to which we shall return. Suffice it to say that the goal of physicists is to reduce to an absolute minimum the number of constants of Nature whose values need ultimately to be determined by observation in their theories. This economy drive is pursued by trying to show that what we previously thought to be independent constants of Nature are in fact related either to each other or to more basic variable quantities.

To the ancients the world was a living organism, but to Newton and his followers it was a unified mechanism—like the interior of a giant watch. Its workings were pristine: precise, mechanical, and mathematical. Once set in motion by the Creator they continued by their own inexorable internal logic. Where the God of the Scholastics was Omega, the ultimate Final Cause; the God of Newton was Alpha, the pre-existent First Cause of all things. This strong belief in the deterministic character of the world owed much to the religious thinking of the times. The Christian view of the one God who laid down the laws of Nature, and 'upheld them by the word of his power', led to a strong faith in the rationality, consistency, and predictability of Nature. However, it should be recognized that these laws had been pursued and found at least in part because a belief in a law-giving Deity existed. Part of Newton's long-standing quarrel with Leibniz was associated with these ideas, which Leibniz thought blasphemous because they gave the Deity no scope to intervene in Nature after the initial act of Creation.

Newton's first universal description of a natural phenomenon was heralded by the supporters of his methods as the uncovering of a fundamental law of Nature—thinking one of the thoughts of God after him. It was the uncovering of such a universal property that gave support to religious apologists like William Whiston, who argued that it witnessed to the unity of Creation, and thereby to a unique Creator. Too remarkable to be viewed in any spirit other than the realist's, it became the mainspring of the new mechanical paradigm of a clockwork Universe running to definite mathematical rules. It saw the foundation of an approach to science that has subsequently characterized English-speaking scientists ever since: the emphasis upon reducing everything to a visualizable mechanical picture. The continental approach has traditionally been more abstract, not requiring a mental mechanical model of a complicated phenomenon before claiming it to be understood. An abstract mathematical description sufficed. The British passion for visualizable models was a psychological feature that would especially exasperate Henri Poincaré at the beginning of the twentieth century. As a distinguished representative of the French appetite for abstract thinking, he claimed that there was no need to reduce everything to the mechanical analogies so beloved of the empirical British scientists. This love of mechanical analogies as a device for explaining complicated abstract physical concepts is a virtue and a vice that still dominates the popularization of science in the English language. It is, in many ways, an inheritance of the Newtonian style: the deduction of great truths about the workings of the Universe using simple mechanical experiments, and the belief in the paradigm that Nature really is a great mechanism—and perhaps the suspicion that God, like Newton, is an Englishman!

Newton's approach to the investigation of Nature brought about the greatest single advance in human knowledge of the Universe's workings that has ever been effected by one man. But the publication of the monumental *Philosophiae naturalis principia mathematica* in 1687 and *Opticks* (in English) in 1704 led to more than a revolution in scientific thinking. It changed the thinking of non-scientists as well. The *Principia* became the first scientific 'cult' book (that is, a book that is read about, but not read), and it created what we might call 'Newtonianism'. This had many consequences, the most interesting of which was the start of the systematic popularization of science through the publication of elementary explanations designed for the lay-person. A vast number of such books were written in the first half of the eighteenth century to satisfy public interest in Newton and his discoveries. By the opening years of that century Newton had established a public reputation unequalled by any British scientist before or since. A Member of Parliament for Cambridge University in 1689 and 1701, in 1705 he was knighted by Queen Anne, the first man of science to be so honoured; the 'ornament of the age', he was President of the Royal Society from 1703 until his death in 1727, and for much of that time he was first Warden and then, from 1696 in an unprecedented succession, the successor to Thomas Neale as Master of the Royal Mint. Remarkably, this institution was running at a loss prior to Newton's stewardship! The practice of 'clipping' (cutting the edges off silver coins), although a capital offence, was widespread, and responsible for the Mint's principal losses since it was giving full exchange value to clipped coins, and so, in effect, giving silver away to unscrupulous members of the public. Newton put an end to this abuse by introducing the idea of milling the edges of coins. They would cease to be legal tender if the milling was found to be absent anywhere around the perimeter. This innovation has remained the Mint's practice ever since, and makes it doubly appropriate that Newton's face was, from 1978 until their withdrawal in 1985, to be found on the reverse of English one pound banknotes (unfortunately the diagram from the *Principia* alongside him is wrongly drawn—the Sun sits at the centre of the planetary ellipse rather than at its focus!). Newton's public service was well rewarded; at his death his personal assets totalled £32,000, a vast fortune in those days. The source of his wealth was twofold. He received a substantial salary of £600 per annum from the Mint, but as Warden he was entitled to a commision of 1s. 5½d. (about 7½ new pence) for every pound of silver that was minted, and after meeting the expenses of the coiners this still supplied an annual supplement to his income of over £1000.

This public side of his life made Newton an establishment figure, and towards the end of his life both he and the Royal Society in general were inevitably the subject of a good deal of satirical writing. Yet unlike so many men of genius, Newton was honoured and esteemed by his contemporaries during his own lifetime, and was buried in a central place of honour in Westminster Abbey along with Chaucer and Shakespeare. Throughout the eighteenth century Newton and his achievements were a subject of fascination amongst London society. His private manner and aloofness, coupled with his pre-eminent intellect, no doubt created an aura of other-worldliness around him. There were queries from awestruck continentals asking if Newton ate and slept like other men. Whereas the intellectuals of the past had arrived at grand conclusions concerning the nature of the Universe by complex and subtle verbal reasoning, Newton was novel in that his great insights were so often arrived at by means of very simple experiments employing mundane objects: the laws of optics from a simple prism, the laws of motion by dropping things to the ground (even apples, perhaps). The general public did not understand these discoveries (which only served to heighten their fascination; a similar effect occurred with Einstein, unlike the case

Figure 2.2 From gravity to levity: a cartoon by one of Newton's contemporaries satirizing his work on gravitation.

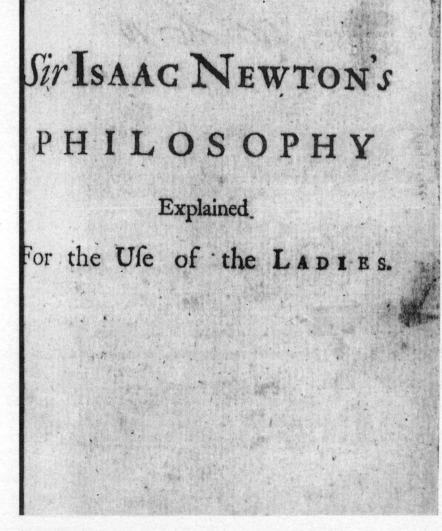

Figure 2.3 Frontispiece of Count Francesco Algarotti's *Sir Isaac Newton's Philosophy Explained for the Use of the Ladies*, translated into English by Elizabeth Carter in 1739. Subsequent editions appeared in 1742, 1752, 1765, and 1772. Voltaire and his mistress considered this popularization to be 'too frivolous, with too many jokes' for their liking. None the less it was translated into six languages and ran through thirty editions to become one of the main vehicles whereby Newton's ideas on light became known to general European audiences.

of Darwin where the public could understand only too well what he had discovered), but they wanted to know what was this new thing called 'Newtonianism'.

There were popular accounts of Newton's work for children, like Newberry's *Newtonian System of Philosophy, Adapted to the Capacities of Young Gentlemen and Ladies* (1761), and for women, who were not as yet regarded as being capable of real study. In the latter category one finds wonderfully incongruous titles like Count Francesco Algarotti's *Sir Isaac Newton's Philosophy for the Use of the Ladies* (1739), which was translated into six languages, or Benjamin Martin's *The Young Gentlemen and Ladies Philosophy* (1759). Beside these lowbrow popular accounts appeared expository works of the highest quality written by leading scientists both for other scientists and the educated public. The most accomplished was Colin Maclaurin's *Account of Sir Isaac Newton's Philosophical Discoveries* (1748), whose author was arguably the most able British scientist and mathematician of his day. On the less reliable side there were eccentric books like John Desaguliers' *The Newtonian System of the World, the Best Model of Government* (1728), attempting to apply Newton's methods wholesale to any and every area of life. All this adulation and public interest in Newton's laws of Nature and his means of discovering them was fuelled by one further important factor which united the interest of intellectuals and lay-persons—religion.

We have already discussed how Aristotle's ideas about final causes led to the development of an anthropocentric teleological interpretation of Nature, by which Man interpreted all good things around him as providentially made for him rather than simply fortuitously used by him. Newton's work gave a new impetus to the traditional Design Argument that the workings of Nature were so wonderfully contrived for our advantage that they must be the result of an all-embracing Divine plan. Whereas the earlier and cruder Design Arguments had argued that natural phenomena were optimal for Mankind's use, and therefore demonstrative of Divine purpose, the Newtonian Design argument appealed to the *universal laws* of Nature themselves as incontrovertible evidence that witnessed to the plan of Nature, and thence to the Planner behind it. The realist view that the Newtonian system of laws described the world as it truly is—a view that Newton himself would not have defended, on account of his provisional attitude towards action at a distance—and did not just provide a useful description, was a natural complement to a belief that the laws of Nature were the edicts of a Divine Lawgiver. The argument was enthusiastically inverted: the Newtonian watch-world required a watchmaker. The earlier Design Arguments were, of course, strongly associated with the Scholastic tradition of 'final causes' which had become anathema to the post-Reformation Protestant orthodoxy of Newton's time. Newton, along with his friend Samuel Clarke, appears to have entertained Unitarian beliefs concerning the Holy Trinity and other fundamental Christian doctrines. An Act of Parliament passed in 1689 debarred those holding such Arian* views from holding certain academic and public offices. Newton was only able to remain as Lucasian Professor at Cambridge because a special dispensation was granted in 1675 which permitted the holder of this chair to remain a Fellow of his college without taking holy orders in the Anglican Church. Newton does not seem to have held his minority religious opinions by default. He researched the subject passionately, tracing the historical basis of the creeds through the history of the early church, and was regarded as something of an authority upon these matters. His voluminous writings on biblical criticism show him to have been the first liberal textual critic. He attempted to lay down 'Rules' for the interpretation of scripture which mirrored those that he had proposed in *Principia* for philosophical reasoning. However, some of this work was regarded as completely bogus even by his otherwise admiring contemporaries and was not fully investigated by scholars until the twentieth century. With regard to Newton's religious views, a telling fact is Conduitt's record that Newton refused the sacrament of the church on his deathbed. Indeed, so shocking was this fact that the witnesses kept it a secret for fifty years after his death, and none of his early biographers record it. By contrast, William Stukeley records a deathbed scene which must have been partially invented by Conduitt and his wife, because it speaks of Newton's final moments as 'truly christian'.

Newton himself did not actively promulgate the Design Argument from the laws of Nature but he does remark, in the preamble to his *Principia*, that when composing it 'he had an eye upon arguments' for belief in a Deity, and he encouraged the use of his work by others who wished to support a Newtonian Design Argument. His famous correspondence with Bentley, which contains so many remarkable scientific insights, arose as a result of Bentley being chosen to deliver the first Boyle Lectures. This lectureship was endowed in the will of Robert Boyle with the express purpose of providing scientific Christian apologetics. Bentley's three sermons provide a classic statement of the Newtonian Design Argument based upon the existence of mathematical and universal laws of Nature, and were prepared with Newton acting as his scientific conscience. The practice of producing examples of the meticulous contrivance of the laws of Nature for human benefit, and the existence of environments tailor-made for the flora and fauna that inhabited them became a major industry which was only stopped dead in its tracks by Darwin's publication of the *Origin of Species* in the

---

*Arius was a fourth-century Libyan theologian who rejected the doctrine of the Trinity, the primary architect of which was Athanasius. Newton's private papers reveal that he questioned almost all the main doctrinal teachings of the Church and persistently maintained that the original Scriptures had been corrupted by the addition of spurious passages during the fourth and fifth centuries, added solely to bolster an invented doctrine of the Trinity. Newton maintained that the former, uncorrupted version of Christianity issued from Arius, rather than Athanasius, and Newton's writings entitled *Notable Corruptions of Scripture* record his somewhat paranoid conspiratorial theories about these corruptions of the early Church. Newton seems to have been the principal recreator of interest in the Arian heresy after it lay dormant for centuries following its condemnation by the doctrinal Councils of the early Christian Church.

mid-nineteenth century. This undermined the claim that the harmony between living creatures and their habitats could only be explained by design, although it was quite irrelevant to the Newtonian Design Argument from the laws of Nature.

For examples of Newton's influence upon religious thinking and the influence of laws of Nature upon 'design' in Nature one has merely to refer to a church hymn-book. But this Newtonian interpretation of Nature was not without its critics. William Blake saw the mechanical world view as the inexorable and depressing machinery of 'dark satanic mills'.

Before leaving Newton we should draw attention to one last issue regarding the concept of laws of Nature which was brought sharply into focus by the popularizers and followers of the Newtonian philosophy: the question of miracles. In Newton's day religious problems of this sort were a big issue, and so we find much discussion as to the compatibility of the Newtonian world, with its God-given laws of Nature, with the possibility of miracles. Indeed, this was precisely why David Hume devoted so much space to undermining the credibility of miracles in his *Dialogues Concerning Natural Religion* (1779). Some took the view that there were two classes of phenomena: one 'natural', governed by the Newtonian laws of Nature, the other 'miraculous'. By observing what does occur they claimed that one could then discern whether one was witnessing a miracle or not. Pieter Van Musschenbroek wrote that the miraculous phenomena 'happen contrary to the laws of Nature'. Although he believed that our knowledge of Nature's laws was incomplete, he did not argue that apparently miraculous phenomena would eventually be regarded as natural when new laws were discovered. Other authors took the view that laws of Nature were simply descriptions of what did occur, and although they witnessed to certain habitual trends, they forbade nothing. John Rowning sums up this view of his contemporaries in this manner:

> Doubtless the Author, both of Matter and of those very Principles by which it acts, can, notwithstanding those Principles, cause it to act differently from what it would do in consequence of them alone, and so by that means produce effects contrary to the common Course of Nature, whenever he shall think proper . . . Upon the whole therefore, to presume, that the ordinary and common Course of Nature is not sometimes altered, is hasty and unwarrantable.

In the immediate post-Newtonian era we find the term 'laws of Nature' explicitly used in texts, usually specially defined in the grand Newtonian manner at the outset, and widely incorporated into religious apologetics. None the less, there persists an uneasy relationship between the appeal to the universality of these laws of Nature as the dictates of God and the problem that they might then become God. Newton foresaw these dilemmas, and we find in Cotes's Preface to the second edition of the *Principia* the following statement about the conflict between the necessity of certain laws of Nature and Divine choice:

Without all doubt this world, so diversified with that variety of forms and motions we find in it, could arise from nothing but the perfectly free will of God directing and presiding over all.

From this fountain it is that those laws, which we call laws of Nature, have flowed, in which there appear many traces indeed of the most wise contrivance, but not the least shadow of necessity . . . He who is presumptuous enough to think that he can find the true principles of physics and the laws of natural things by the force alone of his own mind, and the internal light of his reason, must either suppose the world exists by necessity, and by the same necessity follows the laws proposed; or if the order of Nature was established by the will of God, that himself, a miserable reptile, can tell what was the fittest to be done. All sound and true philosophy is founded on the appearance of things . . . men may call them miracles or occult qualities, but names maliciously given ought to to be a disadvantage to the things themselves, unless these men will say at last that all philosophy ought to be founded in atheism.

The rationality of the world

> The fact that we can describe the motions of the world using
> Newtonian mechanics tells us nothing about the world. The fact
> that we do, does tell us something about the world.
>
> Ludwig Wittgenstein

By the end of the eighteenth century the Design arguments had become part of orthodox religious thinking in Britain; but cogent objections began to be raised, which were also implicit criticisms of the rationality of the new scientific enterprise itself. There were two notable contributors to this sniping who are now regarded, rightly or wrongly, as two of the most important philosophers since the ancients, although in their own day their influence in Britain was essentially nil.

David Hume and Immanuel Kant both took issue with the logical basis of the Design argument. In his famous *Dialogues Concerning Natural Religion* Hume associates it especially with Newton, even though it was a much older notion, and quotes the statement of it given by Maclaurin in his popularization of Newton's work. He refers to it with a touch of irony as 'the religious hypothesis'. It is clear that the Newtonians held a straightforward realist view of the laws of Nature they had found. The world really was a mechanism which obeyed the precise mathematical rules Newton had found (they did not ascribe any significance to Newton's idealizations). Hume did not accept that there was any unique interpretation of the lawfulness of the Universe—even if he were to be convinced that it was lawful. Nor did Hume accept that the Universe had been shown to be a mechanism rather than, say, an organism. Nor did he take for granted the logical connection between cause and effect which Newton assumed. These ideas were set out in Hume's *Dialogues* which was published by an unknown publisher, probably in Edinburgh, three years after his death in 1776. Besides arousing considerable animosity, they stirred the German philosopher Immanuel Kant (1724–1804) to develop a more rigorous analysis of what would now be called the theory of knowledge.

Kant's work really amounted to a questioning of the realist interpretation of Newton's laws of Nature which he had previously accepted uncritically in his earlier excursions into science and astronomy. Like many others of his time, he had believed the Newtonian laws of mechanics and gravitation to be the only logically possible laws of Nature describing motion and gravity. One of his early papers used the existence of the inverse square law to show that space must therefore possess three dimensions. But he then appears to have lost faith in this realist view, and, awakened from his 'dogmatic slumbers' by Hume's writings, he embarked upon a series of critiques aimed at the roots of the scientific method.

Kant argued that it is not possible for us to prove or disprove statements about the real world by reason alone. There are inevitable filters that stand between things as they really are and our perception and understanding of them. These 'categories', as they became known, are inescapable; they are necessary for the process of understanding, and they will create an order where none exists. The things in themselves could be mind-independent, but any order that we perceive is necessarily mind-imposed. This we can recognize as the basis of the idealist position explained in the last chapter, and it suffers from similar weaknesses. Kant used it to argue that the laws of Nature that we find are inevitably our own mental creations, and consequently so are any grand metaphysical conclusions about Design in Nature that we might draw from them. (Kant, incidentally, was extremely sympathetic to the Design Argument, which he found convincing on other grounds. He just wanted to show that it was neither provable nor disprovable by reason alone.) As we mentioned in the last chapter Kant also believed that scientists should proceed with the expectation that the world is purposefully ordered, this being the most effective stance to take. Kant's ideas hark back to Plato's in some respects. We might regard Plato's ideal Forms as being the reality of Nature, which our acts of observation and attempts at representation using mathematical formulae distort into the imperfect versions which we mistakenly claim to be reality. This type of idealism was strongly supported by Jeans and Eddington in the pre-war era of this century. It claims that our observations of the world tell us principally about the categories of our thought.

The most important discussion of Kant's ideas in the context of science was that made by the young Heinrich Hertz in 1900. He drew out, more clearly than Kant had been able to, the distinctions and relationships between the unobserved things in themselves, our sense perceptions of them, and our representations of them.

In practice we proceed by creating mental pictures or mathematical symbols to represent the objects we see in Nature. These images we select to have only one necessary property: that the result of any mental image of a thing is always the same as the mental image of the consequence of that thing in Nature. So, if the effect of some phenomenon P is denoted by $E(P)$ and our mental image of P by $I[P]$, then we require that $I[E(P)]$ be the same as $E(I[P])$. Kantian categories of thought which we use to make sense of the world are those possessing this property, which we shall term *reciprocity*. We have no way of knowing whether our scientific descriptions of Nature are in accord with them in any other respect than the reciprocity property. Newton realized that no further correspondence was necessary to solve the problem of 'how' things behaved. In order to say 'why' one would need to know that the application of the operations $E(..)$ and $I[..]$ possessed other properties as well.

The out-and-out realists who would maintain that reality is precisely described by, say, Newton's law of gravity, and that all the entities mentioned in this law (mass, distance, the constant of gravitation, the mathematical operations involved, and so forth) exist in the same fashion as they appear in the theory, believe that $I[P]$ is identical to P and $I[E(P)]$ is identical to $E(P)$. There is no reason a priori why any such correspondence should be possible, but there is clearly a considerable evolutionary advantage endowed upon any species that does possess it, and *we* certainly would not be discussing the matter if we did not possess it. By contrast, a hyper-idealist who bordered upon solipsism might insist that there is no reason to suppose that the operation I is universal. For all we know, it might differ from one person to the next.

There is a parallel between the requirements on the operations of I and E for reciprocity to hold, and the properties required of encrypting operations employed in developing modern 'public key' codes. It is possible to transmit coded messages between two individuals in such a way that both use different codes which remain unknown to the other, but they are still able to decode any message sent between them. This is far more secure than any procedure that requires extra copies of a single code to be conveyed to the receiver of a message in order that he be able to decode it. For the sake of illustration let us suppose that we wish to send someone a message that has been placed in a box. The analogue of coding the message is the operation of placing our padlock on the box. We would like to be able to send this box to someone in such a way that they can open ('decode') it *without* having to obtain a copy of our padlock key. This sounds impossible; but it is not. We padlock the box and send it to our colleague who then padlocks it with his padlock and returns it to us. We remove our padlock and return it again, whereupon it can be opened by him with his own key. In order to carry out this procedure with two real coding operations they must possess the reciprocity property. If my coding operation is E and yours is I, then acting first with E and then with I has the same effect on the message as acting upon it first with I and then with E.

Hertz argued that the mental images which we can make of things are not determined uniquely by the requirement of reciprocity. This was not known in

Newton's time. But in fact there are many different 'images' of the laws of motion. Subsequently some of these were found by Lagrange, Hamilton, Euler, and Maupertuis. When, as scientists, we attempt to formulate laws of Nature, we will hunt through possible images, and exclude those that fail to satisfy certain criteria. We will throw out those that are logically inconsistent, since they require something to be simultaneously true and false, and hence allow anything to be deduced as true, and we will exclude those images that lead to pictures at variance with our experience of Nature's actual operations. Lastly, we exercise judgement as to which is the most economic in representing the essentials of what is observed with the minimum of superfluous extras. Since we are always dealing with mental images, we must expect that necessarily there will always be found to exist some superfluities. All we can do is minimize their number. This last criterion we can use to modify and tune-up our images, but different individuals can end up preferring different ones, because simplicity and economy are subjective criteria. The chemist may well prefer to see things differently from the physicist.

Having selected the best mental images of what occurs in the world, we now proceed to draw up representations of these images. Mathematics appears to be wonderfully appropriate to the task, and we shall examine the interpretations of this circumstance in Chapter 5.

Kant's motivation for putting forward criticisms of the realist view of the world may have been partly religious. A devout Lutheran, he realized that the mechanical model of the Universe created by the Newtonians was governed by unalterable laws of cause and effect which left no room for free will. His argument that we never observe the 'things in themselves', but only our subjective images and representations of them, allowed him to reconcile the Newtonian view with the existence of free will. Newton's rigid causality governed only the subjective world of the images of observed things, but the true 'things in themselves' need not be totally determined by causality. Despite his self-confessed affinity with the Newtonian Design Argument, Kant argued convincingly that it was a deduction from our images of the laws of Nature with no necessary correspondence to the underlying reality. All the great metaphysical 'Arguments' which use aspects of the laws of Nature—the Design Argument, the Ontological Argument, the Cosmological Argument—are just *arguments*; that is they begin from some *assumptions*, and deduce a conclusion. That conclusion is worth no more and no less than the initial assumptions, and can never be independent of them. They are not disproved by counter-arguments of the Kantian sort, any more than they are proved by those of the Newtonians. Although, for example, the form of Newton's laws of motion excludes any teleological notions, and replaces final causes by initial causes and algorithms for computing the subsequent states which follow from them, one should not draw far-reaching metaphysical conclusions from this image. In 1748 Maupertuis showed that Newton's laws of motion could be derived by the application of a teleological principle. It is possible to define a mathematical quantity, the *action*, which involves the product of mass, velocity, and distance travelled by bodies. Maupertuis's Principle, which we now call the Principle of Least Action, was that

> If there occurs some change in Nature, the amount of action necessary for this change must be as small as possible.

This elegant idea turns out to be equivalent to the Newtonian laws of motion (although it is more powerful in the sense that it can be used to derive the equations of motion in other areas of physics once the appropriate action is identified). But, unlike the formulation of Newton, it is teleological. It says that, of all the paths that could be taken by a body moving from A to B, it actually takes that path for which the associated action is a minimum. This path is therefore determined by both the initial and the final states. Maupertuis attached great metaphysical significance to this result, regarding it as a 'proof of existence of Him who governs the world'. Formerly, arguments of the sort that we lived in the 'best of all possible worlds' were open to the objection that we did not know any other worlds with which to draw such a comparison, but Maupertuis claimed that the other worlds were those in which motion occurred with non-minimal action. Our world was optimal in this well-defined sense, and moreover there existed a teleological aspect to the laws of Nature (in fact, some nineteenth-century commentators interpreted the existence of fossils as relics of the still-born worlds of non-minimal action). What is instructive about these eccentricities is that they exhibit how one image of the laws of Nature, selected to give correspondence only with what is seen, can give completely the opposite metaphysical impression to that obtained from another image which generates exactly the same law of Nature.

Kant and Hume's logical objections to the interpretation of mathematical laws of Nature and the Design Argument that flowed from it fell, for the time being, upon deaf ears. Hume died in 1776, and was regarded in literary circles simply as an impious seeker after fame. Kant's work did not begin to appear in English until about 1796, and was turgidly and unengagingly written. These factors made it easy to ignore their arguments. But the real reason why British scientists did not take such philosophical objections seriously is not hard to find. They had emerged into a period of great success precisely because they had freed themselves from the influences of philosophical discussion, from moots about the meanings of terms rather than the meaning of observations. To have taken these metaphysical objections on board would have been a step backwards into the labyrinth of philosophical disputation, where things unseen dictated the interpretation of things seen. Coupled to this psychological barrier was the fact that experiment and observational data had become the sole adjudicator of disputes. The objections of Kant and Hume made no appeal to observations of the world, whereas the Design Arguments, be they of the Newtonian variety or those drawn from the biological realm where countless examples of 'design' had been meticulously documented, were steeped in observation. This empirical basis appealed very much to the British scientific mind. The Design Argument would be overthrown, not by philosophical objections to its logical soundness, but by the idea of Darwinian evolution. Darwin was able to provide another explanation, itself rooted in detailed observations, for the mass of detailed observations supporting apparent design in the make-up of the natural world. It was because he provided an alternative explanation for the naturalists' observations that he carried the day, not because he undermined the logic of the Design Argument. The Newtonians lived in the post-Copernican era. Although Man was undoubtedly at the centre of their world-view, he was no longer at the centre of their world model. To adopt Kant's approach would have seemed like a step backwards into the pre-Copernican era in which Man was the focus of all things, for idealism assumes for the mind of Man a place at the focus of the Universe, and makes of him a cosmic censor.

Then, as now, scientists placed most trust in philosophical discourses written by scientists. For this reason the most influential work on the philosophy of science written in the first half of the nineteenth century was John Herschel's *Preliminary Discourse on the Study of Natural Philosophy*. Herschel subscribed to observation as the means by which we must learn the laws of Nature. The scientist, he writes, is concerned with 'what are [the] primary qualities originally and unalterably impressed on matter, and . . . the spirit of the laws of nature'. This statement of what amounts to naïve realism illustrates the limited impact of Kantian thinking upon the mainstream of British science of the period. Subsequently, the most vociferous attack upon the uncritical subscription to a grandiose view of the laws of Nature came from the emerging positivist movement, which denied any ontological authority to scientific law, and adopted a rigid operationalist stance. William Jevons and Karl Pearson were the most influential writers representing this view, and they attempted to downgrade the status of laws of Nature by regarding them as approximations and an outgrowth of a purely mental activity. Jevons claimed 'that before a rigorous logical scrutiny the Reign of law will prove to be an unverified hypothesis, the Uniformity of Nature an ambiguous expression, the certainty of our scientific inferences to a great extent a delusion'. He regarded the 'laws of Nature' as propositions, grounded only in probability, about the correlations of events. Furthermore, he interpreted the Second Law of thermodynamics as an argument against assuming the constancy of Nature. The Second Law pointed to a beginning to the Universe, and to a most unusual future in which conditions would be totally unlike the present. Pearson was more radical. He regarded laws of Nature as purely mental responses to sensations about the world. But it would be true to say that the dominant physical scientists of the day did not subscribe to these more subtle views of the laws of Nature. The Faradays, Maxwells, and Kelvins were deeply religious men who adopted without question the view that Nature's laws were real, and that they were imposed upon her by Divine decree.

Darwinian laws

> Natural selection is a mechanism for generating an exceedingly
> high degree of improbability. R. A. Fisher

Until the mid-nineteenth century one had a clear choice regarding the structure
of the world. Either it was a cosmos or a chaos. If the former, then its order must
have a definite source, whilst the latter option flew in the face of everything that
we saw in Nature. The laws of Nature which laid bare the machinery of the world
bore witness to its inner workings, and persuaded both scientists and clergymen
that the Author of those laws of Nature was the essence of order and logic. But
then a new doctrine began to emerge which has influenced our attitude towards
the origin of order in Nature ever since.

In 1813 an expatriate American physician employed at St Thomas's Hospital
in London read an extraordinary paper to the Royal Society. The name of the
physician was William Wells, the title of the paper, *An Account of a White Female,
Part of Whose Skin resembles that of a Negro*. In it Wells proposed what we now
call the process of 'natural selection' as an explanation for the existence of extant
physical characteristics in living things. He derived the hypothesis from his case
study of the adaptation of human skin coloration to climate. He argued, in
contradiction to the prevailing view, that artful design was unnecessary in order
to explain the remarkable adaptation of living things to their environments. If we
could effect adaptation by the artificial selection imposed by breeding then this
adaptation could be achieved 'with equal efficiency, though more slowly, by
nature'. Moreover, Wells appreciated that there was no such thing as the
'uniformity of Nature'; the natural world was in a state of perpetual change, and
the process of adaptation could never be complete. Wells's paper was published
in 1818.

These views were both important and radical. One might have expected them
to have fomented all manner of opposition and public comment. Not so: they
influenced nobody; they were cited by nobody; they attracted neither praise nor
approbation. It is difficult to determine why this was so. Wells was a respected
scientist, a Fellow of the Royal Society, and the winner of the Society's Rumford
medal in 1814 for his classic analysis of dew-drops. It is just possible that by
publishing his paper on natural selection merely as an appendix to his Rumford
prize-essay he actually ensured that it was overlooked, since the essay was widely
cited by philosophers of science as a classic example of the scientific method at
work. Whatever the reasons for Wells's original neglect, he appears eventually in
the later editions of Darwin's *Origin of Species*, acknowledged as the originator of
the idea of natural selection, after Darwin's attention was finally drawn to his
work by an unknown American scientist in 1860.

Darwin and Alfred Wallace went much further than Wells in gathering
evidence for the mechanism of natural selection as an explanation for the

existence of order in the organic world. Through their work a new type of explanation became legitimate. If all possible variants arise at random in a reproducing system, then those variations which most enable the system to reproduce will subsequently survive with greater probability than those which do not. Those reproductions that are best adapted to survive in the environment in which they find themselves will do so more readily that those that are ill-adapted. Hence, time and chance can produce the remarkable match between the living creature and its environment. By this means the spontaneous evolution of order can be explained without recourse to final causes or explicit supernatural design. This evolution through the 'survival of the fittest' completely undermined the traditional argument from design in the biological realm, although it did not undermine those Design Arguments based upon the advantageous character of the laws of Nature themselves. If anything, it reinforced this latter version of the Argument, because now the remarkable contrivances between living things and their habitats was seen to arise as a result of the action of the laws of Nature over aeons of time, rather than from invariances imposed upon the world *ab initio*. Some, like T. H. Huxley (1825–95), attempted to widen the scope of natural selection to encompass the laws of Nature themselves, whilst conservative physicists like Lord Kelvin (1824–1907) opposed evolution for religious reasons, and also argued that there was insufficient time for the process of natural selection to have evolved human life by the present day. Others, notably James Clerk Maxwell (1831–79), restricted their opposition to the attempts to apply the concept of natural selection outside the biological realm by pointing to the invariances of the microscopic world. In particular, Maxwell placed great stress upon the discovery that atoms (which he called 'molecules') were identical, and their properties were not subject to the process of natural selection. He realized that a line could be drawn at some level in the hierarchy of Nature below which natural selection could not supply an explanation for order. That line would need to be drawn above the scale of the atom.

Already in the opening chapter of this book the reader will have detected the influence of the doctrine of natural selection. When faced with the task of explaining the existence of some state of very low a priori probability one now looks to ascertain whether, regardless of starting conditions, this special state is always attained after a long period of time. Moreover, some states will be found to be necessary for the existence of living beings to observe them. Regardless of how improbable their occurrence a priori, we should not be surprised to find them existing today. A priori it might seem extremely improbable that, of all the places where a planet could be situated in this vast Universe, the Earth should be found in orbit around a star. However, such a proximity is no doubt necessary for the evolution of intelligent life-forms.

The Darwinian perspective has further challenging things to say about metaphysical ideas. Take, for example, the Kantian notion that our view of nature is irreducibly conditioned by human categories of thought. The pre-Darwinian Kantian must subscribe to the coincidence that all humans possess, and always have possessed, identical categories of thought. The Darwinian sees the categories of the human mind as a result of the process of natural selection and so fashioned by the physical world—the thing in itself. This view gives realism a new underpinning. For, if we have come to be what we are, both physiologically and psychologically, in response to the adaptive pressures of natural selection, then in many respects we must accurately mirror a physical world that really exists, just as the human ear has evolved in response to the existence of sound and the eye to the existence of light. According to this view any universality in our innate categories of thought could be associated with the universality of the laws of nature.

Fitness is a concept that has been appropriated by thinkers of all disciplines. Indeed, one might apply it to the subject matter of this chapter. Of all the primitive notions that arose about the world, that which ascribes its harmony and regularity to some form of extra-human legislation has been the one that at first had selective advantage for those primitive humans who were able thereby to exploit the regularities of Nature for their well-being. Over the centuries Mankind has become more sophisticated in his curiosity about Nature and in his thinking about thinking, but still he has found certain ideas to have a persistent usefulness that is rich in consequence and utility. One of these ideas remains the concept of laws of Nature. It has consistently adapted to meet the changing intellectual environment, but, as in the biological realm, there is no guarantee of continued survival. The twentieth century was to bring abstractions, ideas, and discoveries about the Universe undreamed of by the actors in our story so far. Their successors were both led by their inheritance of the concept of the laws of Nature, and challenged to refine it to its irreducible minimum by the touchstone of reality.

3

Unseen worlds

It has become increasingly evident, however, that Nature works on a different plan. Her fundamental laws do not govern the world as it appears in our picture in any direct way, but instead they control a substratum of which we cannot form a mental picture without introducing irrelevancies.

Paul Dirac

Mechanism without a mechanism

Scientific concepts are inner pictures.

Heinrich Hertz

Galileo and Newton chose to express laws of Nature in the language of mathematics. They were not the first to realize the importance of using this precise and universal language, but they were the most successful exponents of it. Yet, because they chose to describe the world in this way, they needed to be idealistic to some extent. They had to select what they felt to be the essence of a physical phenomenon. Thus, the collision of a red ball with a blue ball can be described by laws of motion that pay no regard to their colours. It is their masses and speeds that are important. Furthermore, the laws could be most instructively formulated when applied to ideal objects—bodies moving under no forces for example.

The mathematical description very effectively separates the 'whys' from the 'hows'. Newtonian physics was very much a device for telling what the final state of a system would be if you gave its initial state. Although it is often called a mechanical world-view, this is a complete misnomer: the one thing it does *not* do is give any *mechanism* by which changes take place. For this would require knowing why they change. 'It is enough', writes Newton, 'that gravity really does exist, and act according to the laws which we have explained, and abundantly serves to account for all the motions of the celestial bodies and of our sea.' Likewise, when Galileo considers the acceleration of falling bodies, he says that 'the cause of the acceleration of the motion of falling bodies is not a necessary part of the investigation'.

The two particular characteristics of Newton's scheme of natural laws that we wish to focus attention upon are the concentration upon the *visible phenomena* of Nature, and the notion that forces act *instantaneously* between particles that are separated in space. They are both ideas that will gradually fade away. The manner of their departure marks another chapter in the search for the laws of Nature and the entities which they govern.

Force fields

> According to Newton's system, physical reality is characterized by the concepts of space, time, material point and force . . . After Maxwell they conceived physical reality as represented by continuous fields, not mechanically explicable, which are subject to partial differential equations. This change in the conception of reality is the most profound and fruitful one that has come to physics since Newton.
>
> Albert Einstein

The success of Newton's law of gravitation in determining the motion of the tides and the moon, as well as local motions observed on the Earth, in terms of the instantaneous action of an inverse-square law of force led to the prejudice that all Nature's forces depend only on the distance between the bodies concerned, and act along the line joining their centres. Laplace exploited this viewpoint to explain many optical and chemical phenomena in terms of attractive forces. Newton himself held to the older notion that all physical forces were exerted by physical contact, but he could not find a way to implement this belief in the case of gravitational attraction. Certainly, he felt that it would not be possible to explain all phenomena in the way epitomized by his law of gravitation. With regard to the success of that law he wrote in the Preface to the *Principia* that

> I wish we could derive the rest of the phenomena of nature by the same kind of reasoning from mechanical principles; for I am induced by many reasons to suspect that they may all depend upon certain forces by which the particles of bodies, by some cause hitherto unknown, are either mutually impelled towards each other, and cohere in regular figures, or are repelled and recede from each other.

In 1773 the French mathematician Joseph Lagrange noticed that it was possible to express Newton's law of gravitation in terms of the influence of a force-field filling the whole of space in a continuous manner. The mathematical quantity whose rate of change throughout space generated Newton's inverse-square law of force would later be called the gravitational *potential* by the British mathematician George Green in 1828. Laplace was able to find a differential equation giving the continuous variation of this potential outside a mass, and then, to complete the picture, in 1813 Simeon Poisson gave a generalization that determined the potential at any point within a distribution of matter possessing a specified density.

If one were to join up the points where this potential has equal values, the resulting collection of curves—*equipotentials*—gives a contour map of the strength of the force-field's effects upon bodies introduced in its vicinity. Some of the equipotentials of the gravitational force exerted by a single mass are shown in Figure 3.1. We say 'some' because they are continuously distributed all the way out to infinity, with one at every radial distance away from the mass. An infinite number of them could be drawn. The collection of all these lines of force filling space constitutes a description of the *gravitational field*. If a particle is diverted from moving along an equipotential then it feels a force perpendicular to the equipotential at its new position. Thus there can be imagined to exist a continuous field of force as shown in Figure 3.1:

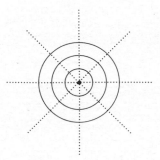

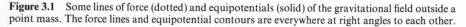

Figure 3.1 Some lines of force (dotted) and equipotentials (solid) of the gravitational field outside a point mass. The force lines and equipotential contours are everywhere at right angles to each other.

The acceleration due to gravity at any point is always along the field line running through that point, and its magnitude is proportional to the density of field lines at the point. The number of field lines ending on a mass is taken to be proportional to the magnitude of the mass.

Consider the famous result that Newton struggled so long to establish before publishing the *Principia*: that the gravitational field outside a sphere is the same as that exerted by an equal point-mass located at the centre of the sphere, if gravity obeys an inverse-square law. In Figure 3.2 we see lines of force emanating from a point mass. Suppose that we superimpose the boundary of a spherical body (dashed). Then, on the outside of the dotted region the lines of force look the same irrespective of whether they are regarded as originating from the point or emanating radially from the surface of the sphere. The same number of force lines intersect the sphere's surface as reach the point placed at its centre, and so they represent the gravitational field of identical masses.

This picture introduces the idea of a continuous influence due to the mass rather than the attraction of point masses. But is it anything more than a helpful picture? After all, the speed with which the gravitational interaction is imagined to travel along the lines of force could still be infinite.

These examples illustrate the introduction of the concept of a force field into the physicists' armoury. The field concept is now a commonplace one, and we recognize that there exist different types of continuous field in Nature. The

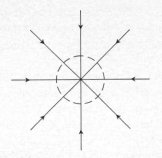

Figure 3.2 The gravitational field lines outside a sphere (boundary dashed) with the same mass as a point at the centre are identical to those that would be seen far from the point mass. The external field of a sphere or a spherical shell is identical to that of a point of equal mass located at its centre.

gravitational field we have just discussed is an example of a *scalar* field. It just varies in magnitude at every point. The height of land above sea-level, or the variation of air temperature or pressure with position are other scalar fields; so is the density of ink on this page. But there exist fields which also specify a direction at every point. These are called *vector fields*. For example, the wind direction at any place, the grain of a piece of wood, or the direction of flow of a liquid are vector fields. The weather map shown in the newspaper each day displays a scalar field of isobars, and usually also a vector field of arrows denoting wind directions. Both of these fields can vary with time as well as from place to place. They can also influence one another in complicated ways. A change in the wind vector field will produce and reflect changes in the temperature and pressure scalar fields in its vicinity.

When physicists identify such fields in Nature they are interested in determining so called *field equations* which specify how a particular field will vary in space and time in terms of the variation of its source. Poisson's field equation determines how the gravitational force field will vary in any distribution of mass in space. It is completely equivalent to Newton's law of gravitation. In describing any force of Nature we attempt to find the underlying field equation (or equations if it is a complicated force) which tell us what type of force field is produced by a given source, and also the equations of motion which specify how particles move when introduced into the field of force. In the case of Newtonian gravitation the equations of motion are simply Newton's three laws of motion. In practice all these equations are partial differential equations.

Electricity and magnetism

Faraday, in his mind's eye, saw lines of force traversing all space, where the mathematicians saw centres of force attracting at a distance; Faraday saw a medium where they saw nothing but distance; Faraday sought the seat of the phenomena in real

> actions going on in the medium, they were satisfied that they had found it in a power of action at a distance impressed on the electric fluids.
>
> James Clerk Maxwell

The idea of lines of force was exploited most successfully by Michael Faraday in the first half of the nineteenth century. Something of a folk hero, Faraday (1791–1867) was born into a poor farrier's family, and received a nominal education in a country school before leaving at the age of 13 to become an apprentice bookbinder. Evidently more of his time was spent reading the books than binding them, and he rapidly became impressively self-educated. He writes that he was 'very fond of experiment and very adverse to trade ... which I thought vicious and sterile'. His one ambition was to be involved in science. As a result, he educated himself so effectively that he was able to compile a comprehensive collection of notes from Sir Humphrey Davy's lectures delivered at the Royal Institution. These he bound up and sent to Davy, together with a request for employment in the Royal Institution. Davy's own account of discussing this application with a friend goes as follows:

> 'What am I to do? Here's a letter from a young man named Faraday; he has been attending my lectures and wants me to give him employment at the Royal Institution—what can I do?'
> 'Do? Put him to work washing bottles; if he is good for anything he will do it directly, if not, he will refuse.'
> 'No, no we must try him with something better than that.'

He was appointed in 1813 as a laboratory assistant to Davy, working for a menial wage. Within three years he was writing research papers, after ten he was a Fellow of the Royal Society, and eventually he rose to become the Director of the Royal Institution's laboratories, the greatest experimental physicist of his day, and a brilliant popularizer of science for the general public. None the less Faraday remained a modest and retiring individual who expressed no interest in the many public honours that were offered to him, and remained at the Royal Institution until his death in 1867.

At the time when Faraday was penning his request for employment to Sir Humphrey Davy, it was known that both electrical and magnetic forces existed and exhibited very similar action-at-a-distance effects to the gravitational force. Two unlike magnetic or electric charges attract one another with a force proportional to the inverse square of their distance apart, whereas two like charges repel with the same dependence on their separation (see Figure 3.3). They appear to differ from gravitational forces only in respect of coming in positive and negative (or 'North' and 'South') polarities, whereas mass is a gravitational 'charge' that is always positive. During the first thirty years of the nineteenth century a series of experiments carried out by André Ampère,

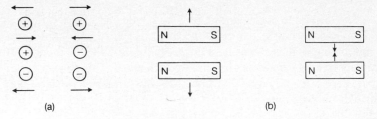

Figure 3.3 (a) Like and unlike electric charges repel and attract respectively. (b) Analogous attraction and repulsion of like and unlike magnetic poles.

Christian Oersted, Jean Biot, and Felix Savart revealed that forces were created by *moving* electric or magnetic charges. This overthrew the prejudice, created by the gravitational force law, that forces of Nature always depend only upon the distance between objects, and act along the line joining their centres.

A moving magnetic pole could accelerate an electric charge, and so create a current flow (see Figure 3.4).

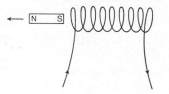

Figure 3.4 The motion of a magnet through a coil causes the acceleration of electric charges and hence induces a flow of electric current.

Furthermore, a moving electric charge could create a magnetic field. This revealed a deep connection between electricity and magnetism: there exist magnetic forces that do not originate from magnetic sources, and electric fields that are generated solely by moving magnets. These discoveries showed the long-known phenomenon of magnetism to be but electricity in motion. The effects of magnets could be reproduced by suitably arranged wires carrying electric currents.

Faraday's fascination with these topical discoveries was turned into critical investigation by the intervention of his close friend Richard Phillips. Phillips edited a journal of philosophy, and was seeking an article describing the history of the ideas and experiments concerned with electricity and magnetism in order that the new discoveries could be evaluated by natural philosophers. He asked Michael Faraday to provide that article.

And so began Faraday's work as a physicist.

Faraday showed that the field picture of magnetic and electric forces gives a revealing picture of the equivalence between the phenomena of electricity and magnetism. In Figure 3.5 are shown some of the magnetic field lines outside a bar magnet. Each field line begins at one pole and ends on the other. If you sprinkle some iron filings over a piece of paper resting on top of a bar magnet you will see the filings arranged along these field lines. By rotating the magnet about its axis Faraday observed the field lines to remain unchanged. This demonstrated to him that they were not individualized lines of stress attached in some way to the magnet like microscopic cords of string. If so, they would have twisted and altered the observed distribution of iron filings.

In Figures 3.6(a) and 3.6(b) are shown the field lines outside two like and two unlike *electric* charges respectively. The collection of all the field lines which would fill the whole of space if they were added to Figure 3.6 constitute the electric field created in each case. The magnetic field created by a current of moving electric charges in a coil, shown in Figure 3.7, has the same external appearance as that produced by the bar magnet field in Figure 3.5.

These simple illustrations indicate that there is a deep connection between electricity and magnetism which is only fully revealed when their sources are set in relative motion, and this symbiotic relationship is best exhibited by the field concept. The true nature of this interrelationship was laid bare in 1865 when the young Scot, James Clerk Maxwell, produced a system of four equations which encapsulated the symbiotic relationship of the electric and magnetic fields, and successfully predicted new phenomena. To a considerable extent our modern technological society rests upon what these equations tell us about the intertwined behaviour of electricity and magnetism. Henceforth they would be known as different manifestations of a single entity: the *electromagnetic field*.

Faraday's investigations undermined the idea that forces acted at a distance in the Newtonian manner. It could now be seen that electric and magnetic forces depended upon the pattern of the field lines, and these lines were everywhere— even inside magnets. The forces were to be found between magnets and charges, not just at them. When he evacuated the region of space between two magnets, Faraday found the field lines were unaffected.

Maxwell's theory of electricity and magnetism gave laws for the behaviour of these natural phenomena which actually went further than the achievement of Newton's laws of gravitation and motion. Newton's laws provided a beautiful encapsulation of everything that was known of these subjects. The motions of the moon and the tides were explained in relation to each other. The regularities of motion were codified, but no fundamentally new unknown properties of the Universe were *predicted* to exist; but Maxwell's equations, while elegantly describing all the intertwined phenomena of electricity and magnetism, also made far-reaching predictions about aspects of the world never before suspected. Some of Maxwell's equations have the form of equations that describe the propagation of waves that, in this context, would have to be interpreted as waves of the electromagnetic force field. The speed with which these waves were predicted to propagate through space was expected to be equal to the value of 300,000 kilometres per second that light had been measured to travel at. Maxwell

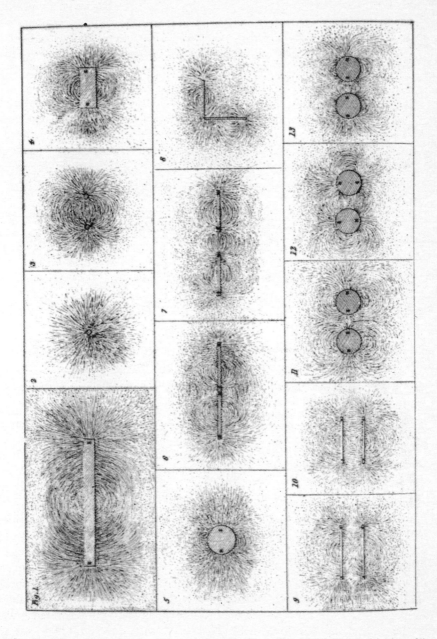

Figure 3.5 Michael Faraday's record of the pattern of magnetic field lines traced by iron filings sprinkled around magnetic poles. This is taken from Faraday's own work *Electricity*.

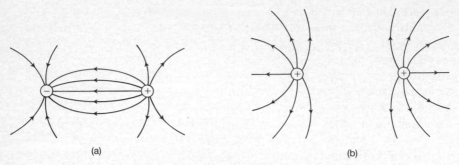

Figure 3.6 Electric field lines in the vicinity of like and unlike electric charges.

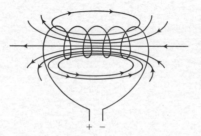

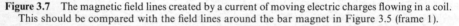

Figure 3.7 The magnetic field lines created by a current of moving electric charges flowing in a coil. This should be compared with the field lines around the bar magnet in Figure 3.5 (frame 1).

concluded that electromagnetic waves were a form of light, which was also known to exhibit the behaviour characteristic of a wave phenomenon, and as a consequence unseen forms of light must exist with a wavelength different from that of the visible region of the spectrum. This dramatic prediction of Maxwell's mathematics was confirmed by Heinrich Hertz who, in 1887, discovered the existence of radio waves having a wavelength nearly a billion times longer than that of the waves of visible light. Later, the entire spectrum of electromagnetic radiation was uncovered.

This discovery revealed that the electromagnetic field was not simply a convenient way of picturing action at a distance, but existed as an entity in its own right. It had predictable and measurable properties, and it mediated the intimate connection between the phenomena of light and electromagnetism.

The Sandemanian world-view

> There is no philosophy in my religion.
>
> Michael Faraday

Faraday's humble background, his lack of mathematical training, and his overriding experimental approach to science are all unusual, but they are in fact all of a piece. The Sandemanian Church, of which Faraday and two generations of his family had been members, was an unusual religious denomination. An eighteenth-century offshoot of the Scottish Presbyterian Church, led at first by John Glas, and then by his son-in-law Robert Sandeman, it eschewed organization and ceremony. It was extremely fundamentalist in its approach to theology and everyday practice. Indeed, it distrusted any form of interpretation of the Bible by theologians, and held fast to a completely literal reading of its text with a minimum of added interpretation. It had no clergymen or pastors, and relied upon the direct revelation of the biblical text to the individual and the appointed elders of a congregation. Strangely, many of its members were, as was once the young Faraday, associated with the business of publishing. Faraday was an elder of this Church in London, and was also active in it as a preacher. He took his religious faith very seriously, and it influenced everything he did and said. His long-time friend and scientific colleague, John Tyndall, who had no religious sympathies himself wrote that: 'I think that a good deal of Faraday's week-day strength and persistency might be referred to his Sunday exercises. He drinks from a fount on Sunday which refreshes his soul for the week.'

Just as Faraday's religious views were biased towards a plain and non-theoretical view of the Bible, so was his reading of the natural world. The mathematical physicists who would produce intricate theoretical descriptions of Nature were to him just like those sapient subtlers of the Word whom the Sandemanians turned aside from. Faraday spoke of his investigations of Nature as the reading of the book of Nature. His reading was literal and simple: an extension of what the Sandemanians called 'plain style'. The only way to know what was in the book of Nature was by experimental scrutiny. Furthermore, he believed that his investigations would reveal the natural laws of God, laws that were a parallel to the moral laws he could 'experimentally' discern by a reading of the Bible. This background goes some way towards illuminating Faraday's attitude towards his fields of force, and to the mathematical work of his illustrious contemporaries. His literal interpretation of God's laws of Nature led him to regard the visual lines of force he saw traced by his magnets and coils as being totally real. His disdain for mathematics speaks of someone with a positive antipathy towards it rather than just a lack of training in its methods. Sandemanians appeared to have had a strong aversion to any use of signs and symbols, because they were regarded as the property of God alone. The use of mathematics was regarded as a transformation of God-given reality into a fallen human representation—an interpretation of the book of Nature by the 'theologians' of science. *Traduttore, traditore.*

These attitudes to symbols are somewhat reminiscent of the ancient reaction to the first actor. One recalls how Thespis stepped from the chorus to impersonate a character in a story that was being recounted in song. By so doing he founded the acting profession, but for a long time the practice was denounced as a dangerous deceptive heresy which usurped and impersonated the position of the gods.

The end of visualization?

> Maxwell's theory consists of Maxwell's equations.
>
> Heinrich Hertz

Faraday's Sandemanian approach to science was practical and totally non-mathematical, and so it should come as no surprise that he found Maxwell's high-powered mathematical description of electromagnetic phenomena quite beyond his grasp; he once asked Maxwell if he could not present his conclusions 'in common language as fully, clearly, and definitely as in mathematical formulae? . . . [thus] . . . translating them out of their hieroglyphics'.

When Maxwell first began to learn about electromagnetism he was a Cambridge undergraduate, mindful of mathematics. He records that he resolved to read Faraday's account before any mathematical ones, having been warned that there 'is a difference between Faraday's way of conceiving phenomena and that of the mathematicians'. However, Maxwell soon realized that Faraday's field lines could be translated into mathematical language. His natural instinct was quite the opposite of Faraday's.

Because of his lack of mathematical training and his aversion to symbolism Faraday's ideas were not taken seriously by many of his gifted contemporaries, and he was unable to understand the theoretical work done by mathematical physicists like Ampère or Maxwell. This made it difficult for his revolutionary ideas to become common currency. The idea of action at a distance was a deep-rooted belief, and for very understandable reasons: it had been fabulously successful in explaining the operation of natural forces. Previous attempts to undermine it had been made by philosophers like Kant and Hume, but their objections had largely been ignored by English scientists. It is also important to recognize just how powerful the image of mechanical explanation had become. Indeed, it was almost a definition of what Victorian physics saw itself to be. As a result, we find that qualities like force fields were still invariably formulated and expressed in terms of mechanical models. Even a mind as abstract as Maxwell's pictured the electromagnetic field as an incompressible fluid moving along tubes formed by the lines of force.

The main difficulty with accepting Faraday's lines of force seems to have been aesthetic—most mathematically-minded scientists just thought them an ugly and cumbersome invention which did nothing to further our understanding of what a magnet actually *is*. The Astronomer Royal thought Faraday's ideas 'senseless'. Fortunately, influential men like Kelvin and Maxwell realized the promise of Faraday's ideas, and Maxwell's work finally raised them up to respectability. Maxwell, in fact, saw the power of the field concept in a more general context. In a private letter to Faraday he expands upon a field view of gravitation:

> The lines of force from the Sun spread out from him and when they come near a

planet curve out from it so that every planet diverts a number depending on its mass from their course and substitutes a system of its own so as to become something like a comet, if lines of force were visible.

Maxwell recognized that Faraday's lines of force were actually equivalent to action at distance, if one incorporated correctly the notion of a potential field of the sort considered by Lagrange, Poisson, and Green. But Maxwell had difficulties of his own. The manner in which he transformed Faraday's physical ideas into mathematics involved abstractions that perplexed far more able mathematicians than Faraday. For some, like Kelvin, Maxwell's electromagnetic equations were unsatisfactory because they lacked anything concrete that could be visualized; they were simply magic recipes that gave the right answers. But they did more than that: they led to new questions. Their success was mysterious enough to make Hertz remark that 'one cannot escape the feeling that these equations have an existence and an intelligence of their own, that they are wiser than we are, wiser even than their discoverers, that we get more out of them than was originally put into them.'

Maxwell had found an expression of natural laws whose outworkings extended far beyond everything that had made their discovery possible. Explanation of the known would be but a secondary role for future laws of Nature to play.

Mathematical modelling

> There is a school of mathematical physicists which objects to the introduction of ideas which do not relate to things which can actually be observed and measured ... I hold that if the introduction of a quantity promotes clearness of thought, then even if at the moment we have no means of determining it with precision, its introduction is not only legitimate but desirable. The immeasureable of today may be the measureable of tomorrow.
>
> J. J. Thomson

The use of analogy to elucidate one area of Nature by comparison with another was first introduced as an explicit methodology by Descartes. Despite opposition from critics who argued that the analogies always differ from the reality they mirror in some (possibly) crucial respect, Descartes went so far as to claim the existence of analogical relations with other areas of science as a *necessary* condition for a particular explanation to be admissible:

> I claim that they [analogies] are the most appropriate way available to the human mind for explaining the truth about questions in physics; to such an extent that, if one assumes something about nature which cannot be explained by any analogy, I think that I have conclusively shown that it is false.

Descartes followed this precept by liberal use of scaled-up models of microscopic physical events. He even used dripping wine vats, tennis balls, and walking-sticks to build up his model of how light undergoes refraction. His statement should perhaps also be taken as evidence of his belief in the unity of things. The necessary existence of analogy reflects a belief in the universality of certain design principles in the machinery of Nature which he expects to reappear in different contexts. A world in which everything is novel would require the invention of a new science to study every phenomenon. It would possess no general laws of Nature; everything would be a law unto itself.

At a less abstract level, the focus upon toy models and analogies facilitates study because it permits visualization. The use of this strategem underwent a new and sophisticated step forward in Maxwell's remarkable hands.

In his explanations of the theory of electromagnetic phenomena Maxwell liked to use analogies and mechanical models of all sorts. For the continental purists this was exasperating. They thought this further evidence of that irritating British prejudice towards producing physical theories which involve only *visualizable* concepts. But Maxwell was not being type-cast in such a role. He made use of mechanical models not just to describe or illustrate the physical phenomena under study, but rather to aid the search for the best possible *mathematical model* of the phenomena. This was the way he had used Faraday's descriptions even when reading them first as a student. Progress had been made in science which supported such an opportunistic approach. Kelvin had discovered that the effect of heat-flow through a material could be described by an inverse-square law of distance, just like the force of gravity. Maxwell saw that despite considerable physical dissimilarity, two natural phenomena could admit the same type of mathematical description, and he proceeded to introduce what we would now call *mathematical modelling*. The practice of producing an exact description of Nature in every conceivable physical detail takes second place to the business of generating a mathematical representation or analogy of it. This description can be arrived at through a sequence of logical or visual steps which need not all have exact counterparts in Nature: they are merely catalysts in the production of a final system of testable equations. Maxwell explains this new approach quite explicitly:

> In order to obtain physical ideas without adopting a physical theory we must make ourselves familiar with the existence of physical analogies. By a physical analogy I mean that partial similarity between the laws of one science and those of another which makes each of them illustrate the other. Thus all the mathematical sciences are founded on relations between physical laws and laws of numbers, so that the aim of exact science is to reduce the problem of nature to the determination of quantities by operations with numbers.

This fascination with mathematical representation was being well supported by the interests of the mathematicians themselves. The theory of potentials and waves was developing into an area of applied mathematics in its own right. A most interesting emphasis in many of these developments was the systematic use of different co-ordinate systems in the mathematical modelling of physical phenomena by partial differential equations.

We are all familiar with the use of co-ordinates in daily life. To arrange a meeting with someone we must specify *four* co-ordinates: the time of meeting, and three others which fix the spatial position uniquely in our three dimensions of space at right angles to one another. These particular position co-ordinates are called rectangular or Cartesian co-ordinates after Descartes. But they are not always convenient. On the Earth's surface we use co-ordinates of latitude and longitude instead. Lamé was the first to develop a systematic theory of the equations of physics presented in arbitrary co-ordinate systems. He was interested in the study of the elastic deformation of materials, and one can see how this leads to the desire for a description of laws of Nature that is more flexible than that traditionally employed for flat undeformed surfaces.

If we start with an object that is very symmetrical, then there will exist a natural co-ordinate system to describe it most economically. But when it is deformed by stress it may lose its symmetry, and the original choice of co-ordinates will render a description that is unnecessarily complicated. Lamé showed how to express the equations of physics in general co-ordinate systems. This step reveals an unspoken faith in the idea that the laws of physics must be written in mathematical form but they must transcend the particular way the mathematical book-keeping is done. The laws of Nature cannot depend upon the way one likes to see them expressed. There should not exist a particular co-ordinate system that simplifies *all* problems. To see the consequences of such a simple idea we will require the assistance of a young Swiss patent officer.

Space and time intertwine

> [There is] a paradox upon which I had already hit at the age of 16. If I pursue a beam of light . . . I should observe such a beam of light as a spatially oscillatory electromagnetic field at rest. However, there seems to be no such thing, whether on the basis of experience or according to Maxwell's equations.
>
> Albert Einstein

The picture of natural laws laid down by Newton but preserved by Maxwell's more sophisticated mathematics was a vista of absolute rules, mathematical in character, which dictate how objects and fields change from place to place and from one moment to the next. The fact that the laws of Nature were represented most expediently in a fashion that highlights how things *change* was later to provoke philosophers into picking upon the features of prediction and falsification as acid tests of whether the proposed changes did in fact correspond to those of experience. But the stage upon which the events of Nature were played out was assumed to be fixed. Newtonian space was an absolutely immutable frame against which to refer all motion, the same for all observers. Likewise, time was a universal linear flow taken to occur at the same rate and in the same way for everyone in the Universe. Space and time were undefined entities, but none the less well known and obvious to everyone. Everyone, perhaps, except Albert Einstein, whose development of the special and general theories of relativity was to alter radically our picture of how the Universe works. The essential new ingredients from his theories and their experimental confirmation teach us a number of revolutionary lessons about the laws of Nature:

1. Only relative motions are involved in the laws of Nature.
2. There does not exist either an absolute space or an absolute time. They are different concepts for each observer in relative motion.
3. There is a maximum speed for the transmission of information.
4. Mass and energy are equivalent.
5. The presence of mass and energy in space and time determines the geometry of space and rate of passage of time.
6. There is no action at a distance involved in gravitation.

These changes in our understanding of Nature's laws were not revolutionary in the sense that they completely overthrew what had been found by Newton. Rather, they described what happened over a far wider range of conditions than did Newton's laws. When the speeds involved in motions were very small compared with that of light, and the strengths of gravity fields very moderate, as they are in everyday life, then Einstein's laws of motion and gravity become indistinguishable from Newton's. However, what they predicted should happen when we examine the world of high-speed motion was bizarre and completely unexpected.

Occasionally one has a curious feeling when sitting in a stationary train. The train seems to be moving out of the platform but suddenly we realize that merely seeing the motion of another train in the opposite direction has temporarily misled us into thinking that *we* were moving. But how do we decide that it is not us who are really moving? Usually, it is because we notice that our carriage is not rattling about, or we see we are not moving relative to something else—the station platform or the view out of the other window perhaps. But what if our train had perfect suspension and there was nothing to be seen out of the windows except for the other train: could we tell who was moving and who was not? Einstein appreciated, like Descartes, Galileo, and Newton before him, that if the trains were moving at constant speed relative to one another, and not accelerating, then none of their passengers could tell which train was moving and which was not. There is no mechanical experiment that passengers in either train can perform which will tell them their absolute speed relative to some universal yardstick. All they can ever know is their *relative* speed with respect to the other train or the ground. They cannot know whether the ground is moving with some other constant speed relative to the Sun or the distant stars. It is not that this information is hidden from them in some way, or that it is impractical to determine it because of the accuracy of scientific instruments. It is unattainable because it does not exist. The trains do not possess any 'absolute' speed because there does not exist an absolute standard of length and time to determine it. This naïve relativity was fully appreciated by Galileo, who described it in terms of a means of transportation more familiar to him. Shut yourself in a ship's cabin below decks: write with a quill pen on some paper or watch some fish in a bowl on your desk. No observations can allow you to ascertain whether the ship is stationary or moving at constant speed. Only when its speed *changes* will the ink flow differently and the fish swim in one preferred direction.

Einstein wanted to go further than his illustrious predecessors with this logic, and ensure that one could not use Maxwell's laws of electromagnetism to discover the difference between a state of rest and one of constant velocity. What must Nature be like in order that Galileo's principle of 'relativity' apply to all the laws of Nature, and not just to the laws of mechanics?

If we take two separate teams of scientists, and place them in separate sealed laboratories moving at constant velocity with respect to each other, then Einstein required them to discover the same laws of physics to hold within their respective laboratories. They might measure quantities to have different values, but the laws or relationships between different quantities would be found the same. This sounds rather different than the picture of the train passengers, but it is in fact the same. The two teams of scientists could not find any way of determining whether their laboratory was moving and the other at rest, or vice versa, because none of the laws of Nature are dependent upon this distinction. This 'thought-experiment' can be re-framed to say something profound about the laws of Nature that is usually called Einstein's *Principle of Special Relativity* (although it is a statement about things that are invariant):

> The laws of physics are the same for all observers moving with uniform relative velocity.

The generalization of what was formerly believed (although not stated in this explicit form) is very slight. The substitution of the word 'physics' for 'mechanics' requires the inclusion of the laws of light and electromagnetism.

This marks another interesting advance in human thinking about laws of Nature. It recognizes explicitly for the first time the need to frame our descriptions of Nature's laws to be independent of ourselves in the role of observers. If it had been possible for the scientists in their sealed laboratories to determine their absolute motion rather than just their relative motion by employing the laws of physics, then those laws would have been different in the different laboratories. They would have possessed an element of subjectivity.

They would have depended upon the state of motion of the observer of them. They could not be universal in the fullest sense.

In order that the Relativity Principle be true, Nature must be constrained in a number of unexpected ways, and yet it is allowed to possess a number of novel freedoms. Maxwell's laws of electromagnetism are found to obey this Principle automatically, but Newton's laws of motion only respect the Relativity Principle so long as the experiments used by the two groups of scientists in relatively moving laboratories do not incorporate light rays or electromagnetic fields. Einstein's theory of special relativity was the revision of Newton's theory to render it compatible with Maxwell's equations and the Relativity Principle. It was the completion of 'classical' physics.

The theory of relativity requires one further invariance assumption in addition to the Relativity Principle:

> The speed of light in vacuum will be found the same by any two observers in uniform relative motion.

One of the consequences of this stipulation is that the velocity of light in empty space is independent of the velocity of its source. No matter how fast an observer moves he will measure the velocity of light in empty space to have the same value: it is a constant of Nature. It is a principle which resolves the dilemma of the 16-year-old Einstein in the epigram at the head of this section.

Newtonian mechanics claimed that if we fire a bullet with speed U from a vehicle moving with speed V relative to the ground then the speed of the bullet relative to the ground is just

$$\text{relative speed} = U + V. \tag{3.1}$$

Einstein's two Principles require that the answer be

$$\text{relative speed} = (U + V)/(1 + UV/c^2) \tag{3.2}$$

where c is the velocity of light. We notice that Newton's law would require different observers to measure different velocities for light. If we fired laser light with speed U equal to that of light in vacuum, c, from a vehicle moving with speed V, then the speed of the laser light measured relative to the ground by the Newtonian equations would be $c + V$, which is not equal to c. The Newtonian result is contrary to what is observed, and conflicts with Maxwell's laws of electromagnetism. It was this problem that had worried the young Einstein. For if he sat on a light ray then, according to Newton's picture of motion, no other light rays would be able to overtake him, so no light would be visible. This he regarded as an absurdity.

The Einstein law (3.2) has the property that if we take $U = c$ then the relative speed is c whatever the value of V. In fact, unlike (3.1), the law (3.2) is consistent with the existence of a cosmic speed limit. If neither U nor V exceed c then the relative speed (3.2) can never exceed c.

Finally, we notice that if we are dealing with velocities, U and V, that are much smaller than that of light, then the term UV/c^2 in the denominator of (3.2) will be very much smaller than 1 and the law (3.2) will be *approximately* the same as Newton's law (3.1). Accordingly, we only expect the peculiar effects of the law (3.2) to be evident when dealing with motions at speeds approaching that of light itself.

The only way that the invariance of the speed of light can be preserved for observers moving with respect to each other at different speeds is if the standards of space and time which they use to determine velocities are not the same. If we measure the length of a rod to be L when it is not in motion relative to us, then we will not find it to have the same length when it is in motion relative to us. If it is moving with a constant velocity V towards us then we will measure its length to be L^* where

$$L^* = L(1 - V^2/c^2)^{1/2}. \tag{3.3}$$

Since V can lie between 0 and c we see that L^* will always be less than L. That is, the rod moving relative to us will be observed to be shorter than when it is not in motion relative to us. As the relative speed of the rod with respect to us approaches that of light, its length is observed to become smaller and smaller. This phenomenon is called *length contraction*. It means that there is no absolute concept of length that exists independent of the observer of it.

Suppose we have two identical rods whose lengths are seen to be exactly the same when we lay one on top of the other so they possess no relative motion with respect to each other. Now set these rods moving towards one another at high speed. If we sit on the first rod then the second rod is moving relative to us, and we will measure its length as it passes to be less than that of the first rod. But if another observer is sitting on the second rod he will see the first rod moving relative to him, and will measure the first rod to be shorter than the second rod. Which rod is *really* the shorter? There is no answer to this question that makes no reference to the motion of the measurer relative to each rod. The concept of length does not have any absolute meaning. The best that we can do is to talk operationally about the length that would be measured by an observer who is not moving relative to the rod. This concept is unambiguous, and is called the *rest length*. It is the greatest length that the rod could be measured to have.

This relativity affects the measurement of time in a similar fashion. If we measure an interval of time recorded by a clock to be T when we are not moving relative to that clock, then when we move with velocity V towards the clock we will find that interval of time to be T^* where

$$T^* = T/(1 - V^2/c^2)^{1/2}. \tag{3.4}$$

Thus T^* is greater than T: moving clocks go slow. This effect of relativity is termed *time dilation*. Again, it means there is no absolute standard of time. The length of my life depends on how fast I am moving relative to the people who want to talk about such a concept. What to one observer is a day will be measured as a thousand years by someone else moving at a high enough speed relative to him. The only unambiguous time is that recorded by a clock that is not in motion with respect to the observer: it is called *proper time*.

Strange as these concepts of space and time are, they are confirmed daily in physics experiments the world over. The simplest confirmation of them is provided by our observation of muons on the Earth's surface. These elementary particles are made by collisions of cosmic rays high in the Earth's atmosphere. If the simple Newtonian picture of space and time were correct, then we should never detect any of these muons on the Earth's surface. Muons are unstable particles that decay on average after only about a one and a half microseconds. Since they are formed nearly 6000 metres above the Earth's atmosphere they would not live long enough to reach the surface even if they travelled at the speed of light. For in their fleeting lifetime they could travel only a fraction of 6000 metres. The fact that we do see them arriving at the Earth's surface is explained by the effects of relativity. To an observer sitting on a moving muon the Earth's surface is moving towards him at a high speed (muons travel at a speed close to $V = 0.999c$), and therefore the distance from the muon to the surface is contracted (according to equation (3.3)) to about 268 metres. The muon can easily travel this distance before it decays, and so cosmic muons are observed at the Earth's surface. One can also view this phenomenon from the point of view of an observer situated on the Earth's surface. It is completely equivalent. To the ground-based observer the muon is approaching at high speed and will therefore have a lifetime that is longer than would be recorded if it were at rest relative to him. The increased lifetime (given by equation (3.4)) allows the muons sufficient time, according to the clock of the ground-based observer, to reach the surface before decaying.

It should be stressed that these counter-intuitive aspects of relative space and time are not just illusions or perspectives, in the way that a body appears to have a different shape when viewed at an angle. The high-speed motion does not break the clocks or physically bend the rules in some way. They are concepts that only exist when referred to a standard of measurement. The muons really do reach the Earth's surface; they would not if space and time were absolute Newtonian concepts.

One might worry about whether it could arise that I observe an event A cause another event B, whereas some other observer could move fast enough relative to me so as to see the event B occur before A. This, fortunately, is forbidden by the rules (3.3) and (3.4): the ordering of events in time is the same for all observers. The sequence of cause and effect is preserved. What is not universal, though, is the concept of simultaneity. If I observe two events to occur simultaneously, then an observer who moves at constant velocity relative to me will not observe those two events to be simultaneous.

It was the relativity of the concept of simultaneity that fuelled the operationalist and instrumentalist philosophies of science that we discussed in Chapter 1. Experimentalists like Percy Bridgman realized that the counter-intuitive properties of space and time revealed by Einstein's special relativity undermined one's faith in everyday concepts. If such an 'obvious' concept as simultaneity had no absolute meaning because it depended completely upon the relative motion of the observer, how could one be sure that many other familiar scientific concepts were not equally ambiguous or meaningless? For this reason, Einstein and Bridgman stressed the necessity for concepts to be unambiguously measureable by a definite procedure before they could be accepted as meaningful.

In the development of special relativity we learn a number of new points of principle about laws of Nature, as well as gaining a new theory of motion. We see how experimentally authenticated laws like those of Newton can turn out to be only approximately correct because the experiments and observations support-ing them scan only a small range of conditions. In this case Newtonian mechanics turned out to be but the slow-motion approximation to special relativity. We have also lost the notion of a universal concept of space and time that is independent of the observer. Instead, each observer determines the nature of those entities relative to that determined by other observers in motion relative to him. The special relativity principles also signal a trend for laws of Nature to be framed in terms of prohibitions on what is possible, rather than direct statements of the effects of certain causes. The principles of special relativity are statements of *invariances* in Nature. They are laws about laws.

Curved space–time

> Our revels now are ended. These our actors,
> As I foretold you, were all spirits, and
> Are melted into air, into thin air;
> And, like the baseless fabric of this vision,
> The cloud-capped towers, the gorgeous palaces,
> The solemn temples, the great globe itself,
> Yea, all which it inherit, shall dissolve,
> And, like this insubstantial pageant faded,
> Leave not a rack behind. We are such stuff
> As dreams are made on.   Shakespeare

Although special relativity removes the absolute character of space and time, and replaces it by the invariance of the speed of light and the invariance of the laws of Nature for all observers in uniform relative motion, it does not propose any radical change in our picture of the nature of space and time itself. But the picture it presented was incomplete. The existence of a cosmic speed limit for the transmission of any signal was incompatible with Newton's law of gravitation. A new description of gravitation was necessary which somehow combined the invariances of special relativity with the success of the simple Newtonian law in describing the dynamics of the solar system. In 1915, ten years after his work on special relativity, Einstein achieved this synthesis. This new theory of gravitation he called *general relativity*. It contains many totally new ingredients that set it apart from all other laws of Nature. It cannot be viewed as a necessary logical extension of previous ideas, as could the special theory of relativity.

Let us think back to our sealed laboratories in which teams of scientists are carrying out experiments to determine the laws of Nature. Einstein noticed that if one of these laboratories were to be falling freely under gravity whilst the other was accelerated at an appropriate rate then so long as the laboratories were sufficiently small the occupants would not be able to determine whether they were inside the accelerating or the freely falling laboratory. It is always possible to neutralize the effects of gravity in a small enough region by applying a suitable acceleration in the opposite direction. However, since the gravity field may be changing in strength from place to place, a different amount of acceleration will be necessary to neutralize gravity at each place. There is no way of removing the gravity field everywhere with the same amount of counter-acceleration; yet, a field of accelerations could neutralize the effects of gravity everywhere. This allows one to imagine that a description of gravity could be arrived at that contains no explicit gravitational forces.

Whereas Newton had pictured gravity as a force acting between different masses all sitting in one absolute frame of space and time, Einstein wanted to do away with the Newtonian idea of space as a fixed arena in which bodies felt mysterious long-range influences called 'gravitational forces', or in which field lines were absolutely placed. In its place would be substituted a large number of local accelerated reference frames for space and time, which experience no forces at all because their gravitational forces are cancelled by their different local accelerations. The law of gravitation would have to show how all these different accelerating motions which neutralize the effect of gravity locally could be woven together to create space and time in the large. This sounds rather odd at first. After all, there is a real difference between sitting next to a large mass and being next to no mass at all. In the first case you feel a gravitational pull, in the second case you don't. In order to reconcile such experiences with the desire to remove the notion of a gravitational force and its peculiar instantaneous action at a distance, Einstein had to give up the usual picture of space and time.

Traditionally, space had been regarded as a sort of cosmic billiard table upon which the motions of matter were played out. Time was the smooth ongoing measure of the linear sequence of these motions. In these rôles space and time were independent of the events that occurred within them. Laws of Nature were laws governing what things happened 'in' time 'on' the fixed billiard table of space. Einstein sought a description in which the presence of matter in space and time would necessarily determine its geometry and temporal flow.

In 1870 William Clifford, an English mathematician, had made a speculative and far-sighted proposal that had attracted little attention. He suggested that space was more like the surface of the countryside than the top of a billiard table. It possessed little hills and valleys, so that it was only flat on the average. Further, he suggested that these undulations need not be static, but could be continually moving from place to place rather like ripples on the surface of the ocean. These undulations in the shape of space were, Clifford suggested, what this phenomenon we call motion really was. The continuous flow of space would be governed by some law of continuity just like the flow of a liquid, he proposed; but none of Clifford's contemporaries appears to have been able to make very much of this strange vision. Maxwell's attention had been drawn to Clifford's ideas by his friend Tait. In a postcard Maxwell dismisses Clifford's picture of space based upon the non-Euclidean geometry of Riemann as a conflict with his notion of absolute space as an arena in which co-ordinates were labels for the fields defined within it. He did not think, as Einstein did, of the field lines as defining the geometry of space. He scribbles hurredly to Tait that,

> the aim of the space-crumplers is to make its curvature uniform everywhere, that is over the whole of space whether that whole is more or less than ∞. The *direction* of the curvature is not related to one of the x y z more than another or to $-x -y -z$ so that as far as I understand we are once more on a pathless sea, starless, windless and poleless.

Einstein was able to develop an idea of this sort in mathematical detail to produce his new theory of gravitation. Instead of having mysterious gravitational forces attracting masses together in a flat space, Einstein proposed that the presence of mass or other forms of energy in space would distort its geometry, as if our billiard table-top were replaced by an elastic sheet. Thus, in places where there are large masses there arise deep pits in the space, whereas in places where there are no masses it will be almost perfectly flat save for some small residue from the indentations elsewhere on the sheet. The flow of time would also be affected by the presence of mass, with clocks going more slowly in the presence of masses.

Newton's first law of motion spoke of bodies feeling no forces moving in straight lines. But there are no such bodies: we expect all bodies to feel the gravitational influence of others, so it would be more natural to adopt the motion that takes place under gravity alone as the baseline, rather than the state of motion under no forces. If the gravitational effects of mass are accounted for by the curved geometry of the space it creates, then the motion that takes place in that geometry when no other forces are applied will be the analogue on a curved surface of a straight line. This 'straight line' is simply the quickest way to travel between two points on a flat surface. We know that if we want to travel most expediently on a surface that is not flat, then the quickest route between two points will not be a straight line. For example, if we want to fly from London to

San Francisco the quickest route over the Earth's curved surface is along the arc of a 'great circle' passing close to the North Pole and down the West coast of North America. These optimal routes on curved surfaces are called *geodesics*.

A body moving under no forces will move along its geodesic route on the rubber sheet of space–time. If there are no masses sitting in space–time then the sheet will be flat, and the body's geodesic path will be a straight line. But if there is a large mass in space–time it will cause a deep depression in the geometry, so that when the moving body passes by this trough it will continue along the geodesic route through the curved geometry just as a mountain stream follows the geodesic route as it meanders down from its source to the sea. The effect of the trough in space is to make the body move towards the large mass, but it does so only by following the geodesic route *locally*, not because of any long-range force emanating from the distant mass. Now there is no distant mass in the old sense, only a distortion of the space–time geometry. Thus, instead of long-range gravitational forces we have bodies taking their marching orders from the local topography of space–time, which is determined by the presence of the masses within it. Just as twists and turns in the route of the mountain stream are dictated by the *local* gradient of the ground it encounters, so the motions of bodies in space–time are determined by the *local* curvature they encounter. There are no gravitational forces; no instantaneous actions at a distance; no absolute space and time independent of the objects that exist within it. In John Wheeler's words: 'Space tells matter how to move. Matter tells space how to curve'. Mass and energy are nothing more than wrinkles in the geometry of space.

Einstein's greatest achievement was to find the extraordinarily complicated set of mathematical equations which tells us how to determine this symbiotic relationship between matter and space–time geometry. These equations are field equations like those of Poisson and Laplace. In every site where the unusual predictions of this picture of curved space–time have so far been checked by observation they have been found correct to high precision.

This last reference to the predictive success has even more significance than usual in the context of Einstein's theory of gravitation. Einstein's theory cannot be deduced from observation alone. It contains steps that need not, as far as we understand, be true. In particular, the class of geometrical types which space and time are distorted into by the presence of mass and energy are taken to be of a particular sort that was first studied systematically by Riemann, a student of Gauss at Göttingen. This is just an hypothesis. It could have led to a contradiction with what is observed. There is no known reason why the world should be expressible as a curved geometry in such a way that, when the curvatures are small, it looks as though it obeys an inverse-square law of force of the Newtonian sort.

An amusing and perceptive résumé of Einstein's recasting of gravitational forces as the curvature of space in contrast to the inverse-square force acting in the Newtonian arena of Absolute Space was provided by George Bernard Shaw, when asked to propose an after-dinner toast to Einstein. According to his secretary, the Shavian 'Irish' account went like this:*

As an Englishman [Newton] postulated a rectangular universe because the English always use the word 'square' to denote honesty, truthfulness, in short: rectitude. Newton knew that the universe consisted of bodies in motion, and that none of them moved in straight lines, nor ever could. But an Englishman was not daunted by the facts. To explain why all the lines in his rectilinear universe were bent, he invented a force called gravitation and then erected a complex British universe and established it as a religion which was devoutly believed in for 300 years. The book of this Newtonian religion was not that oriental magic thing, the Bible. It was that British and matter-of-fact thing, a Bradshaw [English railway timetable]. It gives the stations of all the heavenly bodies, their distances, the rates at which they are travelling, and the hour at which they reach eclipsing points or crash into the earth. Every item is precise, ascertained, absolute and English. Three hundred years after its establishment a young professor rises calmly in the middle of Europe and says to our astronomers: 'gentlemen: if you will observe the next eclipse of the sun carefully, you will be able to explain what is wrong with the perihelion of Mercury.' . . . The young professor smiles and says that gravitation is a very useful hypothesis and gives fairly close results in most cases, but that personally he can do without it. He is asked to explain how, if there is no gravitation, the heavenly bodies do not move in straight lines and run clear out of the universe. He replies that no explanation is necessary because the universe is not rectilinear and exclusively British; it is curvilinear. The Newtonian universe thereupon drops dead and is supplanted by the Einstein universe. Einstein has not challenged the facts of science but the axioms of science, and science has surrendered to the challenge.

The reference to the perihelion of the planet Mercury is made because by 1900 it had been realized that there was a discrepancy between the observed orbit of the planet Mercury and the predictions of Newton's theory of gravitation based upon the simple inverse-square law. Mercury, like all the other planets, moves in an orbit around the Sun that is not a perfectly closed ellipse. If the only force upon the planet was that exerted by the Sun the orbit would be a closed ellipse according to Newton, but in practice it feels small perturbations from all the other bodies in the solar system. The result is an orbit that does not quite close, but very slowly precesses around to form a rosette shape like that shown in Figure 3.8. The angle between the extremities of successive orbits (the 'perihelia') is called the angle of advance of the perihelion.

No matter how they tried astronomers could not account for the magnitude of

---

*Shaw also digressed upon this subject in his play *Too True to be Good*, published in 1934, where he laments that 'The Universe of Isaac Newton, which has been an impregnable citadel of modern civilization for three hundred years, has crumbled like the walls of Jericho before the criticism of Einstein. Newton's universe was the stronghold of rational Determinism . . . Everything was calculable: everything happened because it must: the commandments were erased from the tables of the law; and in their place came the cosmic algebra: the equations of the mathematicians.' The latter point is presumably a reference to the considerable mathematical complexity involved in the formulation of Einstein's theory of general relativity.

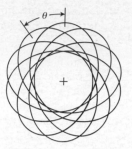

Figure 3.8 A 'rosette' orbit in which the orbit closes after ten circuits. The farthest point of each circuit (the 'perihelion') advances by the angle θ during one orbit as indicated.

Mercury's perihelion advance. The external perturbations failed to explain 43 seconds of arc per century of the actual rate of advance. *Encyclopaedia Britannica* articles at the end of the nineteenth century even suggest *ad hoc* modifications of Newton's inverse-square law which would be able to explain the discrepancy. The greatest success of general relativity was to explain the discrepancy precisely. The effect of the curvature of space due to the Sun's presence is to create an extra 43 seconds of arc per century in the perihelion advance as Mercury follows its geodesic route through curved space–time.

Einstein's theory of gravitation is not just another way of looking at Newton's. Newton's theory gives the wrong description of the behaviour of bodies moving in strong gravitational fields where space–time is significantly curved. Of course, it works well enough in the world of the relatively weak gravity fields which we experience on the Earth's surface. In such a situation Newton's theory is a very good approximation of Einstein's. This is another salutary lesson. We can use a law of Nature for hundreds of years without any adverse results, and build an entire metaphysical view of Nature's mechanical workings upon it, yet find that it is just a little piece of a vast and entirely dissimilar scheme.

In the great step from the Newtonian to the Einsteinian picture of matter and motion there is another interesting feature. Not only did the form of Newton's laws change, but the *meaning* of its fundamental concepts, mass, length, and time would never be the same again. The real extent of this change is masked because the words describing these entities have stayed the same. Most remarkable of all is that Einstein's extraordinary insight, which led to the creation of the general theory of relativity, was not sparked by observations: neither crisis nor adverse experimental facts had made Newton impossible to live with. The anomaly of the perihelion precession was regarded at the time as a minor irritant rather than as a fundamental problem requiring a new theory of gravitation. There was no paradigm shift of the Kuhnian variety (see Chapter 7). The rest of the world's physicists were busy with other completely different problems. It was an insight whose intuitive depth did away with the need for revolutions, and whose novelty still sets it apart from other laws of Nature.

Invariance

> If the universal law of Nature should be discovered, invariance
> principles would become merely mathematical transformations
> which leave the law invariant.
>   Eugene Wigner

During the Stalinist period in the Soviet Union many of Einstein's physical theories were opposed by the organs of the state for political reasons (though physics suffered lightly compared with other subjects, notably biology, at the hands of the ideologists). The subversive feature of special relativity appears to have been the popular notion that it says that 'everything is relative'. A little reflection reveals this journalistic catch-phrase to be meaningless, but faced with political threats, some Soviet physicists astutely emphasized that Einstein's theory could more appropriately be named the 'theory of invariance' rather than the 'theory of relativity'. *Relativity* is an appropriate term to describe the character of things observed in space and time: their lengths, lifetimes, and masses are measurable concepts which are meaningful only by reference to a particular class of observers, who are defined by their motion relative to the object whose properties are being measured. But the reason different observers have different measures of length and time is because two things are *not* relative: the speed of light is measured to have the same value no matter what the state of motion of the observer, and the laws of physics are the same in all laboratories moving with respect to one another at a constant speed. That is, they are *invariant* under a particular type of change. The relativistic laws of Nature are arrived at on the assumption that Nature adheres to certain invariance principles.

Ideally a description of how Nature operates should not depend upon the describer. If it did, then we would be unable to talk to other people about the structure of the world in an unambiguous way. And because there is no reason why the view of any particular observer should have a higher status than that of any other, it is desirable that we should have laws of Nature whose form does not depend on the observer who makes them. This sounds simple and obvious, but it turns out to be far-reaching and profound in its ramifications. There are just so many ways in which the form of the laws of Nature *could* depend upon the state of motion of the observer of them that to require the contrary places many interwoven constraints upon them. They are forced to take on very specific forms. Notice that what is being said here is that the *laws* of Nature found by different observers should be the same. These different observers need not find that all the quantities appearing in those laws take the same values. For example, they will all deduce that momentum is conserved in a collision between two particles, even though they may observe different masses and velocities because of relativity. There is only one observed value, as opposed to law, that they can agree on: the speed of light has the same value for all observers even though the lengths and times they measure in order to see how far light travels in an interval may differ.

The lesson is that observer-independence applies to the relations between events, not to the events themselves.

Einstein regarded the presentation of laws of Nature in a mathematical form that is the same for all observers as a principle of paramount importance. Newton's laws of motion do not have this property. The second law for a body of constant mass, m, is that an impressed force, F, endows it with an acceleration, A, given by

$$F = mA \tag{3.5}$$

But this law will not be the one found by *all* observers. Only a special class of observers, sometimes called *inertial* observers, moving at constant velocities will observe (3.5) to hold. Suppose another observer who is moving with an acceleration A^* relative to the inertial observers carries out measurements of the impressed force and the acceleration it produces. He will find that there exists a law of motion of the form

$$F = m(A + A^*) \tag{3.6}$$

and not (3.5).

The same restriction to a special class of inertial observers applies to the first law of Newton—that bodies acted upon by no forces move at constant velocity. If a satellite is moving through space far from the gravitational effects of any star or planet, and with no source of propulsion, it would be observed to move at constant velocity by an inertial observer. But, if we were to observe the satellite from inside a spinning rocket we would see the path of the satellite spiralling around us. We would conclude wrongly from seeing its path deviate from a straight line that it must be acted upon by external forces. Again, we have encountered the fact that Newton's laws do not apply directly to the observations made by accelerating observers. The law of motion for the satellite must contain additional factors to account for the non-uniform motion or rotation of the observers. These corrections are called Coriolis 'forces', after the nineteenth-century French physicist Gustave de Coriolis. They are fictitious forces used to obtain the Newtonian laws of motion when bodies are viewed from a rotating reference frame.

Einstein regarded this state of affairs as unsatisfactory. Laws of Nature should be expressed in a form that remains the same for *all* observers. This he called the *Principle of Covariance*. In his own words:

> The general laws of Nature are to be expressed by equations which hold good for all systems of co-ordinates, that is, are generally covariant with respect to any substitutions whatever.

This is not a law of Nature in any sense, merely an important guiding principle for their expression. Any law of motion written as a differential equation can be put into a covariant form if one goes to elaborate lengths* that get longer the less satisfactory the proposed law. Einstein struggled for many years to develop his theory of general relativity into a form that was fully covariant. This he achieved with assistance from his friend Marcel Grossman, whom he had known since their student days. Grossman was an able mathematician who introduced Einstein to areas of pure mathematics that were to prove perfect for his purpose.

During the nineteenth century mathematicians had completed the study of a collection of mathematical objects called *tensors* (of which scalars and vectors are particular cases) that are defined by their transformation properties upon changing the co-ordinates they are expressed in. They possess the property that, if any equation involving tensors is true when written in one set of co-ordinates then it must be true *in the same form* in any other choice of co-ordinates. Einstein realized that if the laws of Nature were always written as tensor equations, then if one set of observers found them to have to form, say,

$$S = T \tag{3.7}$$

another set of observers moving in a completely arbitrary way relative to the first set would carry out measurements of the two quantities in the law, and find them to be S^* and T^* where

$$S^* = T^* \tag{3.8}$$

Although S and T differ from S^* and T^*, the form of the law connecting them is the same in (3.7) and (3.8). This is in contrast to the situation with (3.5) and (3.6) where the second set of (non-inertial) observers see a law of a different form. These properties of tensor equations mean that they automatically incorporate the feature that laws of Nature should not depend in any way upon the observers who make them. There is no special set of co-ordinates: no privileged observer for whom all of Nature looks simple. This does not mean that there are not special co-ordinates which simplify the descriptions of particular problems, just that there is no choice that simplifies *all* problems.

These developments are interesting in another respect. Before Einstein's work, scientists had made use of a fairly well-defined mathematical tool-kit. The applications of mathematics did not make use of the 'pure' mathematicians' work. General relativity, by its deployment of tensors and the Riemannian geometry of curved surfaces, was the first physical theory to take up advanced pure mathematics which mathematicians had developed independently of applications. Whereas Newton invented much of his mathematics in order to solve physical problems, Einstein took man-made logical structures, and found them to be beautifully adapted to his needs. If pure mathematics had not taken

---

*If one requires that the laws of Nature possess the stronger property of 'invariance', that is they are covariant *and* all the mathematical objects whose values are unaffected by matter (for example, mathematical derivatives of co-ordinate variables with respect to other co-ordinate variables) are left unchanged, then particular mathematical equations are selected.

the line of development that it did, when it did, Einstein would not have developed his law of gravitation.

Symmetry

> The trouble with facts is that there are so many of them.
>
> Samuel McChord Crothers

There has been a trend in the development of our human codification of Nature's operations. At first we could do no more than list the occurrence of events. The astronomical data we have inherited from the ancients is one of the positive results of this habit. During the early days of this quest attention was paid primarily to the irregularities and eccentricities of Nature, whether they be eclipses or shooting-stars. Later, we find a growing appreciation and exploitation of the regularities of Nature. This led to the notion of natural laws. The next step is to seek the most economical representation of those laws, and to find single principles which give rise to collections of laws. In this way laws come to be treated almost like the events they govern.

Laws of Nature are laws of change. They tell us how bodies will move under various forces or in particular circumstances. If such laws did not dictate the motion of things then they could change in any way at all. The fact that possible changes are not arbitrary means that something is preserved in any change that takes place in accord with the law. We say that there is an *invariance*. To a considerable extent the laws of Nature are just a catalogue of what these unchanging elements of Nature are. Such unchanging properties are called *conserved quantities*, and the statements that these quantities remain unchanged in Nature are called *conservation laws*. The idea carries over to other simple games which have their own microcosm of imposed rules. For instance, chess possesses invariances associated with the 'laws' that govern how each piece is allowed to move: the colour of the square which a bishop sits on is always the same.

This emphasis upon what the laws of Nature leave unchanged rather than upon how any changes do occur is a modern one. It leads ultimately to a view that the laws of Nature are but a catalogue of those things that we can do to the Universe that leave it unchanged. In the next chapter we shall see this approach come to fruition in the study of elementary particles.

The words we have been using to describe a conserved quantity of an invariance are similar to those we would use to convey the idea of a pattern of symmetry. We say that something is symmetrical when it possesses some special type of harmony or unity in the face of superficial diversity. This harmony appeals to the human eye, as is witnessed by the decorative patterns that adorn the walls of (some of) our houses. If we fix attention upon the simplest type of pattern (Figure 3.9), then we can recognize how the harmony is generated: one side of the picture is the mirror image, or reflection, of the other side.

There is a natural connection between this reflection symmetry and the idea of invariance. For if we replace one side of Figure 3.9 by a mirror image of the other side (say, by putting a mirror along a vertical central line) then the Figure will be unchanged. The pattern is said to be invariant under mirror reflections along this axis. It is easy to think of other examples where an invariance with respect to motion is a consequence of a geometrical symmetry. If we rotate a spherical ball about a vertical axis through its centre then it appears the same. There is a rotational invariance. This example is interesting because the same invariance is witnessed if the observer rotates around the stationary ball. This is not true of the example in Figure 3.9, since there does not exist a motion of the observer which mimics the reflection symmetry.

These simple examples display the connection between symmetry and invariance. This is a connection that holds good for any invariance, although the symmetry concerned may be quite subtle. The simplest and most important pairings of symmetries and invariances are given as follows:

Laws of Nature invariant under the operation of:	*Conserved quantity*
Translation in space	Linear momentum
Translation in time	Energy
Rotation in space	Angular momentum

These beautiful relationships reveal the extent to which deep symmetry was unknowingly built into the laws of motion found by Newton. The principal conservation laws are consequences of the fact that the laws of Nature do not depend upon the position, orientation, or time at which they are observed.

So impressed have scientists been with this symbiotic relationship between symmetry and invariance that they have elevated symmetries to a higher station than the traditional equations that tell them how things change. What could be simpler as a law of Nature than the statement that *nothing changes*? An understanding of what symmetries Nature should preserve will enable laws of Nature's changing processes to be deduced. But as we shall see, Nature has not made things quite so simple. It transpires that there exist a number of *almost* symmetries in Nature which we had thought for a long period to be precise before very accurate experiments were carried out. More awkward still, symmetrical laws do not necessarily give rise to events which possess that same symmetry. The laws of motion do not prefer one direction in space over any other, but perch a ball symmetrically on the apex of a cone, and it will surely fall in one direction or the other. All the directions are equally probable, none has any special significance: but this symmetry will be hidden by the particular motion that results in any outcome governed by the law.

The fact that laws of Nature can be identified with *exact* symmetries does not

Figure 3.9 An example of a pattern which possesses reflection symmetry about a vertical line through its centre.

seem to have been adequately dealt with by those, like Cartwright, who believe that the stated equations of physics never apply because they are idealizations which never encompass everything that is happening in a particular situation. However, these equations are but secondary representations of exact invariant principles which are maintained to hold regardless of the number of competing forces at work.

The laws of chance

> Research in physics has shown beyond the shadow of a doubt that in the overwhelming majority of phenomena whose regularity and invariability have led to the formulation of the postulate of causality, the common element underlying the consistency observed is chance.
>
> Erwin Schrödinger

Thermodynamics was the last great corner-stone of nineteenth-century science. Unlike the other laws of Nature we have encountered it does not contain statements about how individual particles will respond to the influence of particular forces. It is not connected with symmetries or invariances. Rather, it tells us about the behaviour of whole collections of particles. And instead of telling us that 'if . . . then . . .' it tells us simply 'if . . . then maybe . . .'.

This introduction does not make thermodynamics seem like a very reliable tool. Suppose one were to run a business using a decision-making procedure which could be no more definite than 'if . . . then maybe . . .'. Surely such a commercial enterprise would proceed rapidly to ruin?

Yet some of the world's most profitable businesses are based upon the pattern produced by a very large number of such 'maybes'. Large insurance companies must successfully exploit the laws of chance in order to stay in business. Their evident prosperity witnesses to the effectiveness with which this can be done. An insurance underwriter cannot predict when your car will be involved in an accident when he signs you up for an insurance policy. Nor is there any law of Nature that would allow him to make such a prediction. But what he can do is determine the *chance* of someone with your sex, age, and driving record being involved in an accident. That is, given a large number of individuals of similar profiles to yourself, what fraction of them having accidents each year, and what is the average insurance claim they make. This he does by looking at the statistics determining the profiles of accident victims. The result is that he can quantify his statement that 'if *you* drive a car for a year then *maybe* you will have an accident'. The more skilfully he can assess his risk by considering very large samples of accident cases, the more reliably he can estimate his likely risk, and charge a realistic premium. He can never be sure that he will not be the loser in the short run, and if his data on accidents of the sort he is insuring against are very limited then his assessment of risk will be poorer. As in the case of insuring satellites on board the space shuttle, he could lose a fortune in short-term claims.

Suppose you stand in the street, and record the heights of all the men who pass by. Gradually you can accumulate the frequencies with which individuals having various heights pass by. Although you can never predict what the height of the next passer-by will be, and may know absolutely nothing about the human physiology that determines our size, what you can predict is something about the fraction of individuals that you find in each height interval. As the number of men recorded gets bigger and bigger, so the relative number will tend to approach a distribution having the characteristic 'bell' shape shown in Figure 3.10. There is no communication between the different passers-by. There is no way that they can correlate their arrivals to make sure you get a particular fraction of tall and short men. Nevertheless, the greater the number of passers-by, the better should be the match of the distribution of their heights to the bell-shaped curve. The reason why we are able to make a statistical prediction as to the distribution of heights when we cannot predict the height of the next passer-by with certainty is a consequence of the fact that the height of each person is totally independent of the height of the others. That is, it is the complete lack of any correlation between the heights of different passers-by that enables us to predict the overall pattern in the distribution of heights. Of course, every now and then a random distribution will

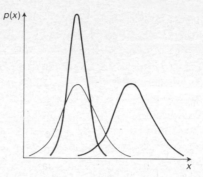

Figure 3.10 The gaussian or 'normal' probability distribution for three different values of its two defining parameters which fix its centre and spread. Each distribution has the same characteristic 'bell' shape and has the same area underneath it. This probability pattern is approached by any large number of independent random events.

throw up some flukes. Two twins with identical heights will come along; a basketball player or a pygmy might pass by occasionally. But in the long run these flukes will be completely outweighed by the randomly run-of-the-mill. The longer the run the closer the agreement with a bell-shaped curve. Not every random process like this will generate the same bell-shaped curve. It will differ in the position of its peak and in its width. The peak will be determined by the typical value (what we might call the 'average'), and the width by the propensity of the sequence of events under study to have a good chance of deviating strongly from the average.

This example serves to illustrate that there can exist laws which are obeyed *statistically*. A high degree of regularity can arise by a concatenation of chance events. These laws are independent of the detailed character of the objects forming the sequence of events (in our example they could have been the heights of passing women, or even passing mice). They will be more accurate when applied to very large collections of objects. They tell us the most probable behaviour of our system, the number of ways that an event can occur in a particular way compared with the total number of possible outcomes of all sorts. The ratio of the two we call the *probability* of that particular event. Thus, the probability of throwing a six with one throw of a die is one in six, and this is the same as the probability of throwing a score of seven with two dice, because in the latter case there are $6 \times 6 = 36$ possible combinations of which only six add to seven: they are the throws where the two dice have the values (6,1), (1,6), (2,5), (5,2), (3,4), or (4,3).

Resort to this type of second-best knowledge that tells us only the chance of something happening seems a poor alternative to the beautiful formulae of Newton, Maxwell, and Einstein. Why bother with it you ask? Unfortunately, it is forced upon us by the complexity of Nature. Newton's laws of motion allow us, in principle, to describe the motions of all the one hundred billion or so stars in our own Milky Way galaxy (the corrections to this description that Einstein's theory would add are tiny because the gravity fields involved are comparatively weak by cosmic standards). However, in practice the problem of their solution is intractable. As yet mathematicians have been unable to solve the equations for the motion of even three masses moving in orbit around each other under the influence of their individual gravitational attractions. Yet astronomical problems present us with systems that involve billions of stars in motion together. Not even the biggest and fastest supercomputers can solve the array of equations governing the path of each star as it is buffeted by the gravitational tugs of the myriad others, first one way then the other. We are in a similar quandary if we attempt to apply our laws of motion to the gas molecules in a typical room. There are so many of them, and their individual motions are so intricately connected to those of others, that knowing the *laws* governing their individual motions is not really of much use. We just cannot solve the equations they give us.

In this cloudy circumstance a compromise is sought, but it turns out to have an unexpected silver lining. Although the problem of three moving stars is too hard for us to solve, that of billions of stars is not billions of times harder still. When very large numbers of objects are involved then statistical trends will start to dominate events. Although we cannot predict the behaviour of every single star precisely, we can predict the probability of finding a star at a particular position moving with a particular speed in a particular direction. Likewise in a gas of air molecules at a particular temperature, we can specify the probability distribution of speeds. The prediction is the so-called Maxwell–Boltzmann distribution shown below.

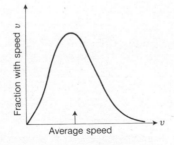

Figure 3.11 The Maxwell–Boltzmann frequency distribution of velocities of the molecules in a gas in statistical equilibrium. The average speed of particles of mass m within a 'gas' possessing this distribution of molecular speeds is $(8kT/\pi m)$ where k is the Boltzmann constant for converting temperatures into energies and T defines the temperature.

A very neat test of this statistical description of the speeds of gas molecules was performed in the 1930s by I. Zartman and C. Ko. They aimed a narrow beam of Bismuth vapour with a temperature of about 800° Celsius towards a rotating drum. The drum had a narrow slit at one point on its boundary, and this was the only place where the vapour could penetrate the walls of the drum. On the inside of the drum, immediately opposite the slit was attached a photographic film; the drum was then set in rotation at about 6000 revolutions per minute (about 133 times faster than the playing speed of a single record). When the narrow slit was facing the incoming stream of vapour the molecules could pass to the inside of the drum and strike the film on the other side of the drum. The faster-moving molecules would reach the film after it had rotated only slightly. By contrast, the more slowly-moving molecules that got through the slit would reach the film only after it has rotated much further. The developed film displayed an exact picture of the distribution of molecular speeds in the Bismuth vapour: it was indeed the distribution predicted by Maxwell and Boltzmann.

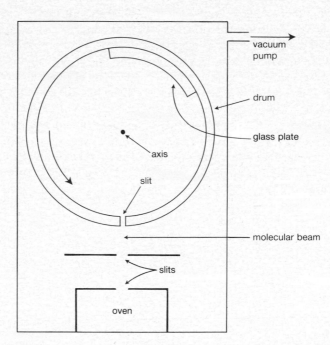

Figure 3.12 The ingenious experimental set-up used by Zartman and Ko to display the spread of molecular velocities in a gas in thermal equilibrium. Molecules are generated with a distribution of velocities in an oven and then collimated into a beam which is fired towards a slit in a rotating drum. The pattern left upon the glass plate on the opposite side of the drum displays the spread of velocities. [Taken from Beiser, A. (1967). *Concepts of modern physics*, p. 247 with kind permission of McGraw-Hill.]

The probabilities we have been discussing in these physical applications have been assumed to be defined unambiguously by analogy with the simple situation we discussed involving the probability of throwing a given total score with two dice. The probability of a particular result is equal to the number of ways in which that result can occur divided by the total number of all possible outcomes. In order for such a definition of probability to be coherent, it must be quite unambiguous what are the basic events which we judge to be equally probable a priori. Only then can we find the correct probability by the calculation of a relative frequency of events. In the case of the dice-throwing the identity of these equally probable elementary events is clear; but there are disingenuous cases where the identification of the equally probable events is by no means unique. There are many possible choices, and the choice that is made will determine the numerical probability that is associated with a particular outcome. A classic example of this impasse is provided by Joseph Bertrand's problem of determining the probability that a randomly chosen chord of a circle is larger than the side of the equilateral inscribed triangle. In Figure 3.13 we have displayed three solutions to the problem. Each gives a different answer! All are correct! Although the problem sounds well-posed, it possesses an ambiguity which allows different elementary events to be attributed equal probabilities. The particular choice one makes will determine whether the answer to Bertrand's problem is one-half, one-third, or one-quarter. Thus the precise assessment of chance cannot be entirely mechanical. Some subjective assessment of the 'global' nature of the problem is necessary in order that the meaning of its solution be clear.

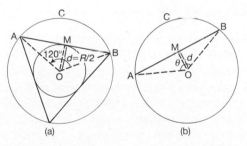

(a)   (b)

Figure 3.13 Bertrand's Paradox. We wish to calculate the probability that a *randomly* drawn chord of a circle is longer than the side of the circle's inscribed equilateral triangle. The two diagrams indicate some of the constructions we can make as aids to calculating the answer. Let us choose the chord to be ACB and mark M the mid-point of the line AB so OM is at right angles to AB. The radius of the circle is R, the distance OM is d. First, we claim that d is equally likely to lie anywhere in value between 0 and R. Clearly, then we see from (a) that a randomly drawn chord will exceed the side of the inscribed triangle only if d is less than $R/2$. The probability of this occurring is therefore $1/2$. This is the first answer to Bertrand's problem. Next, take the view that M is equally likely to lie anywhere within the disc of the circle under consideration. We see that the length of a randomly drawn chord will only be longer than the side of the inscribed triangle if M falls within the interior of a circle of radius $R/2$. The probability required is therefore the ratio of the area πR^2 to $\pi(R/2)^2$; that is $1/4$. Third, we can consider the angle $\widehat{AOB}$ to be equally likely to be found anywhere between 0 and 360 degrees. The desired event will only now occur if the angle $\widehat{AOB}$ lies between 120 and 240 degrees. The required probability is hence $(240-120)/360 = 1/3$. In each case the choice of equally likely events had been made in a different way.

Thermodynamics

> The law that entropy increases—the Second Law of thermo-
> dynamics—holds, I think, the supreme position among the laws
> of Nature. If someone points out to you that your pet theory of
> the universe is in disagreement with Maxwell's equations—then
> so much the worse for Maxwell's equations. If it is found to be
> contradicted by observation—well, these experimentalists do
> bungle things sometimes. But if your theory is found to be
> against the Second Law of thermodynamics I can give you no
> hope; there is nothing for it but to collapse in deepest
> humiliation.
>
> A. S. Eddington

The regularities in the behaviour of large aggregates of particles, whether they be
atoms or stars, gives rise to a 'statistical mechanics' of matter. The beauty of this
approach is that simple statistical quantities, like the average speed of the
particles, determine the macroscopic quantities like temperature that we are
familiar with. The macroscopic consequences of the statistical mechanical
disorder in Nature give rise to three important statistical laws whose scope is very
wide. They apply to physical processes of all sorts as far as we can judge, and they
are known as the three laws of thermodynamics.
The first law states the conservation of energy:

> The total energy in a closed system cannot be changed.

It was first formulated by the German physicist Rudolf Clausius in the first half of
the nineteenth century, following a series of investigations he was carrying out to
determine the properties and limitations of steam engines as sources of power.
What the first law establishes is that, although one can do things within a closed
and perfectly insulated region which changes energy from one form to another—
for example when you wind up your watch you convert chemical sources of
energy within your body into muscular energy of motion, which in turn stores
elastic 'potential' energy in the tightening coil of the watch spring, which is slowly
reconverted into energy of motion as it drives the movement of the watch
hands—you cannot destroy energy. You can only transform it into other forms.
The discrimination between 'kinetic' and 'potential' forms of energy arose in the
1850s. Kelvin first distinguished these forms as 'dynamical' and 'statical', whilst
Rankine called them 'actual or sensible' and 'potential or latent'. The current
terminology was fixed by the usage of Kelvin and Tait. Rankine states explicitly
in 1853 a conservation law that recognizes the exchange of different forms of
energy:

> The law of conservation of energy is already known, viz. that the sum of the actual
> [kinetic] and potential energies in the universe is unchangeable.

Einstein's later discovery that mass and energy are equivalent ($E = mc^2$) does not undermine Clausius's principle of energy conservation; it merely reveals that mass is a disguised repository of huge quantities of energy, and needs to be accounted for in the energy budget when checking the laws of thermodynamics. If I possessed a nuclear-powered watch, then mass would be converted into energy via Einstein's formula in order to move the hands (this design is not recommended). This equivalence between mass and energy means that the conservation of energy is now often termed the conservation of mass-energy.

Clausius's law has all sorts of interesting consequences. It tells you that you cannot create energy out of nothing spontaneously, and it tells you that you cannot build what have become known as perpetual-motion machines, that is, machines which continue to operate indefinitely without any external source of power. In practice any such invention would have to overcome various frictional resistances, using energy to do so. This energy will be converted into waste heat, and because the total quantity of energy is unchanging, the energy of the machine's motion must decrease. Eventually, all its energy of motion will be degraded into heat by the resistances. Of course, engineers design joints and bearings to be as smooth as possible, and we use lubricants so as to reduce the loss of energy by friction. This can defer the demise of a perpetual-motion machine, but it can never alter its inevitability.

Clausius did more than notice that the total energy can be shuffled about into different forms but never created or destroyed. He noticed what is apparent from our description of the operation of real machines. Some very ordered forms of energy, like energy of motion or the chemical energy released when we strike a match, invariably end up as heat energy, but we rarely see the reverse sequence of events. When we apply the brakes on our bicycle the energy of motion is converted into heat within the brake blocks, together with a little sound energy. The heat is itself a motion. An increased temperature of the brake blocks corresponds to an increase in the energy of the molecular motion within the rubber of which they are composed. The only way one can go the other way, and drive the motion of a machine from heat, is if there exists a difference in temperature between two regions within it. In that case the tendency for the temperature to be equalized will create a systematic motion. If a region has uniform temperature then none of its heat energy can be converted back into energy of motion. This indicates that in some way Nature is a one-way street. Energy tends to distribute itself away from large-scale ordered forms like bicycling into the most disordered microscopic form: heat radiation. This degree of disorder Clausius eventually called 'entropy', and it was governed by his famous Second Law of thermodynamics announced in 1865:

> The total entropy of any closed system can never decrease.

It is this law that explains why you cannot make your car go uphill when the engine is off simply by steadily lowering its temperature as the first law, that energy is conserved, would indeed allow if it were not backed up by the second. A lack of familiarity with it by the non-scientist was as shocking to C. P. Snow as the spectacle of a scientist who had never heard of Shakespeare. In his famous lecture, *Two Cultures*, he remarked that

> A good many times I have been present at gatherings of people who, by the standards of traditional culture, are thought highly educated and who have with considerable gusto been expressing their incredulity at the illiteracy of scientists. Once or twice I have been provoked and have asked the company how many of them could describe the second law of thermodynamics. The response was cold; it was also negative.

The Second Law of thermodynamics is probably the most powerful general principle of science. It applies without exception to all the physical processes we have encountered. If it applies to the entire Universe as a whole, then it has fundamental consequences for its future evolution. It is this law of thermodynamics that is responsible for the notion of the 'Heat Death' of the Universe which had such a pessimistic effect upon many philosophers after it was popularized so enthusiastically by Jeans and Eddington in the 1930s. It has also become notoriously misused by non-scientists. The most persistent example is the argument of some American creationists that the Theory of Evolution cannot be true, because it is in conflict with the stipulation of the Second Law that disorder must increase with time. One can see how such a view might arise: evolutionary biologists provide evidence that complex, highly ordered life-forms have developed slowly by a sequence of minute and not improbable evolutionary changes from simpler, less ordered forms by the inevitability of natural selection. This appears to imply that disorder evolves to order in contradiction to the Second Law. Where is the fallacy?

The Second Law was stated for a closed region, that is, for a region that is isolated in that it does not have sources of energy entering through its boundaries. The surface of the Earth where evolution occurs is not an isolated system in this sense. It receives energy from the Sun, whose surface temperature is twenty times higher than that of the Earth. If one takes this into consideration then the total accounting of the evolution of entropy in the evolutionary process must include the heat flow into the terrestrial system as well. When this is accounted for the Second Law is found to hold as required. The fallacy of the original objection is similar to that which one might lodge against a carpenter who has constructed an elegant chair from a pile of old timber. The Second Law does not prevent the carpenter converting disordered wood into a highly ordered piece of craftsmanship, because in the process of its construction the energy he applies with his hands and tools produces a large amount of heat. When this external heat is counted into the entropy budget it is always equivalent to a bigger increase in entropy than the local decrease created by the organization of the pieces of wood. When systems experience throughput of energy from outside, various exotic things can occur, which result in the spontaneous appearance rather than the disruption of ordered structures. (A flame is an example.) Such 'open' systems are the subject of intense study in many areas of contemporary science.

There is a third law of thermodynamics that governs the allowed behaviour of aggregates of particles. This states that

a system cannot be cooled to absolute zero in a finite number of steps.

To understand this we recall that the temperature is a measure of the average speed of the molecules in a gas. It is, therefore, possible to envision a situation whereby so much energy of motion has been extracted from the molecules that they are all at rest. This state is called 'absolute zero', and it corresponds to minus 273 degrees on the centigrade scale. Scientists who deal with the behaviour of gases and liquids at very low temperatures find it convenient to measure temperature on the Kelvin scale. This is denoted by the number of centigrade degrees above absolute zero; thus, the temperature today, 15° centigrade, is 288 degrees on the Kelvin scale.

To summarize these three laws of thermodynamics—the conservation of energy, the non-decrease of entropy, and the unattainability of absolute zero—in a more colloquial fashion we might say that they tell us first that 'you can't win', second that 'you can't even break even', and third 'you can't even get out of the game'! Less seriously still, it has been remarked that capitalism is based upon the false premiss that you can win, and socialism upon the false premiss that you can break even, with the trio of misconceptions completed by mysticism which is based upon the false premiss that you can get out of the game!

It is important to recognize that the laws of thermodynamics are statistical in character. They apply to large aggregates of particles, not to individual molecules. We discern them because we are so much larger than atoms and molecules, and so experience the collective effects of large numbers of them. They represent gross statistical properties of systems of particles which make no reference to the identity of the individual particles or the other forces of Nature that are acting upon them. It is for these reasons that Eddington, in the remarks we quoted at the head of this section, stressed the pre-eminent nature of the laws of thermodynamics in the minds of scientists. This should not be taken as a wilful commitment to maintaining the truth of the laws of thermodynamics in the face of any amount of contrary evidence, but rather an indication of the fact that, because the laws of thermodynamics do not rest upon the detailed nature and properties of the elementary constituents of the world, they need not be shaken if our ideas about the nature of these entities changes.

Untidy desks

If the motion of every particle of matter in the universe were precisely reversed at any instant, the course of nature would be

simply reversed for ever after. The bursting bubble of foam at the foot of a waterfall would reunite and descend into the water; the thermal motions would reconcentrate their energy, and throw the mass up the fall in drops reforming in a close column of ascending water ... living creatures would grow backwards, with conscious knowledge of the future, but no memory of the past, and would become again unborn.
Lord Kelvin

The introduction of statistical laws to describe the behaviour of huge numbers of particles, be they stars or gas molecules, should not be interpreted as indicating that there is anything intrinsically random or uncertain about the motions of these objects. If we were clever enough to solve Newton's equations of motion for all the trillions of mutually interacting bodies in a room full of air molecules, we would find exact trajectories for each particle—but we cannot. Instead, we exploit the fact that the very large number of bodies produces a type of predictable randomicity, and find that the most probable behaviour of the system obeys gratifyingly simple rules. It is characterized by average quantities of the statistical description of motions which possess a macroscopic physical interpretation, like temperature or pressure.

The statistical aspect of the Second Law of thermodynamics is somewhat puzzling at first sight because it tells us something about the way that a particular property of a large aggregate of particles changes with time: it becomes more disordered. One could use this to define the future if you owned a time machine but were unsure of its navigational qualities. However, if we were clever enough we could have solved all Newton's equations to determine the exact motion in time of every particle. Yet Newton's equations and laws of motion possess one simple property: they do not allow you to deduce the future from the past. If you were shown a film of a series of collisions obeying Newton's laws between billiards or pool balls on a table, then you could not, by studying those motions, tell whether you were watching the film as it was actually shot or the same film running backwards. We say that these laws of motion are invariant under the reversal of the direction of time. Both Maxwell's and Einstein's equations possess this same invariance. Here then is the dilemma: if the film of the motions of the particles according to Newton's laws looks the same when run forwards or backwards, how can it give rise to an average property like entropy, which inexorably increases in one direction of time? Are Newton's laws incompatible with the Second Law? Or has the statistical averaging involved in arriving at the Second Law thrown away information? A simple lesson is to be learned from this seeming paradox.

Our everyday experience seems to exemplify the Second Law. Do not our desks and children's bedrooms degenerate effortlessly from a tidy to an untidy state, but never vice versa? The time-reversibility of Newton's laws of motion indicates that fragments of glass should be seen falling into configurations that compose beautiful wine glasses just as commonly as glasses are seen to fall and shatter into fragments. But they are not.

The particular motions we see are not just governed by Newton's laws of change. They also depend upon the starting state, and some starting states are far more probable than others. For instance, the starting state of carelessness that results in us witnessing a wine glass shattering to create a higher entropy state of disordered pieces is so much more probable than the contrived situation in which artfully shaped pieces of glass are all set in motion with exactly the right speeds in exactly the right directions so as to convene to produce a wine glass. Such an eventuality is so absurdly improbable that it has never been seen to occur over the entire history of the human race. There is nothing in the laws of Nature to forbid it, but an extraordinarily special situation would be required for those laws to realize it in practice. In this sense we see how a statistical element is contained in the 'law' of entropy increase.

In the more familiar case of our untidy desk we can understand the one-way trend from bad to worse by a similar appeal to the relative likelihood of different sequences of events. There are so many more ways in which my desk-top could pass from being tidy to being untidy compared with how it could go from being untidy to tidy. The Second Law is statistical in this sense: it embodies the fact that there are more ways of getting from A to B than from B to A when A is more organized than B. Levitation does not violate the laws of Nature—all the molecules in your body could, as a result of their random collisions with each other, all be moving upwards in unison at this very moment—it is just fantastically improbable. Far more improbable in fact than reports of it being simply mistaken.

Demons at work

> The Devil can quote Shakespeare for his own purpose.
>
> G. B. Shaw

A curious problem with statistical laws of Nature was recognized way back in 1867 by Maxwell. It took nearly fifty years to resolve. The problem that Maxwell posed was that of their susceptibility to manipulation by (supposedly) intelligent beings like ourselves:

> if we conceive of a being whose faculties are so sharpened that he can follow every molecule in its course, such a being, whose attributes are still as essentially finite as our own, would be able to do what is at present impossible to us. For we have seen that the molecules in a vessel full of air at uniform temperature are moving with velocities by no means uniform, though the mean velocity of any great number of them, arbitrarily selected, is almost exactly uniform. Now let us suppose that such a vessel is divided into two portions, A and B, by a division in which there is a small

hole, and that a being, who can see the individual molecules, opens and closes this hole, so as to allow only the slower ones to pass from B to A. He will thus, without expenditure of work, raise the temperature of B and lower that of A, in contradiction to the second law of thermodynamics.

The laws of Nature we have encountered so far are of an objective character irrespective of observers. In the Cartesian spirit it has always been possible to separate the observer from the external world. Scientists are naturalists with a perfect hideaway, observing the world without perturbing it. Superficially, special relativity appears to introduce the 'observer' into physics; but this is not the case. True, special relativity tells us that different observers in relative motion disagree about their measurements of mass, length, and time, but they would all agree on the *laws* governing the relationship between sets of masses, lengths, and times, and they would all agree on the description of a particular length, mass, or time that was not moving relative to them. The laws governing the events being witnessed are not conditioned by the nature or motion of the observers in Einstein's theory any more than they were in Newton's. It matters not whether the 'observers' are geiger counters, photographic plates, or human beings.

The discovery of statistical laws appears to undermine this separation of the observer from the observed, and allow the 'observer' to introduce himself as a fifth column into the workings of the microscopic world by use of his intellectual capabilities. For, if the Second Law of thermodynamics is just a manifestation of there being so many more ways of getting from A to B than from B to A, then what if some intelligent agent were preferentially to direct operations through the strait gate that leads from B to A, so that it becomes more probable that events follow this route than the broad road that leads from A to B? Clerk Maxwell gave a striking example of how he thought this manipulation of statistical laws of Nature might occur. It brought to a head the Victorian dilemma of how one was to fit life and human intelligence into the arena of deterministic laws of Nature, although this was not Maxwell's purpose in proposing it. Maxwell believed that the Second Law was an anthropomorphism created by the fact that we are so much larger than the molecules of a gas. If our faculties were acute enough to witness the behaviour of individual atoms, we could make heat flow from cold bodies to hot bodies one molecule at a time. Maxwell's recipe was designed to highlight this feature. Let us unpack his ingredients.

Take a sealed box containing gas in equilibrium, with the box at uniform temperature. Its entropy is at a maximum. The statistical distribution of molecular speeds will have attained the form predicted by Maxwell and Boltzmann, and demonstrated by Zartman and Ko. The attainment of this equilibrium state prevents us from using the gas to drive a machine of any sort. In order to do that we would need a variation in temperature to exist inside the box. If such a variation could be established, then the ensuing molecular motions that would proceed to re-establish the uniformity of the temperature in accordance with the Second Law could be exploited to drive a suitably connected machine.

Maxwell threw a spanner in the works by proposing that we could imagine the existence of a microscopic intelligence, whom Kelvin dubbed 'the sorting demon', making mischief with the gas molecules in our box. Suppose a wall is set up inside the box of gas at constant temperature to partition it into two parts A and B, but with a small connecting trapdoor cut into it. The operation of this trapdoor is placed in the hands of our diabolical colleague, who has been selected for his sharp eye for determining the speeds of the molecules and his dexterity in operating the trapdoor. He is given strict instructions by his boss to open the door only when he sees a faster than average molecule heading towards it from section A, or a slower than average molecule approaching from section B, but otherwise to keep it shut. After a while our demon should have shepherded most of the faster molecules into B and the slower ones into A without exerting forces on any of them. Since the temperature in each half is determined by the average speed of its molecules, the temperature of the two halves will have become unequal with no expenditure of energy. This temperature difference can be used to drive our machine for as long as the demon has the patience to exercise his discrimination over the moving molecules. We have created a perpetual motion machine in contempt of the Second Law of thermodynamics!

Maxwell's demon uses intelligence to influence the statistical distribution of velocities in different parts of the box. In so doing he does not violate the conservation of energy. If this type of fantastic conspiracy is possible, even in principle, then it introduces an unsatisfactory subjectivity to thermodynamic laws. Although the violations are individually microscopic, they will still, if permitted, add up to a large-scale violation of the Second Law. Others, like Tait, to whom Maxwell first suggested the Demon paradox, believed that the Second Law was violated in Nature. Tait argued that a large fraction of the energy emitted by the stars must be leaking out of our Universe into another world governed by different laws of thermodynamics (in fact, it was simply being obscured and absorbed by interstellar dust).

Although Maxwell and his contemporaries did not believe that a 'demon' could do these awkward things, they could not find a good reason why not. Many of them simply contented themselves with the claim that there just weren't any such microscopic demons! This does not solve the problem of course. In order to safeguard the logical consistency of thermodynamics it is necessary to show that there could not be any microscopic demons who do the things that Maxwell's demon does. This turns out to be possible, although not enough was known to show it in Maxwell's day.

The weak link in the demon's enterprise was discovered by Leo Szilard in 1929, and subsequently it has been scrutinized in ever greater detail by others. The problem is that he has got to identify which are the fast and slow particles as they approach the trapdoor, trap them, liberate them into the other compartment, and then reset his apparatus into a state which can identify another molecule's speed as faster or slower than average. In order to do these things he must interact with the molecules in some way, say by shining a light on them and observing how different is the wavelength of the reflected light. The work he must perform in order to discriminate between the fast and slow particles *and then destroy this information in order to repeat the operation from a clean start* always outweighs the work that can be performed by exploiting the temperature difference created by the demonic activity. Maxwell's demon is exorcized. It is not possible for him to create a violation of the Second Law any more than it is possible to show a profit at roulette by always betting on all the numbers. The cost of implementing such a strategy always exceeds the possible winnings.

The eternal return

> To see heat pass from a cold body to a warm one, it will not be
> necessary to have the acute vision, the intelligence, and the
> dexterity of Maxwell's demon; it will suffice to have a little
> patience.
>   Henri Poincaré

The French mathematical physicist Henri Poincaré was much impressed by the question of the real status of the Second Law of thermodynamics. The 'reversibility paradox'—that the increase of entropy with time must arise from Newtonian laws of motion for the individual molecules which possess no preferred direction of time—might be overcome if the Second Law was only a statistical averaging over human dimensions, but was violated microscopically. In 1893 Poincaré developed the conflict between time-symmetric laws of motion and laws of thermodynamics a little further. He showed that it was not necessary to be microscopic in order to see a violation of the Second Law of thermodynamics—you just had to wait long enough. Given any mechanical system of motions possessing finite energy confined to a finite volume it will return infinitely often to a state infinitesimally close to any past state. Even if we start a system in a very ordered, low entropy, state, and it becomes more disordered initially in accord with the Second Law, it must eventually return arbitrarily close to the initial ordered state at some time in the future. The entropy would have to decrease in order for this to be so. The monotonic behaviour of the Universe envisaged by the Second Law could not be deduced from the cyclic consequences of Newton's laws.

There were various responses to this dilemma. The Poincaré recurrence time for a system is fantastically large—$10^{10^{80}}$ years—compared with 4–18 billion years for the oldest objects we have found in the Universe. Having to wait such a long time for a violation of the Second Law to manifest itself is another way of saying that such a violation is extremely improbable. This was quite acceptable to Boltzmann and others who maintained the statistical interpretation of the Second Law. They did not deny that, in principle, any system could give rise to collisions of molecules which made many fragments of glass rush together to form a wine glass in contravention of the Second Law. It was just fantastically improbable.

There were two alternatives. One could hold that the Universe had been created in a highly ordered low entropy state, which then had so many more ways of becoming irregular than staying regular that it is seen to evolve towards irregularity with great probability. As it gets closer to equilibrium in the far future the probability of seeing a recurrence will increase. Alternatively, one could imagine that there existed a steady state of thermal equilibrium in the Universe together, with random fluctuations from place to place that changed with time. On this view, there would be places where the entropy would be increasing and others where it was decreasing, as one would have expected from the time-symmetry of Newton's laws. But if human life can only exist in sites where there is an increase of entropy, then we might explain why we are residing in an entropy-increasing environment. This suggestion was supported by both Boltzmann and Poincaré, and various arguments were framed to show that weird instabilities that occur within the entropy-decreasing worlds would render life impossible. For instance, a slight difference in temperature between two parts of a living creature would progressively increase in an unstable fashion rather than decrease as in our world.

Poincaré continued to hold to the belief that the Second Law was not a statistical artefact of our coarse-grained observations until 1904. He was finally convinced that microscopic violations of the Second Law could occur when they were finally observed to occur in the phenomenon of Brownian motion.

Quantum laws: Nature East of Eden

> There was a time when the newspapers said that only twelve men understood the theory of relativity. I do not believe there ever was such a time. There might have been a time when only one man did, because he was the only guy who caught on, before he wrote his paper. But after people read the paper a lot of people understood the theory of relativity in some way or other, certainly more than twelve. On the other hand, I think I can safely say that nobody understands quantum mechanics.
>
> Richard Feynman

The statistical character of the laws of thermodynamics arises only because of the impossibility of describing the simultaneous higgledy-piggledy motion of huge numbers of particles. The introduction of concepts like 'chance', 'most probable behaviour', or 'random' does not reflect anything intrinsically indeterminate about the collisions of gas molecules or aggregates of stars in space. It reflects our inability to determine.

The existence of such statistical laws does not offer a challenge to the simple realist view of the physical laws. Although the useful constructs like temperature and entropy that feature in these statistical laws exist only as averages or mathematical operations counting the number of configurations in which the system could reside, they are none the less constructed from properties like the speed of the molecules whose existence one has no special reason to question.

During the first quarter of the twentieth century Man's view of the nature of physical reality was to receive a jolt so extraordinary that its reverberations are the subject of deep and unresolved arguments amongst physicists to this day. The problems raised and the possibilities suggested about the underlying character of the laws of Nature challenge all ideas about the nature of reality in a manner that no philosopher could have foreseen. Most remarkably, these disputations surround a physical theory whose experimental predictions are more accurate and more wide-ranging in their technological applications than those of any other scientific development. Upon its forecasts about the microscopic behaviour of matter most of the world's high-precision technology is based. Yet it elevates the observer to a new status, it reveals an intrinsic degree of unpredictability in Nature, and it uncovers something truly extraordinary about reality that is amenable to experimental test. All this is part of the strange but continuing story of the quantum.

Schizophrenic matter

> The poet only asks to get his head into the heavens. It is the logician who seeks to get the heavens into his head. And it is his head that splits.
>
> G. K. Chesterton

Ever since Newton's day it has been known that light rays exhibit an unusual form of 'schizophrenia'. Under some circumstances they behave as though they are tiny particles, while under others they act as though they are waves. But one can see that there are situations where there might exist a real logical difference. It is all very well to say that a particle of a particular mass behaves like a wave of some particular wavelength and vice versa. But if we confine our wave/particle inside a box, and partition the box into two halves, then according to the particle interpretation the particle is either in one half of the box or the other. But in which half is the wave?

The unexpected nature of microscopic particles of matter is best illustrated by considering three situations which were first contrasted in this way by Richard Feynman, one of the physicists who has done most to establish modern quantum theories of matter and light. They illustrate the intrinsic differences between particles and waves.

First, we hit golf balls towards a target through two narrow vertical slits cut in a screen that intervenes between the target and the tee. The slits are wide enough to allow the golf balls to pass through them (see Figure 3.14). The result is quite obvious. The target will have hits scored upon it along two strips where the balls have passed through one or other of the two slits. There will be a narrow scatter of hits around each strip reflecting the fact that the slit has a larger size than a single ball. There will be no hits in the intermediate screened region of the target.

Now carry out this experiment with high intensity sound *waves* transmitted from a loudspeaker towards the two slits (Figure 3.15) The result will be quite different. If we measure the reception of the sound waves at the target by monitoring the noise intensity along the target screen we shall find alternating bands of low and high sound intensity. If we cover up one of the slits and record the new noise pattern at the target and then add this to the pattern we obtain with only the other slit covered we would find that the sum of these two one-slit patterns is *not* the same as that obtained when both slits are open together. When both slits are open simultaneously the phenomenon of wave interference occurs. The reason for the creation of this *interference pattern* is the way in which the waves from the two slits add. The 'arithmetic' of waves is different from that of particles because waves can combine either *constructively*, with peaks matching peaks so as to reinforce their intensity, or *destructively* with peaks matching troughs to produce zero net intensity (Figure 3.16).

Both these experiments are large scale. The first could have been done with cannonballs, the second using water waves impinging upon a divided harbour groyne. Suppose we now see what happens when subatomic particles like neutrons are fired towards the two slits. If we place a photographic film across the target then we find the striking result shown in Figure 3.17. The neutrons behave like the golf balls in the sense that each hit on the target film produces a definite mark. But as more and more neutrons are fired at the screen the individual hits build up a picture that has the characteristics of a wave interference pattern. There are bands where there is a high development of the target, evenly interspersed with underdeveloped bands, each possessing some statistical scatter. Although the neutrons arrive at the target as distinct objects, like the golf balls, the probability that they hit a particular point on the target is determined by a wave intensity. If we close one of the slits then this produces a single wave-intensity distribution with no interference just as in the case of the sound waves. Hence the neutrons manifest particle and wave properties at the same time: they arrive at the target as distinct 'hits', but with an intensity pattern characteristic of a wave.

There are further peculiar aspects of the wave interference pattern produced at the target screen by the neutrons which make it more subtle in nature than the 'ordinary' interference pattern produced by the sound waves. If we fire the neutrons slowly, one at a time, towards the screen so that we can watch the film developing neutron by neutron, and so avoid any obvious interaction between different neutrons which would lead to interference, then we still find the

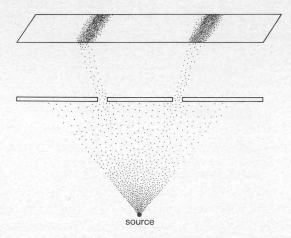

Figure 3.14 The pattern of hits produced by golf balls driven at a target through a screen with two openings. The target is peppered by hits in two narrow bands beyond the two openings.

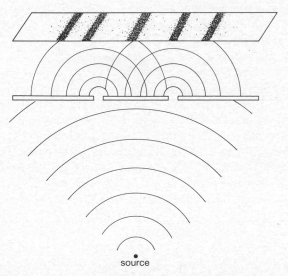

Figure 3.15 The intensity pattern produced by a source of waves directed towards a target through a screen with two openings. The target is highlighted by alternate bands of maximum and minimum intensity.

interference pattern being built up bit by bit. More striking still, we could set up many identical versions of this experiment all over the world and fire just one neutron towards the slits in each of them at a prearranged moment. If we add together the results from all these completely different experiments we would find

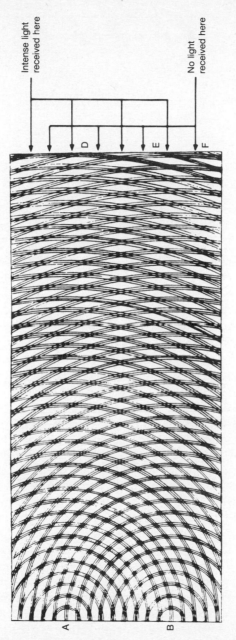

Figure 3.16 The phenomenon of constructive and destructive interference of two waves which is responsible for the intensity pattern shown in Figure 3.15 with two slits at A and B. The figure shows Thomas Young's original drawing. Young played an important role in establishing the wave nature of light during the first twenty years of the nineteenth century and was the first to conduct and analyse the light interference phenomenon in the twin slit experiment shown here.

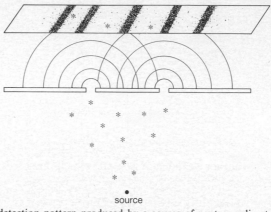

source

Figure 3.17 The detection pattern produced by a source of neutrons directed towards a target through a screen with two openings. The target is highlighted by a detection pattern of the form found for a wave source like that shown in Figure 3.15 even though the detected pattern is built up from individual detection events of the same discrete sort as in the set-up of Figure 3.14.

that the net result would look like the wave interference pattern! The single neutrons seem to be able to interfere with themselves. This is indeed 'one hand clapping'. We could have arranged for the different experiments to be huge distances apart and synchronized the performance of the different experiments so that in the time that it takes for the neutron to get from its source to the target no signal could travel, even at the speed of light, between one experiment and another to cause a correlation of their results in some way. The result is the same: the individual neutron-hits add to form a correlated interference pattern. How does each neutron know which role to play in order to produce the 'right' big picture of wave interference?

An even more perplexing fact about the microscopic two-slit experiment with the neutrons is that any attempt to unravel the wave-particle ambiguity and discover through which slit a particular neutron actually passed *en route* to the target invariably destroys the interference pattern seen at the target. Neutrons are rather delicate. If we set up a photoelectric cell at each slit in the screen to be triggered when a neutron passes through that slit, then we might expect to discover through which slit each individual neutron passes on its way to the target. Unfortunately this ambition can never be realized. The intervention of the light from the photoelectric cells alters the behaviour of the neutrons in a manner that destroys the wavelike result of the experiment. If we are ever able to determine through which slit a neutron goes, then the pattern seen at the target screen is changed from the wave-like one of Figure 3.17 into the particle-like result of the golf balls in Figure 3.14. Whenever we decide to examine whether a neutron is behaving like a particle and determine through which slit it passes, then, and only then, it is found to behave as a particle. If we do not attempt to determine if it is behaving as a particle, then, and only then, it manifests itself as a wave. It is not possible to construct any device which can determine through which slit a neutron has passed without destroying the wave interference pattern at the target.

This is quite unlike anything ever encountered in classical physics. It confronts all philosophical positions regarding the character of the laws of Nature and underlying reality with a totally novel challenge. It appears that the observer of the world plays a crucial role in determining what can be observed, but in a way that is subtly different from the old idealist view that everything is in the eye of the beholder. The naïve realist would hold to the belief that there is an objective world that exists whether we like it or not, and which possesses definite properties that exist independent of any measurement of them. Unfortunately this does not stand up to its first encounter with the quantum laws of Nature. The observed phenomenon together with the act of observation *together* determine what can be observed in the two-slit experiment. This does not mean that one should conclude that everything observed is observer-created, in the sense that the idealist or the solipsist might claim. There is no reason to suspend belief in an underlying reality. It is just that the steps we take to establish it determine what it will be found to be. Reality is contextual. We must also recognize that, at the very least, this reality which dictates goings-on in the micro-world of photons and neutrons is very different from the approximate impression of it that we have assumed from our contact with large objects whose quantum wavelengths are minute, and whose wave-like properties are for all practical purposes indiscernible. It is not that golf balls do not possess wavelike attributes. They do. But golf balls are so large compared with neutrons and the length of their wave attributes so small, that wave interference effects are indiscernible by the human eye.

Intrinsic uncertainty

> The physicists could but make the best of it, and went around
> with woebegone faces sadly complaining that on Mondays,
> Wednesdays and Fridays they must look on light as a wave; on
> Tuesdays, Thursdays and Saturdays, as a particle. On Sundays
> they simply prayed.
> Banesh Hoffmann

The wave-like character of particles was first predicted by the French prince, Louis-Victor de Broglie, in 1924, and confirmed three years later following experiments by Davisson and Germer who, in effect, carried out the neutron two-slit experiment that we described above, and demonstrated the formation of a wavelike interference pattern by the neutrons. De Broglie derived a simple but fundamental formula which gives the wavelength L of the wave-like attribute of a particle of mass m as

$$L = h/mv \qquad (3.9)$$

where v is the velocity of the particles and h is a universal constant, already known to physicists from their studies of radiation as Planck's constant ($h = 6.63 \times 10^{-34}$ Joule seconds). De Broglie's daring suggestion was that this formula applied without exception to all bodies. Roughly speaking, the quantum wave-like character of particles should become significant if their de Broglie wavelength, L, is close to or larger than their actual size, whereas if their physical size greatly exceeds their de Broglie wavelength then their wave-like qualities will be negligible. Because h is such a tiny number we find that everyday objects are far larger than their de Broglie wavelengths, but neutrons and other members of the microscopic world with very small masses are smaller than their de Broglie wavelengths, and hence their behaviour is dominated by the wavelike quantum attributes. This is why golf balls do not produce any noticeable wave-like results at the target. Even though golf balls really do have a de Broglie wavelength, it is fantastically small compared with the size of the ball, and therefore of the slits.

This universality of wave properties suggests that there exists a correlation between the results of an observation of the world and the nature of the measurement being performed. When we observe something, we receive light that has been reflected off the object under study. If we wish to measure its location with very high precision then we shall require light of very short wavelength, so that a single oscillation of the light wave does not miss the object. But, the shorter the wavelength of the light, the higher its frequency of oscillation, the higher its energy, and the greater its perturbation upon the thing being observed.

This limitation upon the precision of observation is a particular reflection of an intrinsic limitation that was first succinctly summarized by Werner Heisenberg in 1927. If one is attempting to measure the position x and the momentum p (the mass multiplied by the velocity) of a particle, then there will always be residual uncertainties in any simultaneous determination of these quantities because of the intervention of our measurement, let us call these uncertainties in the determination of the position and momentum Δx and Δp respectively. Then Heisenberg's Uncertainty Principle is an inequality

$$\Delta x \Delta p \geq h/4\pi \qquad (3.10)$$

Again, the minute universal constant of Planck appears. It means that this uncertainty is far smaller than any that we could discern using the unaided human eye. We will not die in a traffic accident because we were unable to discern the position and speed of approaching cars with sufficient accuracy.

The Uncertainty Principle states that it is impossible *in principle* to measure simultaneously the position and momentum of a particle with any accuracy better than the number $h/4\pi$. At a naïve level this is often interpreted as illustrating only that a measurement inevitably disturbs the state being observed—rather like the dilemma of the photographer trying to take a 'natural' shot, or the bird-watcher trying to remain unnoticed in his hide. This inevitable intervention means that it is impossible to know *exactly* what the state of the object was before the measurement was carried out. For example, if we place an atom at rest in space so that it has zero momentum, then our subsequent attempt to measure its position must (albeit through a number of intermediate stages) involve bouncing light off the atom onto the retina of our eye. But the effect of the light hitting the atom is to change its position, and create an uncertainty in its location which invariably satisfies Heisenberg's relation.

Despite the intuitive reasonableness of this description, it is important to appreciate that Heisenberg's uncertainty is somewhat deeper. The inevitable gap in our knowledge it guarantees is not in any way a reflection of the *imperfection* of our measuring devices (which can now measure very close to the limiting accuracy allowed by the Heisenberg inequality (3.10)). Even if our clocks, scales, and microscopes were *perfectly* accurate, we would still find simultaneous measurements of position and momentum to be limited by the intrinsic uncertainty (3.10). The Uncertainty Principle gives the minimum extent to which the world can be divided into the dualists' conception of the observer and the observed.

This common popular interpretation of the Uncertainty Principle as the result of some sort of disturbance of an otherwise well-defined situation fails to do justice to the full quantum view of the world. It makes no mention of the observer-created reality problem that plagued the interpretation of the two-slit experiment. It has made no explicit use of an observer (a microscope would do just as well), and implies that there really do exist simultaneous values of the momentum and the position, but because of our irreducible clumsiness we can never quite manage to measure them without altering them.

The full quantum theory developed during the first thirty years of this century, and used successfully ever since, preaches a far more radical view: the reason that we cannot measure a position and a momentum simultaneously with arbitrary accuracy is because the concepts of an entity's position and momentum do not simultaneously exist. We have, by our experience with classical physics, been misled into believing that a microscopic phenomenon can be described in terms of concepts like position and momentum which evolved out of our experience with the motion of large objects. In the microscopic quantum domain it is revealed that when we decide to measure one of these attributes precisely, then the state being measured ceases to possess the other property. It should not be imagined that we are just unable to discern them. They cannot coexist. In the two-slit experiment this dichotomy is manifested if we attempt to measure both the momentum and position of a neutron at the same moment in order to discover which slit it is passing through. When this is done, one part of the reality being recorded at the target is destroyed, and the wave pattern whose make-up we are trying to analyse into individual particle paths to the target is transformed into a particle pattern.

Niels Bohr, one of the architects of quantum theory, with whom Heisenberg was working when he discovered the Uncertainty Principle, called quantities like position and momentum which could not be measured simultaneously with arbitrary accuracy *complementary pairs*. There exist other complementary pairs in Nature—time and energy for example. We saw a manifestation of complementarity in the two-slit experiment when we tried to discover through which slit a neutron passes on its way to the wave-like interference pattern. The failure to do this reflects the complementarity of the wave and particle behaviours of the neutrons. One can either realize particle-like behaviour or wave-like behaviour, but not both. The measurement of one sort of behaviour ensures the non-existence of the other.

It should be stressed, therefore, that the quantum view is more radical than the pragmatism of the logical positivists who would maintain that it is meaningless to talk about quantities that cannot be measured. Complementarity asserts that it is not just meaningless to talk about knowing precise values of momentum and position simultaneously: these quantities do not exist. The situation is definite.

But Bohr took the concept of complementarity to have a far wider philosophical significance that its appearance as a rigorous mathematical deduction of quantum physics—his own coat of arms even incorporated the oriental *yin-yang* symbol. He suggested that paradoxical aspects of human experience might fall within the encompass of analogous, but rather vaguer, 'uncertainty principles'. Thus he suggested, for example, that the antithetical concepts of free will and determinism, or vitalism and mechanism, might constitute complementary pairs, so that if one adopts one of these views the other is rendered meaningless. This wider complementarity, like the failure to describe atomic systems precisely, he interpreted as the inadequacy of our primitive classical concepts to capture the nature of reality, and not as a renunciation of the coexistence of any objective reality. These applications of complementarity outside of quantum physics appear at first sight to be rather hopeful extrapolations from the precise quantum idea, but it is more likely that they were the motivation for Bohr's stress upon complementarity in physics. Bohr's father was a physiologist who had introduced this very idea of complementarity into his own subject in order to reconcile the conflict between teleology and mechanism in the description of biological phenomena. He argued that the validity of either view depends upon the question that is being asked about the world. One can see how the presence of complementary descriptions in the biological and human sciences leads to 'uncertainty relations' there also. In order to study the physiology of a living animal it would have to be killed and dissected. The intervention of the observer alters the nature of the specimen irrevocably. This type of observer-dependence also appears to be a major problem in anthropo-logical field studies. It has been claimed in recent years that the classic studies of Pacific Island cultures made by the late Margaret Mead were seriously distorted by the mischievous reaction to her own presence of the peoples she was observing.

Philosophers appear to have paid little attention to Bohr's ideas outside the arena of quantum mechanics. Certainly we are familiar with heuristic examples like trying to obtain a simultaneous appreciation of the detailed technical brushwork and the overall impression of an oil painting: the perspective necessary for one viewpoint excludes the very possibility of the other. Although it is interesting to note the existence of such complementary attributes outside of quantum physics, this has not led to any new knowledge of the problems involved. In the human sciences complementarity appears merely as a new label for an old problem, not as the solution of any problem. But in quantum theory it is a fundamental by-product of a precise theory with detailed observational consequences. It is not just a label. It solves problems, and it reveals new ones.

Niels Bohr's father was a member of a school of biologists who were later named the 'teleomechanists'. They were followers of Kant in his opposition to those biologists who sought a mechanical explanation of every aspect of life. Kant stressed the role of purposeful action by living creatures as a necessary ingredient in any complete description of their behaviour. By analogy, a complete description of a computer requires some reference to the nature or purpose of the program it is running, as well as the purely mechanical workings of the hardware. The two are complementary in the sense that the presence of a complete understanding of the computer at the hardware level does not exclude the need for further, qualitatively different types of information to complete the description. Human beings are undeniably collections of chemicals at the hardware level, but it would be incorrect to conclude from this fact alone that they are 'nothing but' collections of chemicals. Although the teleomechanists made a number of important discoveries in embryology, they faded from prominence because of their error in thinking that the small-scale purposeful activity seen in living organisms was evidence of some intrinsic 'life force'. This view, known as vitalism, is rejected by modern biologists. Yet it was this area of conflict between mechanism and purpose that led Christian Bohr to stress the significance of complementary levels of explanation, and sow the seeds of revolution in his young son's mind. But how curious and direct is the route from Kant to quantum theory.

Waves of chance

Erwin with his psi can do
Calculations quite a few.
But one thing has not been seen
Just what psi really mean.

Felix Bloch

The various strands of the quantum problems we have introduced were brought together by Erwin Schrödinger's discovery of an equation—now known as Schrödinger's equation—governing the behaviour of a quantity called the 'wave function'. It is this equation that replaces the equations of motion of Newton, which govern the positions and velocities of particles influenced by any set of forces. The 'quantum mechanics' that resulted from Schrödinger's description introduces profound new features. Although Schrödinger's equation is as deterministic as Newton's—given the exact specification of the wave function in space at one moment of time we can, in principle, compute its exact specification at any future time—this determinism turns out to be hollow, because we cannot observe the wave function directly. When we carry out a measurement on a physical system all that we can predict about its result using quantum mechanics is the *probability* of obtaining a particular result. From the human standpoint there is a breakdown of determinism in principle. It has nothing to do with fallibilities in our ability to measure with absolute precision. It reflects the intrinsic unpredictability introduced by the intervention of the observer: the irreducible level at which everything in Nature is coupled together. It is not a reflection of our adoption of an expedient averaged mode of description, as in the statistical description of thermodynamics. The exact deterministic laws of quantum mechanics turn out to govern a substratum of the world that is not directly accessible to us. It is unobservable. What *is* observable can only be predicted probabilistically. As the objects under consideration become larger their de Broglie wavelengths become smaller, they are more localized around their average values, and hence described with increasing accuracy by the classical mechanics of Newton, so that the *average* values of quantum observables obey Newton's laws of motion.

The probability distribution of the possible outcomes of a measurement made upon a quantum system in a particular state described by its wave function is determined by taking the squared magnitude of the wave function. This interpretation of the wave function was pointed out in 1926 by Max Born, and differs from what Schrödinger himself had imagined his wave function to describe. If we look back at Figure 3.17, then the quantum description provides a wave function whose squared magnitude produces the probability (or intensity) of neutrons hitting different parts of the target.

It is worth reflecting a little upon the wave nature of the neutron as it is manifested by Schrödinger's wave function. The quantum wave of a particle like the neutron is not a wave that is like a sound or water wave. It is a probability wave: a wave of information. It tells us the probability of finding the effect which we associate with the presence of that entity we have come to call 'a neutron' when we make a measurement at a particular place. We should view it more as a wave of hysteria than a water wave. If a wave of hysteria hits a particular area, then we are more likely to find hysterical behaviour there. Analogously, we are more likely to find a neutron at those places where its wave-function probability is greatest. In atoms, for example, the electron probability peaks at those places that we would have called the radius of the electron's orbit if we had adopted the naïve view of atoms being like mini solar systems, with nuclei surrounded by orbiting electrons at exact radii.

Quantum laws have another striking property which sets them apart from their classical counterparts. In the Newtonian world, identical causes produce identical effects. Alas, this is not so in the quantum world. Like causes will produce like probability distributions for the results of measurements, but two identical measurements will not produce the same result. As we remarked earlier, if we conducted many identical two-slit experiments, and each fired only one neutron at its target, then the results would not be identical. Rather, when added together, they would produce a probability distribution with the characteristic form of a wave-like interference pattern. This is predicted by Schrödinger's equation. There is beautiful agreement between these predictions and the observed outcomes of experiments.

A further reflection about the breakdown of determinism is worthwhile here. We have grown accustomed to the classical brand of determinism, highlighted first by Laplace and Leibniz, in which the future is uniquely determined by the present. An alternative seems almost impossible to imagine. After all, what else is there which could influence the future besides the present? What 'determines' the quantum future?

This new non-deterministic ingredient in our search for the laws of Nature turns out to be two-edged. Whilst it appears to introduce unpredictability everywhere in the microscopic world, it carries with it the key to the stability and predictability of Nature in the large. One of the consequences of the quantum theory of matter—and the reason for its name ('quantum' means amount in Latin)—is that it reveals that the energy of a system cannot take on *any* value. Rather, it can take on only particular discrete values. The energy levels of an atom live on the steps of a ladder rather than the surface of a slide. This is a radical departure from the classical Newtonian view that the energy of a system can take any value. The jumps between the allowed energy values are minutely small, and determined by Planck's universal constant h. Thus, for large and everyday objects the jumps are not noticeable. But when we examine the world on the atomic scale, then the size of the quantum jumps in energy is not that different from the entire energy of the atoms. In such circumstances the fact that the energy of the atom can only reside in a set of very special pigeon-holes has important consequences. For example, take the hydrogen atom: it consists of a proton around which there exists an electron, which is found most probably in a single orbit. Since unlike electric charges attract in the same manner as unlike magnetic poles, why is it that the positively charged proton and the equal but negatively charged electron do not collide and annihilate into a burst of radiation? Why do atoms exist? The Uncertainty Principle provides an answer. As the position of the electron becomes localized closer and closer to the proton, so its complementary attribute—the momentum—becomes larger and larger, and the electron moves so fast that it is impossible to confine the two close together.

Quantization of energy levels is also responsible for the uniformity of Nature. If the orbits of electrons around the central nuclei of atoms were described by the classical mechanics of Newton, then the electron could orbit at *any* radius if its speed were appropriately chosen. This would mean that whenever an electron was bound in orbit around a proton it would, in practice, have a different speed and radius of orbit. The consequence: every time you would have a configuration that is different, and so every hydrogen atom would be different. But in quantum mechanics, the fact that the total energy can only take particular discrete values ensures that when you put one electron into an orbit around one proton there are not an infinite number of possibilities. There is a minimum radius which the electron can take up, in which its wave-function probability is peaked. Once in this ground state, the energy cannot change gradually because of the constant buffeting of the electron by light. It will only change when a perturbation occurs which is large enough to change its energy by a whole quantum. As a result, atoms are stable against the sea of tiny buffetings, and not continually changing their intrinsic properties. The quantization of energy is the reason why all hydrogen atoms are the same. If energy were not quantized every one would be different, and the uniformity of Nature an unrealized idealization.

The nature of quantum reality

> Naïve realism leads to physics, and physics, if true, shows that naïve realism is false. Therefore naïve realism, if true, is false; therefore it is false.
>
> Bertrand Russell

Nothing could appear to be simpler than the picture of a neutron leaving its source, and then being detected by its impact upon a photographic plate. Such is the pre-quantum view of the world. After we have encountered the quantum laws of Nature, the neutron's adventures become deeply perplexing. For we are forced to believe that this entity we call the wave function propagates a neutron probability wave through space, in a way that is precisely determined by Schrödinger's equation. This wave of 'neutron-ness' will be concentrated primarily in a small region, so conveying the impression that the neutron is a localized concentration of something which we call mass. But the neutron wave will still actually be spread everywhere, albeit with a very low intensity far away from the main local concentration, and this reflects the fact that there is a non-zero probability of finding the neutron *anywhere*. When a measurement is made on the system—for example, when the photographic plate is put in place, and hit by the neutron—then the wave function must undergo an instantaneous change if it is still to tell us the probability of finding a neutron at a particular place, because the neutron *is* now at one definite place: the place where we can see the mark on the film. The measurement process, therefore, appears to be in some way distinct from the other physical processes that quantum laws describe. The ongoing quest to interpret and understand the quantum theory, rather than simply exploit its marvellous precision and power to describe correctly the way in which the world works, is at root an argument over whether or not quantum mechanics is a description of *everything* in Nature, including the measurement process, or not.

The 'EPR paradox'

> *Paradox:* a statement that is seemingly contradictory or opposed to common sense and yet is perhaps true.
>
> Webster's Dictionary

Throughout the whole of his amazing early career as a scientist who consistently produced the deepest and most far-reaching insights into the structure of the laws of Nature, Einstein maintained a resolute opposition to the ideas of quantum physics. Although his own discovery of the photoelectric effect paved the way for the idea that energy comes only in discrete quantum packets, he could never accept the element of indeterminacy that was a necessary ingredient of the quantum laws. His oft-quoted refusal to believe that God plays dice with the world reflects his distaste for departures from the traditional realist approach to the laws of Nature. He believed that the task of science was to provide an accurate description of an independent and objective physical reality. Bohr's doctrine of complementarity he saw as a rejection of this fundamental tenet, and merely a reflection of an overly positivistic approach to science, in which only what was measured was allowed to possess any meaning. The degree of his opposition to the quantum picture, and its predictions of probabilities rather than certainties about Nature, is revealed in a private letter to his friend James Franck:

> I can, if the worst comes to the worst, still realize that the Good Lord may have created a world in which there are no natural laws. In short, a chaos. But that there should be statistical laws with definite solutions, for example laws which compel the Good Lord to throw dice in each individual case, I find highly disagreeable.

Einstein made many attempts to undermine Bohr's quantum theory by constructing ingenious thought-experiments designed to exhibit an internal contradiction within the theory. All are generally considered to have failed; but all had a deep and lasting effect upon attempts to interpret the quantum theory. The most profound challenge to Bohr's picture of physical reality came in 1935, and was provided by a scenario composed by Einstein in collaboration with two of his colleagues Boris Podolsky and Nathan Rosen; it has come to be known as the EPR Paradox, after their initials, and appeared under the title '*Can Quantum-mechanical Description of Physical Reality Be Considered Complete?*'. Its aim was to challenge Bohr's interpretation of quantum reality as a revelation that no phenomenon could be said to exist until it was observed: that it is wrong to think that physicists are concerned with discovering what reality *is*, as opposed to what we can say about it. Einstein, by contrast, favoured the traditional view that the nature of things existed independent of whether or not they were observed. To make this traditional view precise, the EPR authors defined a physical quantity as possessing an element of physical reality 'if, without in any way disturbing a system, we can predict with certainty (i.e. with probability equal to unity) the value of a physical quantity'. Einstein and his colleagues believed that Bohr's position could be undermined, and the quantum description of reality revealed as incomplete, if it could be shown that unmeasured quantities would have certain values whether or not they were measured.

The essence of the EPR experiment is to examine the decay of an unstable elementary particle (an electrically neutral one, for example) into two photons. When the decay occurs, the photons will move off in opposite directions with equal momenta in order that the total momentum be conserved. These photons will also both possess an intrinsic *spin*, and if one spins clockwise then the other must spin anticlockwise in order to conserve the total spin of the system during the decay process. By considering the complementary attribute to this spin, EPR were able to produce what they regarded as a paradox which demonstrated that quantum theory could not be a complete description of what was occurring.

When the particle decays we cannot predict which residual photon will be found spinning in the clockwise direction. There is an equal probability for either the clockwise or the anticlockwise direction. But even if we let them fly apart to opposite sides of the Universe, as soon as we measure the spin of one of them we should know, without any shadow of a doubt, that the spin of the other will be found to be equal and opposite to the one we have just measured. This we will know *without* having measured it. Thus, EPR claim that the unmeasured spin of the other photon must be regarded as *real* according to their definition of physical reality, because it is *predictable*. It must, therefore, be related to some element of an observer-independent reality.

The reason why such a set-up attracts the designation 'paradox' is that somehow it appears that the second photon has to 'know' what the measurement on the first photon's spin will be (and that it has been performed), in order to take on the opposite value when it is also measured, even though the second measurement can be performed so soon after the first that there is no time for a light signal to pass between the sites of the two measurements. There is no question, therefore, of the second photon being 'got at' by the first one: they are *incommunicado*.

The new facet of the laws of Nature this experiment reveals is that the inevitable effect of the act of observation can be *non-local*; that is, it has effects upon events that are far away, in the sense of being outside the influence of light signals, and hence of any means of information transfer. The measurement of the spin of the first photon instantaneously brings into being the spin value of the second one. This seems completely contrary to common sense. We recall that one of the reasons why Newton's original theory of gravitation was unsatisfactory was because it demanded instantaneous effects arising on things here as a result of events far away. We should add that this non-local effect has been demonstrated in a series of real EPR experiments by the French physicist Alain Aspect and his colleagues in 1982. They, in effect, changed the initial spin state every ten billionth of a second, and in the meantime made measurements on the equal and opposite spins of the decay products when they were separated by four times the distance that light could travel in the time interval since the initial spin was altered. The predictions of quantum theory were confirmed.

One loophole in the case for complementarity is the possibility that there exists some other set of quantities which are always precisely determined, regardless of the measurement process, and it is these 'hidden variables' that play the primary role in the laws of Nature. In experiments where we find the measurement of position and momentum to be limited simultaneously by Heisenberg's Uncertainty Principle, there may exist some other unknown quantities which we could measure precisely, and from which we could calculate the position and momentum exactly. In terms of these quantities, quantum-mechanical reality might be definite and non-statistical. However, in the 1960s John Bell, a British physicist working at CERN in Geneva, showed that any theory purporting to describe the reality behind experiments like that proposed by Einstein and his colleagues must possess non-local features if a simple arithmetical condition is satisfied. Quantum mechanics meets the Bell condition for there to exist non-local interactions. The flawed idea that the uncertainties of quantum measurements are just manifestations of the observer's ham-fistedness in disturbing the state under study fails the Bell test because it is a local explanation. Quantum ignorance goes deeper than its classical counterpart. The quantum laws of Nature must possess acausal non-local features. No present or future theory that does not contain non-local influences can be consistent with experiments. What is most striking, and quite unique, about Bell's proof is that it demonstrates a necessary logical property of *any* theory of reality, not just a property of its observed manifestations as do all other theorems of mathematical physics: if your view of reality is the right one, then it must be a non-local one.

Yet, remarkably, quantum mechanics does not allow any two measurements to be connected to each other by a faster-than-light signal. The instantaneous non-local influences that must be present in any quantum laws of Nature are not in contradiction to Einstein's theory of special relativity. Einstein's theory places a limit only upon the speed at which information can be transferred. For example, suppose that we assemble thousands of robots in a long line, and program each one to illuminate its monitor at a precisely specified time. We could prearrange the specifications so that each robot lights up after its right-hand neighbour, but so soon after it that to the uninitiated observer it appears as if a faster-than-light signal is propagating from one end of the robot file to the other. But this is not a challenge to Einstein's theory: no information can be transmitted along the line by this faster-than-light phenomenon. If the robots wanted to transfer information to each other along the line they would need to communicate by radio signals, say, from neighbour to neighbour. Since such signals move no faster than the speed of light in empty space, there is no way that information could be transmitted between them at faster-than-light speed. We were able to create a faster-than-light phenomenon by pre-established correlations between independent events. There is no law of Nature against this.

This type of correlation in which there is no transmission of information is what exists in the EPR experiment. We recall that simultaneity is a concept that depends upon the motion of the observer. Three observers could be arranged to move relative to each other and the EPR experiment so that one sees the measurements of the two spins made simultaneously, whilst the others see one or other of the spins observed first. If one really were causally determining the spin of the other by transmitting a signal to it, then we could change the order of cause and effect in Nature by our relative motion with respect to the quantum events being witnessed. The acausal element of quantum reality cannot be exploited for the transmission of faster-than-light communications: EPR is not ESP!

If 'hidden variables' did really exist, then faster-than-light signalling *would* be possible. In a hidden-variables theory of the EPR experiment there exists a hidden quantity which is affected by the measurement of the first spin, and which then propagates an influence to determine the value of the second spin to be measured, even if it is measured simultaneously. This is an unappealing feature of hidden variables, and puts their existence in conflict with Einstein's special relativity. However, the confirmation of special relativity in every high-energy physics experiment ever performed cannot be taken as a refutation of the existence of hidden variables, since we have not been able to make any measurements of hidden variables.

At the time, Bohr's reaction to the original proposal of EPR was to remain unimpressed; he simply did not accept the EPR definition of 'reality'. Rather than retain the conventional way of thinking about reality, and force quantum theory to bend before it, he accepted the results of quantum theory, and regarded its success as a challenge to determine what the consequences are for the traditional realist view of the world. He maintained that it was meaningless to try to distinguish between *reality* and *observed reality*. One could not talk in a meaningful way about some property of a quantum system without reference to the observer: he is part of the phenomenon. Indeed, in a definite sense he creates it. According to Bohr, there does not exist any reality or any laws of Nature independent of the observer. Although the two spinning particles cannot seemingly influence one another, they are connected, he claims, by the presence of an observer who decides to learn about the spin of the second particle by making a measurement on the first. This makes him an inseparable part of the whole system. One cannot think of two spins as independent entities until after the intervention of the observer who separates them by the act of measurement. The result of the first *and* second measurement cannot be predicted before a measurement is performed. Bohr amended EPR's definition of reality to encompass the entire experimental arrangement. The quantum-mechanical description of reality was not incomplete, as EPR suggested, merely non-local, and wider in its compass than EPR had been prepared to countenance.

It is interesting to recall how important the idea of an 'isolated' system was to the development of science. The process of idealizing a particular physical situation to produce an approximation to it in mathematical language that captures the essential features, and ignores external influences, was the secret of the success of Galileo and Newton. Ancient holistic ideas provided no methodology for developing understanding, because they outlawed the concept of cutting Nature up into manageable independent pieces that could be understood individually. The isolation of the observer from the observed was a key element in this divorce. It worked so well because the observers (us), their observing instruments (eyes), and the data-processing machines (brains) are large objects for which the quantum-mechanical influences of observership are imperceptible. If we had been microscopic intelligences we would not have found the idea of an external objective world a useful one. The history of philosophy would have been strikingly different.

It is worth remarking at this juncture that the fashionable association of Eastern mysticism with quantum-mechanical ideas in such popular writings as *The Tao of Physics* and *The Dancing Wu-Li Masters* is based upon a confusion. During the period when we have no instruments to observe microscopic reality, there is no reason to develop a holistic rather than a mechanical view of reality. In such circumstances a failure to separate the observer and the observed leads to a stagnation of inquiry into the natural world, and an inability to extract reliable and useful information about the laws governing its workings. It is the application of a philosophical holism that appears to have contributed to the lack of scientific progress in the Eastern cultures with a strong mystic tradition. The more recent discovery of an holistic ingredient in the microscopic world of quantum physics, and the observation, retrospectively, of its connection with mystic intuitions about the world, would never have occurred in any culture subscribing to these intuitions about the world *ab initio*.

In quantum theory we face the problems that confront the necessary abandonment of such a view. The observer cannot be isolated in any sense from what he is observing; it becomes very difficult to say what, in fact, the 'external world' is actually external to. Bell's demonstration of the inevitability of non-local influences seems to place no bound on the extent of the 'complete' system of apparatus plus observers on which Bohr laid such continual stress. How do we tell what distant observations are *not* affecting a state?

Bohr's once-revolutionary view has become known as the 'Copenhagen interpretation' of quantum mechanics, because it was conceived by Bohr when working in Copenhagen where he had gathered around him a school composed of many of the greatest physicists of his day. Whilst others worked furiously to achieve remarkable successes with the new quantum theory, and explain many of the basic properties of matter, Bohr thought deeply about the interpretation and meaning of the quantum laws of Nature. The Copenhagen interpretation is what most quantum physicists would subscribe to if pressed to opt for an interpretation of this most successful of all scientific theories—although they might be outnumbered by those with no interest in such a question. It maintains that there is no deep reality for us to discover in the traditional sense, only a description of it. The reality that we observe is determined by the act of observation. It really exists when it has been measured—it is not an illusion—but there is no sense in which we can say that it exists in the absence of an act of observation. We must recognize that 'things' like photons and neutrons cannot be 'real' in the same way that we think that chairs and tables are real. They are more like shadows: arising from a combination of light and the observer's situation. Shadows are real enough, but they do not exist in the sense that a book does. The element of reality that Einstein associated with the photon because it possesses a predictable property can mean nothing more concrete than the occasional manner in which it manifests its measurable attributes.

One might call this 'quantum idealism' or even 'quantum positivism', depending upon one's perspective. According to this interpretation, the complementarity epitomized by the uncertainty relations reflects the inability of the classical concepts like position and momentum to coexist and capture the quantum world. The reason they are found wanting in the task is because, like all 'classical' concepts, they assume properties to be defined in the absence of an act of observation, and such a world does not really exist according to Bohr. It is meaningless to talk of a particle having an unmeasured spin, or to think of the reality of the measured spin as created from some other underlying reservoir of reality. If Bohr had been asked to interpret the wave-like pattern formed by the neutrons in the two-slit experiment, and the particle-like pattern created when one attempts to determine through which slit the neutron has passed, he would have pointed to it as an example of how the type of measurement that an observer decides to carry out determines whether the neutron will be found to be a wave or a particle phenomenon. It makes no sense, he would have argued, to talk about what it 'really' is, or to think that measurement somehow disturbs it from exhibiting what it really is. No such underlying reality exists.

That crazy mixed-up cat

Like the waters of a river
That in the swift flow of the stream
A great rock divides,
Though our ways seem to have parted
I know that in the end we shall meet.
Twelfth-Century Japanese verse

Many physicists do not find 'Copenhagen' quite so wonderful. It still leaves embarrassing gaps in our understanding of the quantum world. We are asked to accept that a measurement has non-local influences. But the most awkward problem for the Copenhagen Interpretation is more basic: what is a measurement? We recall that, according to Bohr, when a 'measurement' is made the infinitely dispersed wave function must 'collapse' to a definite, but unpredictable, state at a particular place—that place where a particle is found to have hit the film in the two-slit experiment for example. The transition from the quantum to the classical picture is sudden and mysterious. Is the wave function of a neutron collapsed by the record made on the film playing the role of an inanimate observer, or is it collapsed by the observation of a physicist who looks at the combined system of the neutron wave function and the film interacting together? Where and when does the wave function ultimately collapse? The problem for Bohr's interpretation is that it does not really claim to describe what quantum states and measuring devices are: only the nature of their interrelationship. This is rather puzzling. For there appears to be nothing special about measuring devices like geiger counters and sheets of photographic film. They are themselves, at root, composed of quantum particles like neutrons. Bohr's interpretation must regard a particular class of physical processes called 'measurements' as something special, because they collapse wave functions instantaneously. Yet, it does not seem possible to find a good reason for regarding these 'measurement' processes as being special in any way. However, it is none the less clear that the process of measurement has a number of properties that are quite the opposite of those possessed by the deterministic evolution of the quantum wave function in accord with Schrödinger's equation, in the absence of a measurement reduction. Whereas the latter is deterministic, linear, continuous, local, and recognizes no preferred direction of time, the measurement process is quasi-random, non-linear, discontinuous, non-local, and irreversible. No two phenomena could be more dissimilar.

The ambiguity between the quantum evolution of wave functions and the definite act of measurement was beautifully captured by an imaginary experiment conceived by Schrödinger in 1935. He wrote that he was motivated by reading the article of Einstein, Podolsky, and Rosen which appeared earlier that year. Like Einstein and his colleagues, Schrödinger never accepted the Copenhagen interpretation of the quantum theory, and regarded his imaginary experiment as a 'ridiculous case' which challenged its rationality. The 'Cat Paradox' reads like this.

Suppose a cat is confined in a sealed room along with a geiger counter sitting beside an occasional source of radioactivity. If the geiger counter records one of these (for all practical purposes) random decays within an hour, then it triggers the release of poisonous gas into the room which quickly kills the cat. If no atom decays in the allowed time then the cat survives. The experiment ends when we look into the room after an hour to see if the cat is alive or dead. Schrödinger claims that, according to the Copenhagen interpretation of quantum mechanics, before we look inside the room the cat possesses a wave function which describes it as existing in some mixture of the definite states 'dead' and 'alive' in which it can be found after the act of looking at the cat determines what state it is in. When and where does the mixed-up, half-dead cat state change from being neither dead nor alive into one or the other? Who collapses the cat's wave function; is it the cat, the geiger counter, or the physicist? Or does quantum theory simply not apply to 'large' complicated objects, even though they are composed of smaller ones to which it does apply?

A last bizarre twist was added to the 'Cat Paradox' by the Indiana physicist F. J. Belinfante, who pointed out that, according to the quantum formalism, you can bring dead cats back to life simply by determining precisely the complementary attribute to the property 'dead cat' or 'live cat': a sequence of these measurements will sooner or later actually measure the cat's state to be alive!

Quantum ailurophobia

> When I play with my cat, who knows whether she is not amusing
> herself with me more than I with her?   Montaigne

The point that Schrödinger was trying to make by proposing the Cat Paradox was that quantum theory is not a complete description of physical reality. It is our knowledge of the cat that we should regard as being in the mixed state rather than the cat itself. Bohr, however, would have argued that there is no such thing as 'the cat itself': 'our knowledge of the cat' is the only reality that there is. Furthermore, the large scale of Schrödinger's example tends to undermine its importance. We might object that we wouldn't know a cat in a mixed state if we saw one; although lasers can now be made to exhibit mixed states for us, they are not macroscopic.

Whilst some notable physicists have argued that these conceptual problems witness to the dissolution of the traditional concepts of the 'observer' and the 'observed', or even the classical concept of logic, others, most notably Eugene Wigner and John von Neumann, have argued for the surprising position that only special types of measuring device—those possessing human-level conscious-ness—can collapse wave functions. They maintain that it is not possible to formulate the laws of quantum mechanics in a completely consistent fashion, unless explicit use is made of the fact that the final repositories of the acts of observation are conscious beings. Such a position requires us to believe that human consciousness is something that behaves in a manner that is quite distinct from everything else we have encountered in Nature. It also leaves us wondering what will happen if intelligent observers ever die out.

There is no doubt that the brain is made of ordinary matter, and *aficionados* of Artificial Intelligence (the artifical intelligentsia?) would argue that there is no barrier of principle to the construction of a fully conscious intelligent robot. One could imagine the performance of a sequence of neurosurgical operations which gradually replace the bio-circuits in your brain with silicon chips, one by one. Would you ever suddenly cease to be human during this series of replacements, any more than if your vital organs were sequentially replaced by perfect mechanical transplants? But quantum theory unsettles naïve arguments for the inevitably of Artificial Intelligence. These arguments rest upon the premiss that the brain is a mechanism of a biochemical sort, which can be duplicated in its effects by a super-computer. If we analyse what we know so far about the human brain we do indeed find it to be explicable in biochemical and computational terms, down to a particular level. But if we keep on going the quest for determinism must run into quantum uncertainty. Until we understand how quantum computers operate, our understanding of how that special brand of software called 'consciousness', which the brain employs, is related to the hardware of the nervous system will remain incomplete. Paradoxically, as we try to analyse an observer like you or me into smaller and smaller mechanical pieces, eventually we come upon a level of quantum structure that requires an observer to endow it with meaning.

A most unusual property of human-level intelligence is its propensity for self-contemplation. We do not have to learn by trial and error. We can simulate the results of future actions, and act accordingly with foresight. We do not need to be at the mercy of natural selection unless we want to be. This ability allows us to create a mental world of abstractions: music, dreams, mental images, logical systems, sets of rules. All of these differ in kind from the concrete physical objects around us, and the emotional and thought-filled world of our conscious experience. This latter, personal world mediates between the objective world of 'things' out there, and the abstract ideas and theories we have about them.

A most amazing facet of observer-created reality was suggested by John Wheeler. He points out that all the astronomical observations that we make involve the present-day reception of light rays which emanated from distant stars and galaxies billions of years ago. The sea of low-temperature microwaves we observe all over the sky are the dying embers of the Big Bang from which our expanding Universe sprang. At the most fundamental level, all the photons we detect from distant stars are quantum waves whose wave functions are collapsed to classical certainty by the detectors and the astronomers who observe them. Does this mean that in some sense *we* bring these astronomical objects, and even the whole Universe, into being when we observe them today?

The most awkward problem of the consciousness-created reality school of opinion is its kinship with a form of 'quantum solipsism', in the face of the unanimity of our observations of the world. How is it that all these different human consciousnesses which disagree so vehemently about everything from wallpaper to welfare payments are in complete agreement (as far as we can ascertain) about the observed facts of quantum reality, about the results of the two-slit experiment, about what the stars look like, and the difference between dead cats and live cats? The most awkward problem for Bohr's observer-created reality, on the other hand, is articulating how it actually differs from consciousness-created reality except in name, because of the special status it gives to a class of processes called 'measurements'.

How many worlds do we need?

> I think that it is quite likely that at some future time we may get an improved quantum mechanics in which there will be a return to determinism and which will therefore justify the Einstein view. But such a return to determinism could only be made at the expense of giving up some other basic idea which we now assume without question.
>
> Paul Dirac

The quantum laws of Nature arose from the study of physical reality; they produced staggeringly accurate descriptions of atomic structure; but, in the hands of Bohr they led to the conclusion that the reality from which they were born did not exist in a naïve realist's sense. The creation of reality by the act of measurement has produced two awkward problems: what is a measurement, and how does the dispersed wave function collapse instantaneously to produce a single definite location when the position of particle is measured? In 1957, Hugh Everett III, a graduate student of John Wheeler's at Princeton University, took a radical but logical view of the problem. What happens, he asked, if you take the quantum formalism at face value? Suppose that it describes everything—even the measurement process. What if the wave function *never* collapses? The result is the nearest we can get to a realist interpretation of quantum mechanics, but the price to be paid is considerable. In order to maintain this realist position self-consistently we must dramatically revise our picture of reality.

According to Everett, when a measurement takes place the observer's state splits into each of the options that populate the set of possible outcomes in the Copenhagen Interpretation. Whereas Bohr's picture leads to a breakdown of determinism because one can only predict the *probability* of a particular measured result on the state being realized, Everett insists that complete determinism is maintained: each of the possible results really occurs. None has a status any different to the others. All are equally real, as 'real' as anything we experience or measure in the laboratory. Each possible sequence of paths through the sequences of allowed outcomes is disjoint from the others. We experience a history composed of just one of them.

This extraordinary 'Many Worlds' interpretation of quantum mechanics is taken seriously for only one reason: it follows undeniably from the mathematical formalism if quantum mechanics is held to apply to everything in Nature. To avoid its conclusions one must assume that quantum mechanics does not describe the measurement process.

We can see that the Many Worlds interpretation is deterministic in principle. As the wave function evolves forward in time, we can predict all the future states that could arise by the splitting of the observer's state. But paradoxically, we cannot establish the past evolution of our state because we cannot know all the other branch paths that our state has not followed during its evolution. We can predict the effects of quantum causes, but not find the causes of quantum effects. The logical appeal of the Many Worlds interpretation to its supporters (who include amongst their ranks some of the world's greatest quantum physicists) is the avoidance of any mention of 'consciousness', or 'observers', or scholastic distinctions between systems and measuring devices. Presumably, all interactions between elementary particles must create branchings of their universes, and not just those interactions that are 'observed' in Bohr's sense. Seen in this light, the Bohr interpretation appears slightly pre-Copernican because it gives us a special status that Everett denies us. Indeed, it is a form of vitalism.

Everett's interpretation of the Cat Paradox is simple. As a result of the experiment there exist two equally real states of the observer: one sees a live cat in his box, the other sees a dead cat in his box. Both situations are equally real. We experience only one of them. The Many Worlds interpretation regards the wave function as providing a description of the definite state of the neutron as it moves towards a target film in the two-slit experiment. The wave function splits into all possibilities, instead of just collapsing into one. Instead of acts of measurement there just exist correlations between different states. For Everett, the idea of wave-function collapse was simply a consequence of our inability to interact with the totality of quantum reality, confined as we are to meandering along a single branch of our ever-splitting schizophrenic reality.

This scenario has extraordinary consequences: like a world designed by Borges, everything that logically can happen does happen. There are worlds in which we never die. The evolution of life in the Universe, no matter how improbable, must occur. Each of us will continue to exist for as long as there is space and time, for even if we die in this world there is another where we do not, *ad infinitum*. How ironic that the evolution of consciousness is only inevitable in the one interpretation of quantum mechanics that does not require it in order to create quantum reality. It is also true that the Many Worlds interpretation of quantum reality is the only doctrine on offer which requires no non-local interactions: Bell's theorem does not apply. But the existence of the many worlds could hardly be called a local phenomenon.

Simple as the Everett interpretation is, it requires just as major a suspension of disbelief in order to swallow an unlimited multiplicity of real worlds as it does to swallow Bohr's claim that there is no real world at all. It poses puzzling problems: why should conscious minds be aware of only one of Everett's branches? Can branches that once split ever reconvene with observable consequences? Can we interact with the 'other worlds'? The reason Everett's view plays a growing role in fundamental physics is because cosmologists have begun to join forces with elementary-particle physicists. Their common goal is a quantum theory of gravitation. This synthesis appears at present a most unholy marriage. Quantum theory lacks the sophisticated property of general relativity that the presence of mass and energy in space produces its local geometry and the rate of flow of time. General relativity lacks the intrinsic uncertainty and complementarity of the quantum world. How unusual any new unified theory of these two great physical theories might prove to be can be gauged by pondering the consequences of their marriage. General relativity tells us that it is the presence and motion of masses in space and time that determine the structure of space and the rate of flow of time. Yet quantum theory tells us we can never know the position and motion of any mass in space and time simultaneously with arbitrary precision. How, then, can we know the space and time continuum into which we want to introduce the masses? We are caught in a vicious circle. Despite this problem, preliminary attempts have been made to produce hybrid theories of quantum gravity. The principal application of this new synthesis will be to provide a quantum theory of the entire universe in the way that we can use Einstein's general theory of relativity at present to give a non-quantum cosmological picture. It is here that the Everett interpretation becomes essential, for without it we are left asking the question 'who or what collapses the wave function of the universe?'—some 'Ultimate Observer' at the world's end, or outside the Universe of space and time altogether? It seems more natural to some to believe that quantum wave functions never collapse. It is no coincidence that all the main supporters of the Many Worlds interpretation of quantum reality are involved in quantum cosmology.

Another deep problem of principle in any application of the quantum theory to the Universe as a whole is the unusual status of time in the quantum theory. The steady-state solution of Schrödinger's equation for the wave function of the Universe does not contain a quantity that we could call 'time'. Unlike in the old pre-quantum theories of motion, there seems to be no natural way of introducing the concept of time into the picture. All through the applications of quantum mechanics to the world time appears as a peculiar concept in a way that does not characterize pre-quantum laws of Nature. There exist versions of Heisenberg's Uncertainty Principle which tell us that simultaneous measurement of energy and time are impossible with accuracy better than Planck's universal constant. But this only places a limit upon measures of time that are defined intrinsically by the state being observed. There is no limit to the accuracy with which a simultaneous measurement of energy and time could be made if some external standard of time were employed to measure it.

The quantum legislature

> If I get the impression that Nature itself makes the decisive choice what possibility to realize, where quantum theory says that more than one outcome is possible, then I am ascribing personality to Nature, that is to something that is always everywhere. Omnipresent eternal personality which is omnipotent in taking the decisions that are left undetermined by physical law is exactly what in the language of religion is called God.
>
> F. J. Belinfante

The laws that have been found to describe the quantum world are fantastic in many senses of that word. They are the most accurate and precise tools we have ever found for the successful description and prediction of the workings of Nature. In some cases the agreement between the theory's predictions and what we measure are good to better than one part in a billion. Yet when those laws are explored down to their foundations they force us towards possible views of reality that are totally at variance with all our intuitions and common-sense notions about reality. They reveal a remarkable depth of structure and novelty that lay latent within a mathematical formalism developed with other, more pragmatic, ends in view.

The profundity of our ignorance about the interpretation of quantum theory might be thought to presage a disaster in our ability to use it successfully. Nothing could be further from the truth. As yet, the radically different quantum ontologies have not led to a different prediction for the result of any experiment. The expectation of non-locality has been confirmed experimentally by Aspect's experiments. The Copenhagen and the Many Worlds interpretations seem totally at variance with each other, yet it is a widely held prejudice that they are experimentally indistinguishable. However, this view may prove to be false. Recently David Deutsch has proposed that one day it will be possible to design quantum computers whose operation will depend on whether the Many Worlds interpretation is true or false. Quantum computers are a fascinating prospect because of their intrinsically parallel information-processing capability. If the quantum computation resulting from one hour of running is equivalent to two computer-hours of non-quantum serial processing, then Deutsch argues that we must conclude that these computations must have been performed in the other worlds. The novelty of his detailed proposal is that it exploits the property that it is possible to design a quantum computer which splits reality into different 'worlds' and subsequently recombines, and which *remembers that it has split*. This property ensures that any quantum computer will be able to carry out computations faster than its non-quantum counterpart. This distinguishes the Bohr and Everett interpretations.

The most obvious difficulty with this exciting idea is that, although a reconstitution of the quantum system may be possible on paper, it is still necessary to get the required state prepared and extracted at the end of a practical demonstration. At some level the experiment becomes limited by the non-quantum problems of getting the information in and out of the quantum computer without altering it in dramatic and irretrievable ways.

There is one last response to the dilemmas of quantum reality which is in accord with the attitude of many of the 'old masters' responsible for its development: quantum mechanics may not be true. Quantum mechanics has never been found to disagree with experiment—indeed, by this criterion it is the most spectacularly accurate theory of Nature that we have ever possessed. It leads to a prediction that the magnetic moment of the electron is

$$1.001\ 159\ 652\ 46$$

and the latest experiments find a value of

$$1.001\ 159\ 652\ 21$$

with uncertainties in the last two digits in each case. This is accuracy to within one part in ten billion! Clearly, if quantum theory is 'wrong' then it fails either in an environment that is quite foreign to our experience, or it is in some way inappropriate when systems of large size or complexity are involved because new higher-level laws of organization come into play spontaneously. This is the classic situation in which venerated laws of Nature are found to be inadequate. Newton's laws of motion and gravitation remained undoubted for nearly two hundred and fifty years before they were revealed to be but first approximations to more complicated laws when large gravitational fields and high-speed motion was subjected to very accurate observation.

Einstein, Podolsky, and Rosen questioned not the correctness of quantum mechanics as a description of observation, but its *completeness* as a description of all aspects of reality. Schrödinger did not believe that his equation could be applied to things that possessed the vast complexity of a cat. It is possible that he was right. The Schrödinger equation, with its linear, time-reversible evolution of the wave function, may be but an approximation to a more complicated *non-linear* equation whose solutions are irreversible, and so possess the essential features of the measurement process. In the future it may be possible to scrutinize this possibility by employing the discovery of quantum behaviour in relatively large systems of atoms at very low temperatures. Going to larger systems still, we suspect that the quantum theory will experience some unusual modification when it is synthesized with the theory of gravitation. These modifications will have deep consequences for our understanding of both cosmology and elementary particle physics. It is to the far reaches of inner and outer space that our story now turns.

4

Inner space and outer space

The true, strong and sound mind is the mind that can embrace
equally great things and small.
Samuel Johnson

Setting the scene

Physics originally began as a descriptive macrophysics, containing an enormous number of empirical laws with no apparent
connections. In the beginning of a science, scientists may be very
proud to have discovered hundreds of laws. But, as the laws
proliferate, they become unhappy with this state of affairs; they
begin to search for underlying principles.
Rudolf Carnap

The development of relativity and the quantum theory during the early decades
of this century laid the foundation for a deeper understanding of the
astronomical realm of the Universe and the microscopic world of subatomic
elementary particles. These theories can be seen, with hindsight, to have
effected a radical change in our understanding of Nature's laws. Quantum
theory revealed that the deepest laws of the micro-world govern strange and
unobservable things. It marks the end of visualization and 'common sense' as
trustworthy guides to the frontiers of knowledge. No longer could we have
confidence in the Victorian belief that everything can be pictured in terms of
simple mechanical models—atoms do not behave like little cricket balls, nor
does space lie flat and true like the top of a billiard table. The complementarity
manifested in quantum laws reflects the inability of our classical concepts
to accommodate the richness and subtlety of the world, and removes the
Cartesian divide that insulates the observer from the observed. Naïve realism is
dead.

Relativity showed us that the bedrock of our experience—the space and time
within which we find ourselves immersed and swept along—is as malleable as
anything else we know. Space and time are not just fundamental categories
within which we must organize our experience: they are influenced by those
experiences. The rate of flow of time; the geometry of space; both are determined
locally by the material within the Universe. We cannot distinguish between the

curvature of space and the masses that effect it. They are equivalent ways of describing the same phenomenon.

Traditionally, the domain where the effects of Einstein's picture of space-time are of practical importance is the astronomical one. It is only over large distances, and in the vicinity of the largest aggregates of matter, that the subtle effects of Einstein's elastic space-time show themselves to an extent that lies within the sensitivity of our present-day instruments to detect. By contrast, the quantum theory is of practical significance in the world of the very small, where the de Broglie wavelength of particles is comparable to or larger than their physical size. Quantum theory is an inessential detail to our description of just about everything larger than small numbers of atoms. Throughout most of our experience, and even that of experimental physicists, the domains over which general relativity and the quantum theory legislate with influence are curiously disjoint. We live in the in-between world where things are too small and move too sluggishly for the effects of relativity to be overt, yet are too big for their quantum ambiguity to be perceptible. We are fortunate to stand on this middle rung in the ladder of Nature; the physical world seems simpler for the bourgeoisie who inhabit this in-between world betwixt the 'devil' of the quantum world and the 'deep blue sea' of curved space. The complications that arise there are those spawned by the collective behaviour of many atoms and molecules. One of the most appealing of these complications is the helical structure of carbon, hydrogen, nitrogen, oxygen, and phosphorus which we call the DNA molecule. From its subtleties has arisen the phenomenon we call 'life'. Its structure is not simple. Were it so, we would be too simple to know it.

The units of measurement that we have adopted and inherited bear witness to our parochial considerations. Many are anthropomorphic—the foot, the yard, the cubit, the span—and arise from the dimensions of parts of the human body. Others, like the day or the year, are dictated by the periodic motions of the Earth. Our modern metric units, like the gram or the centimetre, originate from their handy description of everyday quantities. When told that the Universe is 15,000,000,000 years old, or that the nucleus of a hydrogen atom weighs about 1,000,000,000,000,000,000,000,000 times less than a gram, it is natural to feel stupefied, but the significance of these huge numbers is their revelation of the peculiarity of the human position straddling the immensities of outer space and the microcosm of Nature's innermost workings. It is not that the size of the Universe or the atomic nucleus is peculiar; rather it is our own intermediate situation that makes both ends of reality seem so far away. There are about as many stars in the Universe as there are atoms in a sugar cube. But the cutting-edge of modern research into the ultimate laws of Nature has sought out a situation in which the laws of the very large and the very small must collide. In this environment we may have our only opportunity of seeing just how complex, or how simple, are the most basic laws of Nature. It is this collision course of inner and outer space, and the laws that govern them, that we shall be concerned with in this chapter. It will take us to the frontiers of modern physics, and reveal new challenges to any simple description of Nature and Nature's laws. Our story begins with the Universe in the small.

A world within the world

If that this thing we call the world
By chance on atoms was begot
Which though in ceaseless motion whirled
Yet weary not
How doth it prove
Thou art so fair and I in love.
                                        John Hall

The idea that matter is composed of indivisible microscopic units is an old one that originated amongst the Greeks as part of their rationalization of natural philosophy, but it then subsided until resurrected by William of Occam in the Middle Ages. In ancient times there neither was, nor could have been, any experimental support for such an idea. 'Atomism' arose as a philosophy of Nature within a wider view of things, rather than a scientific theory in our sense. Greeks like Aristotle, who opposed 'atomism', equated it with a blind desire to abnegate the governance of Nature in favour of pure chance. Atoms required no divine assistance. They had existed from a past eternity. Their motion did not derive in any way from inherent properties of themselves: they possessed no final causes. Such 'accidental' properties were, like material causes, of secondary status in the Aristotelian view. The essential features determining motion and material composition were the formal and final causes. Aristotle was also convinced that human knowledge must be grounded in sensations. His thoroughgoing empiricism recoiled from the atomist assumption that reality is grounded in unobservable microscopic entities.

   For centuries the term 'atomism' was equated with atheism and the worst form of materialism, and aroused the sort of passions that 'evolution' still evokes in the southern states of America. It suggested chance and disorder, the very state that the Creator had done away with. It leans towards encouraging an extreme form of reductionism wherein everything is reduced to atoms, and as a result is judged to be 'nothing but' atoms. Such a view found its most eloquent statement in Lucretius' *On the Nature of Things*, wherein the mortality of the mind, body, and soul of man was the conclusion to be drawn from their atomic composition. The other atheistic association of the early atomic doctrine was its claim for an infinity of worlds, each formed by the random coalescence of atoms. Such a cosmogonic extravagance appeared to diminish the magnificence of the created order in our own world.

Atomism was not an idea that emerged with force in the East. It arose briefly in the writings of Hui Shih, a philosopher of the fourth-century BC with a penchant for paradoxes of the infinite reminiscent of Zeno, which introduce the notion of the 'Small Unit': the smallest entity in Nature which contains nothing within itself. Such ideas had little survival value in the Chinese intellectual environment, and, like Zeno's, were soon submerged by criticism. Today, physicists would not regard mathematical paradoxes of infinite divisibility as even relevant to the issue. The distinction between mathematical and physical divisibility was first considered and appreciated by the twelfth-century Arab philosopher Averroes, who considered the concept of infinite divisibility to be unphysical because it could not be realized in practice.

The holistic bias in oriental thought favoured a view which envisaged the basic elements of Nature to be wave-like, antithetical influences, from whose conflation visible objects and structures would effortlessly emerge. The Chinese viewed the Universe as a seamless whole; although there existed individual objects and pieces to that whole, they had meaning only through their spontaneous and harmonious participation as individual chords in the great symphony of Nature. Atomism would have been an offence against their faith in the ultimate continuity and homogeneity of Nature. By contrast, the Arab and Indian cultures were sympathetic to ideas having much in common with Greek atomism. In Islam there were often considerations of a theological character motivating adherence to an atomic view of matter. In particular, atomism was believed to be a necessary property of the world in order that God could bring all things systematically to account at the Last Judgement (it is interesting to note that the modern Cantorian notion of a *countable* infinity, rather than finiteness, is what is actually required for this systematic accounting).

The Greek idea that matter was composed of indivisible basic particles first arose in the fifth century BC in order to explain how it could be that things undergo small changes, and yet maintain their essential identities. Everything material and spiritual, even the gods themselves, were supposedly made out of 'atoms', and the Greek word '*atomos*' means indivisible or uncuttable (by the Middle Ages this word became latinized, and meant only the smallest unit of time). Later, the desire to equate the indivisible 'atoms' of matter with the infinitesimal geometrical points introduced by the Pythagoreans spawned controversy, and led to Zeno's well-known paradoxes of the infinite with which he sought to show that Nature possessed an underlying continuity in space, rather than the discontinuity implied by the existence of discrete atoms in space.

Not all of the early Greek atomists required their atoms to be necessarily minute in size. They just needed them to be indivisible—a property that did not exclude the possibility that they possessed a structure of some sort. Nor were they assumed to be identical. They were imagined to exist in a variety of shapes and sizes, but those of one variety could not be transmuted into those of any other. Always in motion, they were not necessarily concentrated together, but separated by a 'void'. The properties that these atoms possessed were regarded as the only 'real' features of matter. Others, like colours, tastes, or smells, which arise in large aggregates of atoms, were regarded as secondary constructions of the human senses. So, for example, the fact that some substances possessed higher density than others could be explained by their containing a larger number of atoms, and their resistance to fracture by the distribution of voids within them between the atoms, and so on. Such 'explanations' were not really based upon observational data, but were used to confirm it. The division between atoms and void was significant in that it attributed existence to empty space for the first time. This was necessary in order to explain motion. For, if all being consisted of atoms, then because atoms could only influence one another by contact, they would have to pass from being into non-being in order to move. There was no analogue of the notion of an influential force whereby a particle could exert an effect on others without contact. That is a comparatively modern idea that even Newton was uncomfortable at having to employ, and only did so in the spirit of the instrumentalist.

The atoms themselves were made of some mysterious primary material, and from their determined but apparently random motions the four elements of earth, air, fire, and water were believed to have condensed out of vortices spun by the wakes of atomic collisions. It is curious that, although the original picture of Democritus and Leucippus portrayed the atomic motions as deterministic, the later Epicurean atomists had to modify this idea to preserve their reverence for the freedom of the human will. To this end they introduced the idea that there exists an intrinsic randomizing factor, called 'swerve', which influences the motions of the atoms in an unpredictable fashion. The greater the total 'swerve' content of an entity, the more spiritual and less material was its nature. At the pinnacle of this spiritual hierarchy of capriciousness was the seat of the will: the human soul.

Despite his opposition to the atomist doctrine of reality, Aristotle proposed a theory of 'minima' for living things. He claimed that no body could be dismembered into an infinite number of parts. Each will possess smallest parts, but they will be determined by the nature of the type of matter involved, and unlike atoms, these minima were subject to change and decay. Thus, the minimal parts were those below which a further division would cause a loss of identity. Curiously, Aristotle's ideas seemed to be preserved and perfected during the Middle Ages, and were incorporated into the labyrinth of Scholastic philosophy even when the atomists' views were opposed. In this way the idea, if not the vocabulary, of the atomists was preserved.

Although the atomists' views might now appear to have been far more promising a basis for science than their more metaphysical competitors, they did not become a significant influence in the ancient world; that place was taken by the Aristotelian system. After the Middle Ages the belief in the 'atomic' divisibility of matter was rehabilitated by European scientists like Copernicus, Sennert, Basso,

Van Goole, Gassendi, Huygens, and Boyle, following the rediscovery and reprinting of ancient atomist texts in the early fifteenth century. Again, the adoption of the atomic view was primarily an outgrowth of a particular philosophical view.* It was merely convenient to view Nature as composed of solid, imperceptible building blocks. The atoms played no part in any definite theory, nor were they observable (in Gassendi's day the smallest observed objects would have been about one thousandth of a millimetre across). They were necessary to allow one to see why bodies possessing identical size, shape, and motion could none the less be manifestly different. Francis Bacon regarded them as extremely useful in 'the exposition of Nature', regardless of whether they existed or not. Against this instrumentalist view was ranged the contrary Cartesian doctrine of the infinite divisibility of matter. Faced with little hard evidence either way, an important impetus to the atomist view was undoubtedly provided by the publication of Hooke's *Micrographia* in 1665. In its pages, the intricate images of tiny life-forms revealed by the recent invention of the microscope were open to public view for the first time. This helped to displace the popular sixteenth-century view of 'atoms' as a deep fantasy beyond human ken— a view well illustrated by Shakespeare's line in *As You Like It* that 'It is as easy to count atomies as to resolve the propositions of a lover'. Two years after the appearance of Hooke's opus, Robert Boyle published his influential work *Sceptical Chymist* which marks the beginning of modern chemistry. Boyle argued that only by laying hold of atomist ideas could the occult and teleological influence of the alchemists be removed from the subject. From now on chemicals were to react by definite dynamical laws, rather than by strange affinities and incantations: oratory was at last shooed out of the laboratory. This approach was only completed in the late eighteenth century when Antoine Lavoisier finally rid chemistry of the phlogiston concept, which regarded heat as a material substance whose migration in and out of matter had been invoked to account for chemical changes without any use of the concept of smallest particles. In physics Huygens attempted to explain natural phenomena in terms of changes in the speed and location of microscopic atoms, subject only to the conservation of extrinsic properties like energy and momentum. It is significant that, whereas past

---

*The traditional Greek form of atomism could not be revived as a sensible explanation for the obvious rationality of Nature until its intrinsic concepts of randomness and chaos became adequately harnessed. Such a harness only became available when the picture of laws divinely imposed upon matter from without completely displaced the Aristotelian notion of innate tendencies and potentialities within matter. Only then could one fall back upon the idea of the Cosmic Law-giver controlling the superficially random motions of the atoms to produce order. By this side-step the traditional atheistic associations could be removed from the atomic doctrine and it was ready for embrace by the proponents of the mechanical world-view.

Gassendi, who was a French priest, introduced God explicitly as the source of atomic motions, whilst Walter Charleton, his influential English popularizer, reversed the logic to deduce the existence of God from the fact that atoms did evidently give rise to ordered phenomena. Charleton's promulgation of the new atomist picture also influenced Newton during his early Cambridge years. Newton's early notebook records his study of Charleton's atomist views.

atomists have looked to universal properties intrinsic to the atoms themselves for an explanation of the regularities of matter and motion, Huygens sought an explanation of those regularities in the existence of conservation *laws* which the atoms obeyed.

Newton placed limits on the size of the smallest natural objects of about one hundred thousandth of a centimetre, from his determination of the thickness of fine soap films. He ignored past prejudice that belief in atoms was ungodly in some way, and stressed the Deistic vision of a world composed of atoms that had been brought into existence and set in motion by an act of God. In his *Opticks* Newton conjectures that

> it seems probable to me, that God in the Beginning form'd Matter in solid, massy, hard, impenetrable, movable Particles, of such Size and Figures, and with such other Properties, and in such Proportion to Space, as most conduced to the End for which he form'd them; and that these primitive Particles being Solids, are incomparably harder than any porous Bodies compounded by them; even so very hard, as never to wear or break in pieces; no ordinary Power being able to divide what God himself made one in the first Creation.

Newton was of the opinion that these solid basic components of matter were held together 'by very strong attractions'. The systematic revelation of these very strong attractions, and the world of elementary particles which they orchestrate, has been the great success story of twentieth-century physics.

Inevitably, Leibniz was in disagreement with Newton over the atomic doctrine. We saw in Chapter 2 something of the importance that Leibniz attached to the *continuity* of Nature. For this reason he found it shocking to contemplate a discontinuity marked by atoms of finite size which, if they formed the basis of all microscopic motion, would also require discontinuities in their velocities at the moments of collision with other atoms. Consequently, Leibniz maintained that matter was infinitely divisible and the Universe infinitely extensive, so completing the continuous spectrum of Nature from the infinitesimally small to the infinitely great. While not denying that there might be evidence for very small particles of matter, Leibniz believed that they would each be found to contain 'a world full of an infinity of different creatures'. The idea that atoms mark an end to the subdivision of matter he believed to be a depraved and seductive illusion:

> Atoms are the effect of the weakness of our imagination, for it likes to rest and therefore hurries to arrive at a conclusion in subdivisions or analyses; this is not the case in Nature, which comes from the infinite and goes to the infinite. Atoms satisfy only the imagination, but they shock the higher reason.

Whereas disciples of Aristotle opposed the existence of actual or achieved infinities and infinitesimals in principle, Leibniz welcomed them, and indeed expected to find them everywhere in Nature as a witness to the infinite perfection of her Creator.

Despite our familiarity with the concept of atoms as building blocks of matter, this idea did not come easily even to early twentieth-century scientists. In the opening years of the nineteenth century the idea had been used to good effect by pioneering chemists like John Dalton, Amedeo Avogadro, and Stanislao Cannizzaro. Dalton was born in 1766, and spent most of his scientific career in Manchester. His first scientific work, the discovery of colour-blindness (from which he suffered), sprang from his own deduction that he did not see colours in the same way as others (in France colour-blindness is still called *le Daltonisme*). Following his successful work on vision Dalton turned his interests to meteorology, and even took the time to publish a work on the 'Elements of English Grammar', before having his curiosity widened by a public lecture and demonstration of the chemistry of gases. His chemistry displayed a deep commitment to atomism, and was inspired by reading Newton and the Greek atomists. It brought about two lasting innovations in the study of chemistry. Dalton introduced the practice of constructing 'Lego-like' models of compound atoms by joining together coloured wooden blocks, and he was the first to use a comprehensive written system of chemical symbols to codify combinations of atoms. He introduced his atomic theory between 1808 and 1810, primarily in a major work entitled *A New System of Chemical Philosophy*, in which he argued that all matter was composed of indivisible atoms of finite size. These atoms were unchangeable and identical; they could be rearranged into different combinations by chemical reactions, but never created or destroyed. This picture thereby provided a simple explanation for the conservation of mass in chemical processes which had been demonstrated and proposed earlier by Lavoisier. It had a further great success in establishing the law of multiple proportions for chemical reactions. Dalton's significance lies primarily in the fact that he supposed atoms to have very definite properties which he elucidated by observations of how chemical changes took place, rather than as a result of some traditional philosophical inclination of his own. His theory was based upon the observed facts, although not grounded entirely upon them, because of the unavoidable historical influence of the atomist tradition. Dalton distinguished between atoms and 'compound atoms', the latter being what we would call molecules.

The molecule was the smallest amount of a chemical compound which could exist. A water molecule could be split into hydrogen and oxygen atoms, but it would lose its identity as water in the process. The atoms of hydrogen and oxygen, on the other hand he believed to be indivisible. By the end of the century J. J. Thomson would reveal that atoms could also be divided into parts. Dalton believed that atoms and molecules were in close static contact. This was disputed by his Italian contemporary Amedeo Avogadro, who maintained that molecules were separated by distances which were very large compared with their sizes—a view reminiscent of the ancients' atoms in the void.

As we have already seen, the early thermodynamicists like Maxwell and Boltzmann implicitly gave credence to the atomic idea by treating gases as though they were mechanical systems of microscopic masses. Maxwell, who had a penchant for putting his scientific ideas into semi-serious verse, explained the historical process which culminated in this view as follows:

> In the very beginning of science,
> the parsons, who managed things then,
> Being handy with hammer and chisel,
> made gods in the likeness of men;
> Till Commerce arose and at length
> some men of exceptional power
> Supplanted both demons and gods by
> the atoms, which last to this hour.

Even pragmatists like Faraday were attracted by the notion of small particles of matter, despite the worry that they might be no more than a heuristic. Maxwell laid great stress upon the idea of 'molecules' (by which he referred to what we would term atoms). He made accurate estimates of their maximum possible sizes, and regarded them as universal and immutable—vital elements in his belief in the design of Nature, and an indication that Darwinian evolution was limited in its extent. The spectroscopic properties of atoms led Maxwell and like-minded % —proponents of the atomic view to expect that the indivisible atoms would possess some form of internal structure, so that each variety would respond in a different way to being perturbed—a view also in essence not dissimilar to that of the ancients.

None the less, the strong influence of empiricist and positivist philosophy at the end of the nineteenth century carried with it a notable body of eminent scientists, men like Mach, Ostwald, and Duhem, who opposed the atomic doctrine on philosophical grounds. In the absence of direct observational evidence for the existence of atoms, they regarded the introduction of the entire concept as a theoretical device which should find no place in a philosophy of Nature founded upon observed fact alone. For Duhem, this instrumentalist doctrine played a key role in maintaining his religious and scientific views in peaceful coexistence. By arguing that all scientific 'laws' offered useful descriptions of observed facts which merely 'saved the appearances', he could claim that true reality was only touched on by the Catholic faith in things unseen. Wilhelm Ostwald contrived to write an entire chemistry text which studiously avoided all mention of the atomic concept, despite its having now assumed fundamental importance within the subject. A statement he made in 1895 during a lecture on his anti-atomist philosophy is instructive because it displays his opposition to all alternatives to it which involve entities that do not reveal themselves directly to the senses:

Thou shalt not make unto thee any graven image or any likeness of anything! Our task is not to see the world in a more or less clouded mirror, but to see it as directly as the constitution of our being will possibly allow. The task of science is to put realities, demonstrable and measurable quantities, into definite relation with one another . . . it cannot be fulfilled by applying some hypothetical model, but solely by the evidence of interdependent relations between measurable quantities.

Ostwald developed thermodynamics and physical chemistry into a combined approach to Nature based upon 'energetics'. He regarded energy as the only physical reality (he even called his house 'Energie'), with matter just a secondary derivative from it. He also opposed any form of mechanical explanation, and rejected the kinetic theory of gases developed by Maxwell and Boltzmann. Eventually, his views were plainly discredited by Planck's demonstration that the Second Law of thermodynamics could not be explained solely by use of the energy concept.

Even in 1906 these opponents still maintained that the atom was nothing more than a symbol for representing facts. They would not attribute reality to anything that had not been seen. Later, when direct observational evidence of atoms became available, they generally converted to the atomic view—with the exception of Ernst Mach, who never accepted the atomic doctrine. He retained his instrumentalist opinion of the atomic doctrine, firm in the belief that successful theories were merely the fittest of the competitors vying to match the facts. Mach, like Ostwald, also denied the pre-eminence of mechanical explanation. He argued that the dominant role of mechanical models and modes of explanation was an historical accident rather than a necessary prerequisite for a mature natural philosophy. The Machian view of reality that underpinned his opposition to unobserved and unobservable concepts was founded upon the precept that only our sensations are real. This 'sensationalism' was strongly opposed by convinced atomists like Boltzmann, who maintained that the type of purely 'energetic' explanation for chemical and gas dynamics preferred by the sensationalists was equally reliant upon unobservable elements. It involved appeal to the flow of energy governed by thermodynamics and differential equations—all concepts possessing unobservable aspects—and made full use of a manifold of mental pictures. Sensationalists claimed to believe only what they saw, but in fact they were rather better at believing than seeing. For, if one does not believe in atoms because of direct visual evidence, why believe that stars are huge astronomical bodies? As far as the senses are concerned they are microscopic pinpricks of light about the same size as small pieces of matter; but neither Mach nor anyone else would treat them as 'really' being the same size as fireflies, or other local sources of light of the same *apparent* size. Hence, some other consideration is being employed to add to the information available to the senses alone in placing the stars so far away despite their small visual appearance. No observations or other varieties of sense data can avoid the association of contextual elements that weave them into the coherent picture of Nature, nor can they be fitted into that picture if there is no theoretical understanding of the intrinsic biases inherent in the processes by which observational data are collected. There simply are no bare facts.

One can see that an over-enthusiastic adherence to the positivist dogma was likely to become a considerable handicap for a physicist in the early years of the twentieth century. This suspicion has led to the intriguing suggestion that there may exist a connection between the sudden demise of French science in the early years of the twentieth century and its resistance to the atomic picture of matter. As the focus of science turned upon the microscopic world, and found there things that defied visualization, so a greater premium was placed upon the flexibility of mind of a Maxwell than upon the rigid formality of a Duhem or a Mach.

To the modern scientist these metaphysical disputations seem extraordinary. We are used to the 'facts' of science setting the pace for natural philosophy to follow. But nineteenth-century philosophers did not assume such a secondary role. They did not take the data and the theories of the scientists as the starting point of their philosophical speculations about science and Nature in order to seek out the basic metaphysical problems. They had already inherited those problems in their traditional forms along with the basic philosophical stances of the Kantians, Hegelians, and positivists. The appropriateness of a particular view for resolving scientific questions would be judged primarily by its success in resolving these traditional philosophical issues.

Dissecting the atom

> We now know, it is true, that the often expressed skepticism with regard to the reality of atoms was exaggerated; for, indeed, the wonderful development of the art of experimentation has enabled us to study the effects of individual atoms.
>
> Niels Bohr

From the middle of the nineteenth century onwards there grew up a belief amongst those of the atomist persuasion that there existed a close relationship between atoms and electricity. Faraday had demonstrated this connection most strikingly. Common salt was believed to be an electrically neutral combination of sodium and chlorine: one negatively charged atom of chlorine combined with one positively charged atom of sodium to produce a neutral molecule of sodium chloride. Faraday showed that when a positive and a negative electrical terminal were immersed together in some salt water to complete a circuit, then the positive atoms of sodium were attracted to the negative terminal whilst the negative atoms of chlorine migrated to the positive terminal. Moreover, the total amount of each type of atom that accumulated at the terminals was found to increase in direct proportion to the quantity of electricity applied.

This type of phenomenon indicated that single atoms and molecules, although individually neutral, must be able to shed or attach themselves to electrical charges in some fashion. These transitory atoms that have in some way lost or gained additional electric charge were called *ions* (after the Greek verb 'to wander'). By 1833 Faraday was led from these observations to the idea that atomism extends to electrical charges as well as the components of matter. Furthermore, these basic electric charges appeared in some way to be part of the atom, or at least an adjunct to it.

As early as 1848 Weber had suggested that the electrical properties of metals might be explained by the motion of a collection of negative and positive charges within them, but it was not until 1874 that the Irish physicist George Johnson Stoney suggested that there must exist a basic unit of electric charge to explain experiments of the sort conducted by Faraday. Two decades later he proposed that this basic charge be called the 'electron' (a direct transliteration of the Greek for 'amber'—a material that was often employed in electrostatics experiments because it readily displays a negative static electrical charge when rubbed with fur), and predicted its numerical value. Just six years later the existence of the electron was demonstrated experimentally by J. J. Thomson in Cambridge. Thomson then tried unsuccessfully to rechristen it the 'corpuscle', but it is Johnstone Stoney's terminology that has stuck.

There were several imaginative attempts to build a specific picture of an atom of matter as a result of this discovery. The attempts all bear witness to the realization that atoms were not the indivisible pieces of structureless matter once envisaged, but possessed an internal structure all of their own. The crudest early picture, proposed by Lenard, envisaged the atom possessing thousands of pieces each possessing equal positive and negative charges, held in balance by the electrical forces of attraction and repulsion. A little-known Japanese physicist, Nagaoko, suggested a central positive charge surrounded by Saturnian rings of negative charge, whilst the most influential view was that of Thomson himself, who proposed a homogeneous sphere of positive charge populated with electrons, like a raisin pudding. He even allowed for the possibility that the electrons might maintain a stable configuration by moving in circles or oscillating back and forth about a state of equilibrium. By 1910 Ernest Rutherford and his collaborators at Manchester had interrogated an atom by bombarding it with heavier particles. By examining the manner in which the incident particles were deflected from atoms they were able to deduce that the bulk of the mass in the atom was concentrated in a tiny 'nucleus' of positive electric charge, with the electrons of negative charge distributed around it at assorted distances. Later, Rutherford suggested that the nucleus was composed of massive particles called 'protons', each of which was to carry an equal and opposite charge to the electron, but have a mass about 1836 times greater, together with hypothetical particles of similar mass to the proton but possessing no electric charge. These, he suggested, should be called 'neutrons'. In 1932, twelve years after Rutherford's prediction, neutrons were discovered by James Chadwick.

There followed a spectacular development of the now standard picture of atoms composed of massive nuclei no more that 10^{-13} centimetres in diameter, surrounded by a collection of orbiting electrons extending 10^{-8} centimetres from the central nucleus, and whose detailed arrangement creates the chemistry of the atom in question. When these pictures were merged with Bohr's principle of energy quantization they produced a model of extraordinary precision, able to predict the detailed properties of all types of atom. The subjects of chemistry and physics were now simply defined. The chemists were interested in the arrangement of the electrons in atoms: the subtleties of the electron patterns determined how atoms could link together into molecules and compounds. Physicists focused their attention upon the structure of the nucleus and the properties of its constituent particles.

In the period from the 1920s until the present, the physics of elementary particles has assumed major importance. It has become the most fundamental, the most expensive, and the most enticing area of physical science, and continues to attract the most able mathematical physicists. Experiments have long outgrown the domain of a single scientist. They have become huge international collaborations involving hundreds of scientists, technicians, and computer experts. The leaders of such enterprises require at once the skills of the physicist, the economist, the politician (some say even of the mafia boss!), and the personnel manager in order to succeed. The Italian Nobel laureate Carlo Rubbia, who has a reputation for combining all these attributes, has gone so far as to compare the achievements of modern experimental particle physics with those of the Renaissance. In centuries to come the unravelling of the elementary constituents of matter may be seen as the greatest achievement of our civilization; our particle accelerators great standing-stones and secular cathedrals to the human spirit.

Brave new world

> An explicit scientific world view may arise by a higher specialization of the same basic grammatical patterns that fathered the naïve and implicit view. Thus the world view of modern science arises by higher specialization of the basic.
>
> Benjamin Lee Whorf

When the twentieth century began we knew of only two types of natural force: gravitation and the intertwined influence of electricity and magnetism. This knowledge has come to be labelled 'classical physics' to distinguish it from the quantum revolution that followed it. Some groups at the time believed that classical physics constituted all that there was to know about the world. This over-confident negativism is something that resurfaces every so often when scientists have had a successful period. People worry that they might be so successful that the subject could be finished and the book closed. In the year 1880 the director of the Prussian patent office asked his government to close his department on the grounds that there was no conceivable thing left to be invented! In 1894, Albert Michelson, one of the leading physicists of the day, claimed that

> The more important fundamental laws and facts of physical science have all been discovered, and these are now so firmly established that the possibility of their ever being supplanted in consequence of new discoveries is exceedingly remote . . . Our future discoveries must be looked for in the sixth place of decimals.

To some extent such pronouncements witness merely to the parochial view of physics at the time. It was perceived to have a very limited domain of relevance, and to be quite disjoint from chemistry and the life sciences. None the less, scientists like Michelson were oblivious of the huge gaps in physics at the time. Little or nothing was known about the properties of materials, the nature of light, or the nature of the heavenly bodies. There was a simmering conflict between the physicists' view that the Sun was powered by heat generated by its gradual contraction under the force of gravity, and the requirement that it had existed long enough to allow Darwinian evolution to occur. The seeds of a revolution were beginning to poke their heads through the ground, but few paid much heed. Within fifteen years of Michelson's statement would come Einstein and Bohr, bringing with them relativity and the quantum, turning the world upside down.

Since those days when the way of the world was fashioned by electricity, gravity, and magnetism, our experimental investigations have identified two further basic forces of Nature: the 'strong', or 'nuclear', force which is responsible for binding atomic nuclei together, and the 'weak' force which creates some types of radioactivity. Whereas the forces of gravity and electromagnetism have no restrictions on the distance over which their influence is significantly felt, the strong force has a range roughly equal to the size of the smallest atomic nucleus, about 10^{-13} of a centimetre, and the weak force's range is one hundred times shorter still. The details of how these influences work cannot be elucidated with apples and magnets. They require the creation of special circumstances and environments in which the signatures of the microscopic forces reveal themselves to sensitive devices. But their effects are by no means hidden from us. They include the atom bomb and the Sun.

It was soon discovered that the proton, the neutron, and the electron are not alone. The elementary particle club did not appear to be terribly exclusive. A steady flow of new discoveries continued: in 1936 the muon, in 1947 the uncharged pion, in 1950 the charged pion, and in 1956 the long-sought neutrino predicted by Pauli in 1931. During the 1950s and 1960s the construction of the first machines able to accelerate protons to very high energies, and smash them into pieces by collisions one with another, gave rise to a population explosion of elementary particles. In the end several hundred such particles were found. What had begun as an exclusive club had burgeoned into an all-comers bazaar.

Incestuous matter?

> So, Nat'ralists observe, a Flea
> Hath smaller Fleas that on him prey,
> And these have smaller Fleas to bite 'em
> And so proceed ad infinitum.
>
> Jonathan Swift

From the debris surviving the collisions between protons at higher and higher energies during the 1960s there flowered for a time two radically different philosophies. The first, in accord with the ancient atomist intuition, held to the view that matter was divisible into particles like protons and neutrons, and that these in turn contained a small number of elementary constituents which might in their turn contain smaller components. Eventually one would encounter the 'bottom line' to this divisibility, and *the* elementary particles, finite in number, out of which all matter would be additively constructed, would be revealed. The second view, in complete contrast, owed nothing to the atomist tradition. It harks back to the oriental intuition for holistic interaction between all the elements of Nature and Leibniz's prejudice for the infinite divisibility and continuity of matter. It maintained that matter is indefinitely divisible into an infinite number of elementary particles which are all composed of each other in a well-defined sense. This incestuous arrangement became known as the 'bootstrap theory' in celebration of the eccentric Baron von Munchausen who claimed to have lifted himself up by his own bootstraps! It differs from the first picture in the following ways: if we take a box of material particles, and heat them up, then the first picture predicts that eventually the material within the box will break up into its most basic constituents, and these will continue to rise inexorably in energy and temperature as the box is heated. The second picture leads to the conclusion that, as the box is heated, so energy is channelled into producing more and more different types of an infinity of possible particles having higher and higher mass, and so requiring greater and greater energies to produce them, and as a consequence the temperature need not continue to rise. One of the striking predictions of the early bootstrap theory was that there must exist a maximum possible temperature of about a trillion degrees. We know now that this type of bootstrap theory is in conflict with the facts, and much higher temperatures than its suggested limiting temperature are attained in particle accelerations, but it is still possible that there does exist a bootstrap structure to matter at a more microscopic level than has been probed so far, and with a vastly higher maximum temperature.

The very entertainment of the notion that 'everything is made of everything else' sounds like a mystic aberration, but it was in tune with what was seen in particle physics experiments. The smashing of protons was seen to produce yet more protons rather than collections of internal constituents. Given three different elementary particles 'Tom', 'Dick', and 'Harry', one could reconcile what was seen with an intransitive bootstrap picture which had 'Tom' made of 'Dick' and 'Harry', and 'Dick' made of 'Tom' and 'Harry', and 'Harry' made of 'Tom' and 'Dick'. Such a dilemma appears to be inevitable when one considers what must happen when an attempt is made to split a proton or a neutron into pieces. All such pieces must be bound together by nuclear forces that are so strong that when the binding is broken the energy released is sufficient to make new particles, in accord with Einstein's famous mass-energy equivalence $E = mc^2$. The concept of an elementary particle becomes a little nebulous in such circumstances. It is akin to trying to break down a bar magnet into more elementary single magnetic poles by cutting it in half. Instead of making single North and South magnetic poles we produce two magnets (see Figure 4.1) each with a pair of magnetic poles. In some sense magnets are made of other magnets.

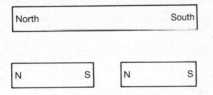

Figure 4.1 The breaking of a single magnet produces two magnets, not two isolated magnetic poles.

The bootstrap picture gradually faded away as evidence emerged for a very definite type of substructure within protons and neutrons.

Quarks

Three quarks for Muster Mark!
Sure he hasn't got much of a bark
And sure any he has it's all beside the mark.
But O, Wreneagle Almighty, wouldn't un be
a sky of a lark.
To see that old buzzard whooping about
for uns in the dark
And he hunting around for uns speckled
trousers around by Palmerstown Park?

James Joyce

During the 1960s theoreticians began to make sense of the pretenders to elementarity. Murray Gell-Mann and George Zweig showed that a beautifully simple picture could explain all the known particles. They proposed that particles like the proton, the neutron, and the mesons, which feel the strong nuclear force, are not elementary particles at all, but possess internal constituents. These Gell-Mann called 'quarks' and Zweig 'aces'. It is Gell-Mann's literary allusion to James Joyce's 'three quarks for Muster Mark' that has stuck (in Joyce's *Finnegans Wake* the three quarks appear to be the three children of Mr Finn. On some occasions the concept of Mr Finn is represented by himself, but on others that role is played by the three young quarks).

We can divide the subatomic particles into three groups: baryons, mesons, and leptons. The leptons ('*lepton*' means 'light' or 'little one' in Greek—it was the 'widow's mite' in the famous New Testament gospel story) feel the influence of gravity along with the weak and electromagnetic forces. The baryons ('*barys*' is the Greek for 'heavy') feel all the forces, but possess a conserved quantity called baryon number. The mesons also feel all the forces but do not possess baryon number. The proton and the neutron are baryons; the electron, the muon, and the neutrino are leptons; whilst the pions are mesons. For some time after its discovery the muon was referred to as the mu-meson because it was mistakenly believed to be a meson.

In recent years evidence has steadily accumulated to confirm the idea of Zweig and Gell-Mann that protons and neutrons are not elementary particles. In scattering experiments they behave as though they contain three microscopic constituents, which reveal themselves through the pattern of scattering that takes place when protons and neutrons are bombarded. But no one has ever seen one of these quarks of which all matter is made. Why not?

Quarks carry two types of charge. Beside the ordinary electric charges carried by electrons and protons they possess another variety which has been termed 'colour'—although it has nothing whatsoever to do with the hue of things. Like electric charge, this colour charge appears to be conserved in Nature. That is, when particles interact and transform their identities in collisions, it is possible to redistribute the colour charge but not to alter the total amount of it when the final book-keeping is performed. Now, just as particles possessing electric charges feel an electromagnetic force of attraction or repulsion between them, so particles possessing the colour attribute exert forces upon one another. This 'colour force' between the quarks that constitute nuclear particles is the origin of the nuclear force seen between the nucleons, just as the electromagnetic forces between protons and electrons give rise to secondary chemical forces between molecules and compounds. Nuclear power is the process by which we can extract energy from the colour force between quarks. It is as real as you and I. Yet the colour force is distinctly odd. Whereas the forces of gravity, electricity, and magnetism all weaken in strength as the inverse square of the distance between the particles involved, the colour force gets stronger as the distance between quarks increases. It is like an elastic force. And it keeps quarks bound within baryons and mesons.

Quantum fields

> The laws of Nature give a fundamental role to certain entities. We are not really sure what they are, but at the present level of understanding they seem to be the elementary quantum fields.
>
> Steven Weinberg

The basic characteristics of the four known forces of Nature are listed in Table 4.1.

Table 4.1 The four basic forces of Nature.

Force	Range	Relative Strength	Influences	Carrier Particle
Gravitation	∞	10^{-39}	Everything	Graviton
Electromagnetic force	∞	10^{-2}	Electrically-charged particles	Photon
Weak force	10^{-15} cm	10^{-5}	Leptons and hadrons	W^+, W^-, Z bosons
Strong force (colour force)	10^{-13} cm	1	Colour-charged particles	Gluons

We have found that all these forces are best described by a merger of the classical concept of a continuous 'field' of force introduced by Faraday, with the quantum ambiguity of Nature. The offspring of this union is the *quantum field*. We recall that the classical field of Faraday was an unspecified 'something' filling all of space, and which possessed a collection of attributes that could be described by numbers. These numbers give its strength and direction at each point of space. The electric, magnetic, and gravitational fields provided the classical examples. If, like Lord Kelvin, one insists also upon a picture, then that of a flowing liquid filling all space is a good analogy for certain fields. Each field of force is viewed as a different such fluid. Quantum fields manifest the probabilistic element introduced by our measurement of them in accord with Heisenberg's Uncertainty Principle. They can be viewed as a turbulent flow of liquid in which the chaotic fluctuations get larger as one examines the fluid with a magnifying glass on a finer and finer scale. Yet, over a larger finite region the microscopic fluid turbulence averages out to produce a relatively smooth flow in the same way that the ocean surface appears smooth from a distance. Quantum field theories are mathematical descriptions of the average properties of force fields over finite regions of space during finite intervals of time. They predict how those fields will change, and how they interact one with another. Whereas a classical scalar field will be specified by an intensity at each point, a quantum field will be specified by a set of probabilities; they will tell us the likelihood that the field will be observed to take a particular value at a certain time and place. All our successful laws of Nature's elementary constituents are quantum field theories.

Quantum field theories differ radically from their classical predecessors. The most important manifestation of this difference is the picture they present of the vacuum. To the pre-quantum student of Nature nothing could be simpler than the vacuum—it's nothing, pure and simple. But the quantum view of the world forbids us the sort of precise knowledge that would allow us to say that a box contains so definite a thing as 'nothing'. A measurement introduces radiation. Electromagnetic radiation is everywhere. The quantum vacuum is alive with activity. It is distinguished solely by being the lowest energy state in which the entire system can reside.

We picture the quantum vacuum as a sea of continuously appearing and disappearing pairs of oppositely charged particles (see Figure 4.2). Each of these pairs is individually unobservable, in accord with Heisenberg's Uncertainty Principle, if it possesses an energy ΔE and exists for a time Δt such that

$$\Delta E \Delta t < h/4\pi.$$

These are called 'virtual' particle pairs. Although individually unobservable, the collective effects of virtual particles produce tiny but measurable changes to the energy levels of atoms. Later, we shall see that the existence of an active quantum vacuum has important consequences for the structure of the world.

The image of virtual particles also dictates the picture we have of how forces of Nature act. Whereas Newton was forced to adopt a model in which forces acted instantaneously over great distances by some unknown means, the quantum view requires that an interaction be mediated by the exchange of a 'carrier' particle. In the case of the electromagnetic interaction between electrically charged particles this carrier particle is the photon of light; the strong nuclear interaction between quarks is mediated by particles called gluons, and the weak interaction by particles called W and Z bosons, which were recently detected in colliding beams of high-energy particles at CERN in Geneva; finally, the carrier of the gravitational interaction is called the graviton. Heisenberg's Uncertainty Principle gives the relation between the mass of the carrier particle and the range of the force it mediates. The range is roughly equal to the de Broglie wavelength of the carrier particle, and hence is inversely proportional to the mass of the appropriate carrier particle. Thus, the massless photon and graviton mediate the infinite range forces of electromagnetism and gravity respectively. The very short

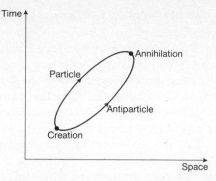

Figure 4.2 The quantum vacuum: this space–time diagram illustrates the spontaneous production of a particle–antiparticle pair. The pair subsequently annihilates after a time interval so short compared with the mass-energy required for their creation that the process is unobservable by the constraint of the Heisenberg Uncertainty Principle. Such an unobservable process is called a virtual process. Under certain circumstances the reannihilation can be prevented by the imposition of some external force field in which case the virtual particles become detectable. The quantum vacuum is envisaged to consist of an infinite sea of virtual processes. This picture is none the less subject to direct experimental test.

range of the strong and weak interactions are reflections of the very high mass of their carrier particles.

The fundamental legislation of inner space

> I am inclined to think that scientific discovery is impossible without faith in ideas which are of a purely speculative kind and sometimes quite hazy; a faith which is quite unwarranted from the scientific point of view.
>
> Karl Popper

Physicists have deduced mathematical laws which describe how the different forces of Nature behave at a microscopic level. Each theory is internally consistent, useful for practical work in elementary particle physics, and by any pragmatic criterion 'successful', but particle physicists are far from content. There remain a number of unanswered questions which illustrate modern prejudices and expectations concerning the laws of Nature:

(i) Are there simple principles which dictate what forces and particles exist in Nature and how they interact with one another?

(ii) Is there a single unified law of Nature: a 'theory of everything'?

(iii) Why do the forces of Nature have the strengths they do?

(iv) What are the ultimate constituents of matter—the most elementary of elementary particles?

(v) What are the ultimate principles which we should employ to discern laws of Nature?

The fact that such questions are asked is evidence of our deep faith in the simplicity and unity of Nature. Physicists believe that the degree of rationality already discovered in the make-up of Nature is a strong evidence for a deeper and all-encompassing rationality governing its most basic entities. The burning question remains—what are those basic entities? On what do the laws of Nature act?

The development of the scientific laws we have followed so far, be they Newtonian, relativistic, or quantum in character, has displayed a common imperative. Rules are sought out which dictate or describe how physical phenomena will behave elsewhere and in the future, if they can be specified here and now. The extension of this approach into the world of elementary particles would appear to call for mathematical rules that tell us how the most basic bits of matter behave. But this is only half of the truth. The study of elementary particle physics wants to do something extra that no other physical science aims at. It wants the laws of elementary particle behaviour to dictate what elementary particles there are. The student of Newtonian mechanics has no interest in cataloguing all the objects in Nature which obey Newton's laws of motion. He has learnt the lesson of Galileo, that the specific identity of bodies does not affect their response to forces. Moreover, every macroscopic body he encounters is different. The particle physicist, by contrast, has found that elementary particles fall into a finite number of *identical* sorts. Everywhere we have looked in Nature there exists an entity called the electron, defined by a small number of properties like its mass, electric charge, spin, and so forth. Every electron appears to be identical to every other electron: 'once you've seen one electron you've seen them all'. This is the most persuasive of all testimonies to the 'unity of Nature' (John Wheeler and Richard Feynman were once so struck by this situation that they suggested there may exist only *one* electron, whose influence can be propagated forwards and backwards in time in an infinite number of ways to give the illusion of there being many different identical electrons!).

If we accept that the most basic elements of matter come in collections of *identical* members, then next we would like to know whether there exist laws which restrict the number of different types of particle. To some extent this has been achieved.

Modern theories of elementary particles and of their interactions with one another are members of a breed called *gauge theories*. The first such theory was Maxwell's theory of electromagnetism. Today's best working descriptions of all the known forces of nature—gravity, electromagnetism, and the weak and the strong forces—are all gauge theories. Such theories are predicated entirely upon symmetries. We recall from Chapter 3 how the presence of a geometrical symmetry is always associated with the conservation or permanence of a particular attribute of the system under study in the face of general changes. For each symmetry there exists an associated conserved quantity. This property is maintained even when the symmetries involved are more esoteric than simple geometrical translations and rotations. These additional symmetries are called *internal* symmetries, and correspond to invariances under various re-labellings of the identities of particles—for example, swopping the identities of all the protons and neutrons in the Universe. Like the geometrical symmetries, these internal symmetries also give rise to conservation laws—the conservation of electric charge arises in this way.

Gauge symmetries are rather different. In particular, they do not lead to conserved quantities in Nature, but the requirement that a particular gauge symmetry be maintained by, say, the electromagnetic interactions in Nature places very powerful mathematical restrictions upon its detailed structure. So, for example, the requirement of the gauge symmetry of Maxwell's equations and their quantum edition forbids the existence of a photon possessing a mass, and it dictates the precise way in which electrically charged particles interact with light. But the power of such considerations is even deeper than this. The simplest geometrical symmetries we mentioned in Chapter 3 were all examples of what are called *global* symmetries. Suppose we take some object, your hand will do, and transform its position in some way so that every point of it is transformed in an identical fashion. The result is simple: the hand just moves bodily to a new place in space. It looks the same in every respect. But physicists wish the laws of Nature to be unchanged under far more general changes. It is unnatural to expect an invariance under changes that are required to be the same everywhere. If a change occurs on the other side of the Universe, how can electromagnetic phenomena here and now know instantaneously how they must behave in order to keep in step, and maintain the global symmetry? The finite speed of light seems to forbid it. What we would require to avoid this difficulty is invariance of the law of Nature under changes which could be different, and completely arbitrary, at every place in time. This is called a *local* gauge symmetry. Whereas the geometrical and internal symmetries were invariances that existed when you changed things in the same way everywhere and everywhen, local gauge symmetry requires invariance after you have done different things in different places at different times. At first sight this appears an impossibility. If every bit of my hand is allowed to move in space in a different and arbitrary manner, then the form of the hand cannot be maintained. It will be dispersed into bits and pieces going their separate ways. There is only one way in which the form of the hand can remain intact and unaltered in the presence of such unbridled changes: forces must necessarily exist with a form that constrains how different parts of the hand can move. Trivially, we might imagine elastic bands placed around our fingers to prevent their dispersal in the illustration given above. What this means is that the imposition of a more general, or *local*, gauge symmetry actually dictates what forces must exist between the particles involved. Einstein's general theory of relativity is a local gauge theory of this sort. The laws of motion that Einstein framed to account for uniform motion at constant speed constitute the theory of special relativity. They have the same form for all observers in constant relative motion. The only way in which the form of the laws of motion can remain the same for all observers in arbitrary accelerated motion relative to one another—in which case they feel forces—is if the gravitational force field exists to cancel them out.

The reason that elementary particle physicists are fascinated by gauge theories can be traced to these remarkable properties. Given a particular symmetry, this can be used to generate a complete theory of one or more of Nature's interactions in a way that contains very rigid constraints upon the types of particle that can exist within its domain, and the forces that can act between them. The gauge age has reduced laws of the micro-world to symmetries in a systematic and powerful fashion. Whereas the laws of classical physics were recognized only in retrospect as being associated with the maintenance of particular symmetries, those of modern quantum physics can be derived *ab initio* from the belief in the universality of symmetry.

These considerations appear, for the first time, to present physicists with a trustworthy guide to extending our knowledge of Nature's laws without having primary recourse to observation. If gauge symmetry is the master-key to the basic workings of Nature then it appears that we possess a systematic method for generating new laws of Nature which can then be put to the test of observation and experiment. However, this is some way from the true state of affairs. In particular, although gauge theories prescribe what type of particle is allowed, they do not specify how many varieties of each allowed particle exist. So, for example, we have evidence for three types of neutrino and their associated charged particle (electron, muon, and tau), but the stipulation of the gauge symmetry of the associated interactions does not tell us how many other undiscovered neutrino types there may be. Nor does gauge symmetry tell us the particular masses of the elementary particles and the precise strength of their interactions. It tells us that certain things are proportional to others, but it does not fix the values of the proportionality factors. It tells us something of the pattern of the laws of Nature, but not the population of particle types, nor the values of the constants of Nature. These defects are not ones which could bring the known gauge theories into conflict with observation. These are questions that are beyond the scope of the current edition of these theories to answer. They reflect the fact that these gauge theories are not the ultimate descriptions of Nature. This we might suspect from the simple fact that there exist so many of them. The successful working gauge theories of the different forces of Nature are all distinct, self-contained, and insulated from each other. We believe that the basic laws of Nature are not like this. Rather, there should exist some deeper description in which the many interactions of Nature appear as different manifestations of a single force. This quest is an old dream, but has only begun to be investigated in earnest since the late 1970s. It is important to recognize that contemporary particle physicists are dissatisfied with the current editions of their theories, even though each appears to work perfectly as a *description* of its own area of jurisdiction. The search for bigger and better theories is not motivated by the contradiction between the *status quo* and experiment. Rather, there is a philosophical dissatisfaction with the scope and lack of universality of the existing theories. They are evidently incomplete, and lack unifying connections with each other. The engine of progress here is not a shattered Kuhnian paradigm, but a desire to join together what Man has put asunder.

Before we turn to the question of whether there exists a sweeping unification of all the laws of Nature—a single all-encompassing symmetry—it is well to be aware of one profound fact about the symmetries possessed by the laws of Nature. It is this fact that makes the deduction of natural laws from observations of the world so difficult.

When we say that Nature possesses a certain symmetry, what we mean in practice is that there exists some quantity which possesses that symmetry, expressed in mathematical language, and from which we can generate a collection of differential equations which will also possess this symmetry. For all practical purposes these equations are our laws of Nature governing the entities under study. However, what we see in the real world are particular events and interactions described by particular *solutions* of the equations embodying those laws. But the solutions of the equations need not display the symmetry possessed by the equations themselves. Consider a simple example. If we support a pencil balanced on its point, then the law of gravity which governs how it falls when released is symmetrical in the sense that it does not favour one direction in space over another. The mathematical equations of Newton or Einstein make no explicit mention of direction in space. However, the pencil must always fall in *some* direction when released. In other words, the symmetry of the law is broken in particular realizations of it.

With a little thought this is obvious. I am sitting at one particular place in the Universe at this moment, and so are you. But we saw that the laws of Nature possess an invariance that makes them the same to observers at every place. This symmetry gives rise to the conservation of linear momentum. The only way in which such a space-blind set of laws can be reconciled with the fact that they permit me (and you) to reside at particular places in the Universe is if the outworkings of symmetrical laws do not necessarily reflect the same invariance principles. Another example we might take is the fact that human beings have hearts on the left of their bodies. There is no deep symmetry in Nature that appears to require this. We could all have our hearts on the right of our chests, like our mirror images, and our bodies would function identically. Presumably, the first humans to evolve had this 'left-heartedness', and the effect then snowballed through hereditary reinforcement. We could not deduce anything about the left-handedness of the fundamental laws of Nature from the fact of our own left-heartedness.

Hence, our study of Nature is difficult. Everywhere we look in the Universe we witness the particular asymmetrical outworkings of underlying laws which may be symmetrical. The true symmetries are hidden. They govern laws, not their consequences; we are able to observe only the consequences of broken symmetry.

When a gauge symmetry is broken in a particular realization something unusual occurs. The carrier particle of the force of Nature that is required to implement the local gauge invariance will take on a mass. This mechanism is believed to be the origin of mass. There is, in effect, an underlying energy field called the Higgs field (after Peter Higgs, an Edinburgh physicist) permeating every point of space, and this induces the presence of a mass for the carrier particle when the symmetry is broken. Some gauge symmetries, for example those that give rise to the electromagnetic and gravitational force fields, are not broken, and hence their respective carrier particles, the photon and the graviton, remain massless. But the gauge symmetry giving rise to the weak force is broken, and the carrier particles, the W and Z bosons, thereby aquire masses.

Unification

> Physics is littered with the corpses of dead unified field theories.
>
> Freeman Dyson

The dream of a single, all-encompassing, law of Nature spanning the whole of fundamental physics, showing the dissimilar forces of Nature to be but different manifestations of one underlying influence, has risen many great physicists from their slumber of contentment with the *status quo*. Eddington and Einstein are the most famous pioneers of a grandiose 'theory of everything' uniting electricity, magnetism, and gravity, from whose synthesis would be revealed, behind the inner logic, the peculiar values of the constants of Nature. Although, with hindsight, we see that these imaginative pioneers were searching in the wrong place for this Holy Grail of physics, and attacked the problem at a time when we lacked even the knowledge of what it was that needed to be unified, they created a vision that has remained as the ultimate theoretical challenge in physics, awaiting another Einstein to take it up successfully.

Since 1975 there has again been serious work on the problem of unifying the known laws of Nature into a single 'Grand Unified Theory', as it has become known. Spurred on by the evident exploitation of gauge symmetry by Nature, theoreticians have attempted to embed the known symmetries separately respected by the strong, electromagnetic, and weak forces into a larger all-encompassing symmetry. The form that this unifying symmetry can take is constrained in a number of important ways, in order that the sub-symmetries fit within it in a logically consistent fashion. Roughly speaking, it must be to mathematical symmetry operations what a prime number (like seven or eleven) is in arithmetic—one must not be able to factor it exactly into pairs of sub-symmetries. These requirements generate a collection of candidate symmetries, each of which must then be fleshed out with a physical outworking to see if it produces a correct description of the world.

At first it seems strange that any such unification of different natural forces can even be contemplated. How can one expect to unify gravity and electromagnet-ism with the weak and strong nuclear forces, when each of these forces have such different strengths and ranges. Gravity is 10^{39} times weaker than electromagnet-ism! Moreover, the heterogeneity goes farther: the different forces have been found to act upon different and exclusive classes of elementary particles. When Einstein and Eddington sought unified theories of Nature, the richness of the elementary particle world lay unexplored. When its complexity was revealed there was no rush to create new unified descriptions because of these basic problems: the different forces of Nature appear as qualitatively and quantita-tively different as any four things could be.

Despite these obstacles, unification is gradually proving possible, piece by piece. We possess a unified theory of the weak and electromagnetic forces whose predictions were beautifully confirmed by experiments at CERN in recent years. The next step is to incorporate the strong interaction into this synthesis. The final step—the addition of gravity—is the most formidable conundrum engaging the attentions of theoretical physicists today.

The problem of merging together interactions which appear to have such different strengths has been overcome by our understanding of the properties of the quantum vacuum. In the pre-quantum picture of electromagnetism two negatively charged electrons would repel each other (like two similar magnetic poles) with a force that is a measure of the intrinsic strength of the electromagnetic force of Nature, and which does not depend upon the temperature of the environment in which the interaction takes place. But place a charged particle in the quantum vacuum and things become dramatically different. The sea of virtual particle pairs contains electrons and their antiparticles (positrons) continually appearing and re-annihilating into radia-tion. When a real negatively charged particle is introduced into this sea it tends to attract all the oppositely charged members of the virtual particle sea, and so becomes surrounded by a shielding cloud of virtual particles with positive charges. Hence, any approaching charged particle will be repelled rather weakly if it approaches with low energy, since it will only reach the outer sea of shielding charges, whereas a rapidly moving particle will penetrate the shielding cloud and get close enough to the centre to feel the full negative charge within. Thus, the higher the energy of the incoming particles, the stronger the electromagnetic interaction it feels. A helpful analogy is to imagine two solid billiard balls surrounded by thick layers of woollen padding. If these fluffy objects are

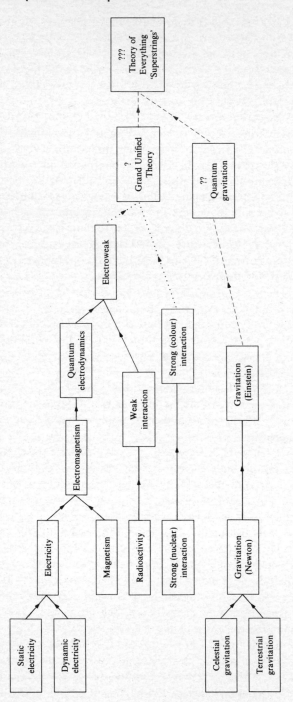

propelled towards each other, then the force with which they rebound is strongly dependent upon the speed of collision. At slow speeds there is only a weak interaction between their woolly exteriors, whereas at high speed the hard cores meet and rebound strongly. Likewise, the effective strength of the electromagnetic interaction between elementary particles depends upon the energy and temperature of the environment within which it is measured. As the temperature rises the interaction becomes *stronger*.

When the same reasoning is applied to the strong and weak forces of Nature something more complicated occurs. The strong interaction occurs between particles which carry the attribute called 'colour' charge which we introduced earlier. Where the quantum theory of electromagnetism is a description of how electrically charged particles interact with light, the quantum theory of the colour force describes how quarks and their gluon carrier particles interact with each other.

The interaction of two quarks sitting in a quantum vacuum is rather different from that between two electrons. A single quark will be surrounded by a sea of continuously appearing and annihilating virtual pairs of quarks and antiquarks, and virtual pairs of gluons and antigluons. But the gluons are unlike the carrier particles of the electromagnetic force which appeared along with the virtual electrons and positrons. Photons do not possess electric charge, but gluons do carry the colour charge whose influences they mediate. Whereas the virtual antiquarks tend to attract to the quark in the vacuum, so shielding its colour force, the gluons congregate preferentially around quarks of the same colour charge, and thus tend to smear out the central colour charge over a wider volume. This trend opposes the segregation of the colour charge created by the virtual antiquarks. Who wins? If there are fewer than eight different varieties of quark, then it is the colour smearing by the gluons which prevails. In this case, the higher the energy of the environment so the higher the energy of one quark as it approaches another, and the closer they will be able to get to each other's centre. But the closer they get, the more spread out they find the repelling colour charge to be, and the *weaker* the resulting interaction. This effect is completely opposite to that which occurred with the electromagnetic force, and it means that the strong interaction will become progressively weaker at higher energies of interaction. This trend is called *asymptotic freedom*, to stress the fact that, as the energies get higher, the interactions get progressively weaker, so that the particles behave asymptotically as though they were free from the influence of any forces. Similar considerations to those we have just given show that the weak force changes in effective strength as the energy at which it is measured increases.

Thus, the strengths of the electromagnetic, weak, and strong forces depend upon the energy at which they are measured. At the ambient temperatures where life is possible the strengths of these three interactions look very different and distinctly un-unified. But as the temperature rises the strong interaction weakens, while the other two interactions strengthen at different rates. The weak and electromagnetic forces become of equal strength at energies of about 90 GeV (giga-electron-volts), but they should only become similar in strength to the strong interaction at an energy of about 10^{15} GeV—far beyond the reach of direct experiments on Earth.

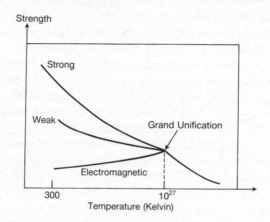

Figure 4.3 The variation of the 'effective' strength of the strong, weak, and electromagnetic forces of Nature with the energy at which they are measured. The variations are produced by the properties of the quantum vacuum.

Thus we see how the low-temperature environment which physicists necessarily inhabit biases the picture they have had of the forces of Nature. The quantitative problem of unifying forces of apparently different strengths is resolved by the properties of the quantum vacuum. But this still leaves the qualitative problem. According to our conventional wisdom, quarks carry a *conserved* type of charge called 'colour'; leptons (like the electron and the muon) do not. So experience had convinced us that there is no way in which a quark can transform into a lepton and vice versa. In order to unify the forces of Nature governing the interactions of quarks and leptons, it would be necessary for there to exist some undiscovered type of elementary particle in Nature which could mediate interactions between quarks and leptons. These particles should appear in profusion only at the very high energies at which the unification takes place. Thus, they must have a rest mass energy roughly equal to the unification energy. This state of affairs makes it very unlikely that the effects of this unification would be significantly manifested at the low energies of everyday life—even that of particle physicists. A similar logic was applied to the well-studied case of the weak and electromagnetic unification which occurs at energies close to 90 GeV. For this unification to be manifested there must exist three new elementary particles with masses very close to the electromagnetic–weak unification energy. These three particles were discovered a few years ago—with masses just as predicted—and are called the W^+, W^-, and Z bosons. The analogous heavy bosons associated with the grand unification at 10^{15} GeV can only be detected by their indirect effects, if our present ideas are correct. Their most striking consequence should be that the proton is unstable and decays into leptons, mesons, and light. One can think of this as a new type of radioactivity. It is expected that the proton will have an average decay time exceeding 10^{31} years. This lifetime corresponds to about one proton in the reader's body decaying in his or her human life-span. This is an extraordinarily long time; but the huge number of protons in large bodies means that it is practical to detect such decays from aggregates of matter containing a thousand tons of water or iron buried deep underground. There, surrounded by particle detectors, and shielded from the effects of confusing cosmic rays by the Earth's crust, these tell-tale decays could be detected. Despite a number of experimental searches, this characteristic signature of grand unification has yet to be seen in more than one experiment, and so far we can only conclude with certainty that the proton has an average lifetime exceeding 10^{32} years. 'Grand Unification' may or may not be true.

Richard Feynman has some pithy comments regarding the direction of these quests for symmetry in Nature, and the significance of the finds already made. We have good gauge theories of the different interactions of Nature; they possess a compelling similarity in structure. But why, asks Feynman, are all the gauge theories of physics so similar in structure?

There are a number of possibilities. The first is the limited imagination of physicists; when we see a new phenomenon we try to fit it into the framework we already have—until we have made enough experiments, we don't know that it doesn't work. So when some fool physicist gives a lecture at UCLA [the local rival institution to Feynman's!] in 1983 and says, 'This is the way it works, and look how wonderfully similar the theories are', it's not because Nature is *really* similar; it's because the physicists have only been able to think of the same damn thing, over and over again.

Another possibility is that it *is* the same damn thing over and over again—that Nature has only one way of doing things, and She repeats her story from time to time.

Another possibility is that things look similar because they are aspects of the same thing—some larger picture underneath, from which things can be broken into parts that look different, like fingers on the same hand.

This quest for a unification of all the forces of Nature has continued to develop in a technical fashion, but has yet to arrive at an experimentally verified symmetry that does the trick. It is not our intention to provide a detailed description of these developments, but merely to indicate new aspects of the laws of Nature that are suspected. The most striking new development has been the dusting-off and revamping of an old idea.

A new dimension

> The idea of achieving [a unified field theory] by means of a five-dimensional cylinder world never dawned on me . . . I like your idea enormously.
>
> Albert Einstein [to Theodor Kaluza]

In 1919, soon after Einstein's revolutionary general theory of relativity had been developed, Einstein received a letter from Theodor Kaluza, an unknown mathematician at the University of Königsberg where Immanuel Kant had lived and worked. Kaluza noticed that if Einstein's theory of gravitation were formulated for a universe possessing one more dimension of space than our own, then it became mathematically identical to Einstein's theory for our three-dimensional world, together with the electromagnetic field of Maxwell. By adding an extra spatial dimension to the world one could reduce electromagnetism plus gravity to gravity alone. The abstract symmetries that gave rise to the conserved quantities of electromagnetism became simply the conditions for the four-dimensional gravity theory to be generally covariant; that is, for the equations to have the same form in all co-ordinate systems after the manner that Einstein had demanded of all physical laws.

This idea seemed so outlandish to Einstein that he kept Kaluza's paper for two years before giving it his final *imprimatur*. But eventually, in 1921, it was published by the Prussian Academy. Few took this idea seriously. Indeed, even fewer were in a position to do so; the mathematics used in Einstein's theory of general relativity in just three space dimensions was beyond the ready grasp of most scientists of the time. The addition of extra dimensions was an even worse abstraction. Furthermore, Kaluza offered no observational support or possible test of his theory. It was just a pretty mathematical idea. Today physicists are rather susceptible to pretty mathematical ideas. In 1921 this susceptibility was not quite so well developed.

In order to make sense of Kaluza's idea one had to explain where the extra dimension had gone. Why do we not experience it? Why does the Universe appear three-dimensional if it is in reality four-dimensional?

Kaluza had suggested that the extra dimension of space might differ from the other three in some fundamental way. In 1926, the Swedish mathematician Oskar Klein developed this suggestion in a way that connected it with physics. Klein showed that if it mediated the connection between electromagnetism and gravity, then it was natural for the extra dimension to have a minute extent—maybe as small as 10^{-33} centimetres—so we would be unaware of it both in everyday experience and all scientific experiments carried out hitherto. We would be in the position of flat insects spending our lives walking over the flat Earth unaware of the dimension of height extending above us. Although our experience of the surface of our 'Flatland' could extend far and wide, that of the orthogonal third dimension could be confined to a tiny, imperceptible extent.

Klein's picture of the four-dimensional world wraps up the additional space dimension so that what we have been calling a point in space is really a tiny circle, and what we view as a line is a narrow tube.

After half a century of dormancy the speculations of Kaluza and Klein have been enthusiastically resuscitated in recent years. At first it was suspected that just by adding even more spatial dimensions to make a total of ten one might be able to extend the trick, and reduce the weak and strong nuclear forces along with their generating gauge symmetries to pure gravity in a higher-dimensional realm.

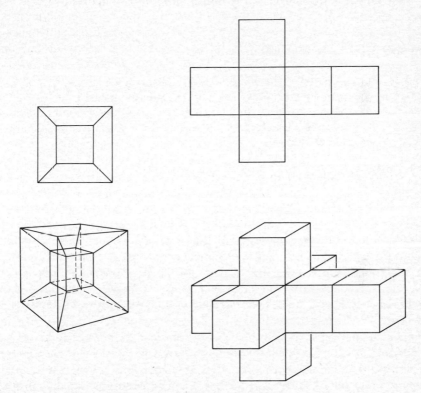

Figure 4.4 (a) If a three-dimensional cuboid is viewed in projection in a mirror then it appears as two rectangles joined by four lines joining the corners of the two rectangles. We cannot imagine what a four-dimensional version of a cuboid would look like but if we were to view it projected in a mirror it would look like a three-dimensional figure which we can visualize (b). It would appear as two cuboids joined by twelve surfaces, one emanating from each edge of the interior cuboid and ending on an edge of the external cuboid. If we unravel the one-dimensional boundary of a two-dimensional square it produces a line composed of four equal line segments. If we unravel the two-dimensional boundary of three-dimensional cube then we produce a cross of six equal squares (c). Correspondingly if we were to unravel the three-dimensional boundary of an (unvisualizable) four-dimensional 'cube' it would look like a criss-cross of eight equal cubes (d).

It turns out that things cannot be quite as simple as that, but the idea that there may exist additional spatial dimensions is now taken seriously. It may well prove that the laws of Nature display their most basic and symmetrical structure in more than three dimensions. We may glimpse only the three-dimensional shadow of the higher-dimensional laws (Figure 4.4). This possibility introduces a new degree of difficulty to the problem of unravelling the laws of Nature. It also creates interesting cosmological problems of its own. What determines the total number of dimensions of space? Why are three of them large and observable by us, while all the others remain curled-up to an infinitesimal extent?

Why are there three dimensions of space?

> In two dimensions any two lines are almost bound to meet sooner or later; but in three dimensions, and still more in four dimensions, two lines can and usually do miss one another altogether, and the observation that they do meet is a genuine addition to knowledge.
>
> A. S. Eddington

The fact that we experience three dimensions of space is intimately connected to the laws of Nature that exist in those three dimensions. This connection was first recognized by Immanuel Kant during the early 'pre-critical' phase of his career, when he was still an ardent Newtonian who believed that the form of the law of gravitation found by Newton was unique in just the way that the geometry of Euclid was believed to be unique. He claimed that the ubiquity of the inverse-square law in Nature was the reason for space possessing three spatial dimensions. Kepler also appreciated the connection between spatial geometry and the inverse-square law of light intensity fall-off, pointing out the differences expected in two- and three-dimensional spaces. (Today we would turn this argument around, and say that three dimensions lead to inverse-square laws for basic force fields.) During the twentieth century there have been several investigations into the connection between dimensions and laws of Nature, which have been very revealing. We can see how it is the three-dimensionality of space that allows stable orbital motions to occur under the influence of central forces of attraction like gravity, electricity, and magnetism. Such stable configurations enable a whole range of structures, from atoms to solar systems, to exist in Nature. Furthermore, the propagation of waves is very delicately conditioned by the dimension of the space in which they move. Only if the dimension is an odd number, like 3, 5, 7, . . . do waves travel only at the fundamental wave speed. In the case of an even number, parts of the wave travel at all speeds less than or equal to the fundamental speed. The result is that sharp signals cannot be propagated. Some parts of the signal arrive at different times to others. If we restrict our attention to the odd-dimensional spaces, we find that when the dimension exceeds three, the wave signals distort as they travel. Only in three dimensions can waves propagate in an undistorted and reverberation-free fashion. This has important ramifications in all the areas of Nature where wave propagation lies at the heart of the Universe's workings. Quantum waves, brain waves, gravitational waves, electromagnetic waves—the form of the laws that govern them and the natural phenomena are all inextricably linked to the dimension of space. And only in three dimensions is the number of dimensions equal to the number of axes about which one can perform different rotations. This simple fact dictates the entire form of the laws of electromagnetism.

These special properties of three space dimensions make it very difficult to envisage how chemical life of any sort could have emerged if the Universe had existed with more than three dimensions of space having large extent. But one can be less anthropocentric in judging the special features of three dimensions. Mathematicians have repeatedly found that spaces with three and four dimensions have unusual and often unique properties not shared by other dimensions. In fact, the study of these low-dimensional spaces exists as an area of mathematical study separate from that of spaces with more than four dimensions. In the future we may discover that the special properties of three and four dimensional spaces tell us important things about the three-dimensional universe of space and the four-dimensional universe of space-time which we experience. If there really exist more than three dimensions to space, but these extra dimensions are coiled up to an infinitesimal extent, then we need to understand why it is that three and only three of them have been allowed to grow large. For it is this state of affairs that dictates much of the form of the laws of Nature. As yet we do not know whether there had to exist only three large dimensions of space, but we begin to suspect that only large three-dimensional worlds would be cognizable to complex beings. If the Universe did once possess many more dimensions of space, we need to determine whether there existed a random element in the dimensional gymnastics that resulted in three dimensions going their separate ways to become the space in which we live. Could there have been four or five dimensions that escaped, or is there some deep property of Nature which ensured that there had to be three?

What are the ultimate building blocks of matter?

> *Examiner:* What is Electricity?
> *Candidate:* Oh, Sir, I'm sure I have learn't what it is—I'm sure I *did* know—but I've forgotten.
> *Examiner:* How very unfortunate. Only two persons have ever known what electricity is, the Author of Nature and yourself. Now one of them has forgotten.
>
> Oxford University science viva, *c.* 1890

The atomistic ideas that have been adopted by twentieth-century physicists, and transformed into precise mathematics, have resulted in a collection of symmetries which allow certain types of particle to exist and to interact with others in very particular ways. These symmetries are in some sense what the forces of Nature 'are'. This is either a consequence of the intrinsic mathematical personality of Nature or, as Kant would have claimed, just the inevitability of a representation of Nature being required which is intelligible to our brain's intuition. Setting a resolution of this higher-order question to one side for the moment, we see that we must learn precisely what the most elementary objects in Nature are if the role of symmetry is to be placed in its proper perspective. In short: we need to know what the laws of Nature govern. What are the basic elements of the Universe?

Habitually, the modern physicist has assumed that these basic elements must be 'points', that is, entities which have zero size. The most basic things whose behaviour is orchestrated by gauge symmetries are usually taken to be the quarks and leptons. They are assumed to be point particles. In practice, this idea means nothing more than the operational fact that when they collide with other particles they rebound as if they are points. It is possible that our limited experience of the elementary particle world has yet to uncover the fact that quarks and leptons contain further constituents. Whether or not they do, it is undeniable that, until very recently, the overriding assumption that points in space were the most basic entities held sway. The theories which enshrine this assumption in mathematics are the quantum field theories. But, successful as these theories are in describing different interactions of Nature, they appear to be beset by an endemic 'disease' which has blocked attempts to carry out the programme of unifying all the forces of Nature. It has been found that attempts to calculate some quantities that we observe in Nature give rise to infinite answers in these theories. Only by carrying out some sleight of hand can these infinities be removed. Moreover, in some cases not even this can be done. When calculations are made they produce numerical answers which are given by the sum of an infinite series of terms. What one requires of a 'good' theory is that this infinite sum should contain terms which get successively smaller so quickly that the series sum gets closer and closer to a definite value as more and more terms are added. A simple example is the series

$$1/1^2 + 1/2^2 + 1/3^2 + 1/4^2 + \ldots \textit{ ad infinitum}$$

If you keep on calculating these terms and adding them up you will find the answer getting closer and closer to 1.64. In fact the infinite sum of terms equals

$$\pi^2/6 = 1.644934 \ldots$$

However, if we take the series

$$1/1 + 1/2 + 1/3 + 1/4 + 1/5 + 1/6 + \ldots \textit{ ad infinitum}$$

you will find that the running sum never settles down to approach a definite number. It just keeps increasing steadily. The sum of an infinite number of terms following the pattern given leads to an *infinite* answer, even though each successive term in the series gets smaller. Such a series is called *divergent*. It is quite easy to check that the last series diverges just by grouping its terms in successive batches of two, four, eight, sixteen, and so on:

$$1/1 + 1/2 + [1/3 + 1/4] + [1/5 + 1/6 + 1/7 + 1/8] + [1/9 + \ldots + 1/16] + [1/17 + \ldots + 1/32] + \ldots$$

Each bracketed term totals to more than 1/2. There will be an infinite number of these sub-totals, and so the grand total must be an infinite number of halves, that is: *infinity*. Clearly, a physical theory which yields answers like this has something wrong with it. It is rather like having a piece of financial accounting software that contains a serious bug.

In 1984 two physicists, Michael Green of the University of London and John Schwarz of the California Institute of Technology, made a remarkable discovery. They showed that if one relinquishes the idea that points are the basic elements of Nature, and replaces them by lines, or '*strings*' as they have become known, then all the diseases of the traditional quantum field theories are miraculously cured, and all the possible symmetries of Nature can be included neatly in only two possible cases—worlds with nine dimensions of space and one of time, and worlds with twenty-five dimensions of space and one of time. Some of these nice properties were appreciated ten years earlier, but the idea of more than three spatial dimensions was not then taken seriously. It was prior to the revival of Kaluza and Klein's ideas. As a result, this line of inquiry was abandoned by the main stream of research in elementary particle theory. Only the real diehards, like Green and Schwarz, persisted with strings as others rushed to explore the success of the gauge theories. Eventually it was found that another of the early problems with a string theory of Nature—its prediction of 'tachyons', particles which move faster than light and render the description of Nature inconsistent—disappeared if the property of 'supersymmetry' was incorporated into their formulation. This was a desirable step, because gauge theorists had developed this mathematical symmetry in order to unify elementary particles of different intrinsic spin. In effect, it unified matter and radiation into a single theory. It is still not known whether supersymmetry is a reality. Most particle theorists act as though it is true, but we will only know when particle accelerators discover the extra elementary particles whose existence it predicts. Thus strings became 'super-strings', and embodied all the possible symmetries that they could possess.

One interesting detail is worthy of note at this stage. We have just recounted how the form of the force laws and wave equations of Nature owe their forms and properties to the dimension of space. The special properties of super-symmetric string theories do not pick out particular dimensions of *space* as being preferred; they select particular dimensions of *space-time*. However, in practice it is always assumed that there is only one dimension of time, because otherwise weird problems with causality could arise, and energy would not be conserved because the laws of motion would cease to be invariant under translations in time (see

Chapter 3). Only because of this assumption do we conclude that ten or twenty-six dimensional superstring theories require nine and twenty-five dimensions of space. These considerations serve to impress upon our minds once again how different time really is from space physically, despite the formal mathematical similarity between the two quantities that mathematical theories of physics create, because of the way that they use spatial position and temporal locations as labels to help identify events uniquely.

At first this introduction of 'superstrings' rather than points sounds peculiar when we know that particles behave very much like points at the energies of accelerator experiments. However, the strings which form the building blocks of Nature possess a tension which varies with the energy of the environment, and as the temperature falls the tension increases and contracts the linear string down to a point. But the real attraction of the string picture is that one can get so much out of so little. When the known elementary particles of matter are regarded as points, then one is faced with a large collection of them—embarrassingly large in view of the many physicists. The properties of each one of these particles requires a different explanation. But a single string possesses many possible energies of vibration. The goal of the string picture of reality is to attribute each force and elementary particle species of Nature to a different vibrational state of a single string. The lowest-energy vibration should be associated with gravity, the weakest force, whilst the more energetic excitations of the string may give rise to the other forces and particles.

open string   closed string

Figure 4.5 An open and a closed superstring.

Strings can come in two varieties: open or closed. Each of these varieties can be either 'disoriented' or 'oriented'. In the former case waves propagating around the string have the same properties regardless of their direction of travel around the string, while in the oriented case those properties differ. At present, the most suggestive structure for an ultimate description of the forces of Nature appears to be an oriented closed string, called the 'heterotic string' by the four Princeton physicists (now known as the 'Princeton string quartet'!) who introduced it.

Strings can interact with each other in clearly defined and intuitively obvious ways. An open string can break into two open strings, or pinch off a loop to leave a closed and an open string, or simply close up and form a closed loop. A closed string can fission into two closed loops. Finally, two open strings can intersect and separate off to form two different open strings.

At present this picture forms the focus of interest in the physics of elementary particles. It provides the first compelling candidate for a theory of everything that physicists have ever had. Miraculously, it may even be uniquely specified by the requirement that the theory possess no infinities. As yet there are no experimental facts for or against it. In coming years the formidable mathematical defences of its basic equations will be breached, and predictions will surely be forthcoming. Only then will we know whether the unique prescription it offers is a theory of everything or a theory of nothing.

The inner space credo

> No, no! You have merely painted what is! Anyone can paint
> what is; the real secret is to paint what isn't!   Oscar Mandel

What have we found from our brief exposure to the elementary particle world? The laws of Nature that govern there are found to be precise and pristine in a way that we do not witness in the complicated everyday world, where a multitude of simple events interfere to obscure the basic features. In the elementary-particle world symmetry—whether preserved or broken—is the guiding principle, because there exists a deep connection between the laws of Nature and the identity of the objects they govern. The search for the laws of elementary particle processes appears Platonic in its approach. Symmetries are systematically explored in the belief that Nature will employ the biggest and best. The laws of Nature are reduced to the list of things that can be done to the world without changing its observable features. This philosophy leads to systems of mathematical equations which then constitute 'the theory'. However, it is often easier to find a theory than to solve the equations which constitute it. Although the theory may possess an elegant symmetry, this need not be manifested by any of the *solutions* of that theory, and it is these that tell us what we should see in Nature. Again, we find the profoundest logic reserved for the constitution of things behind the scenes. Nature uses symmetry most deeply to dictate the laws of Nature, not to specify the form of individual things. These symmetries are all, at root, safeguards which allow the form of the laws of Nature to look the same no matter what the state of motion and individual perspective of the observer. This view, which Einstein pursued with innovative genius, has proven the most fruitful interpretation of the use of symmetry in Nature. Furthermore, we are probably justified in suspecting that the symmetries manifested by the version of the laws of elementary particle behaviour that we have at present are at least some of the properties possessed of the truly fundamental laws that we have yet to uncover. At worst, they form part of the truth.

The successful assumption that Nature employs symmetry maximally has led to two types of attitude towards the forms of unknown laws of Nature. On the one hand there is the 'totalitarian' approach that everything which is not explicitly forbidden by the dictates of symmetry must be a compulsory part of the laws of

Nature. On the other, there is the more 'liberal' philosophy that everything which is not necessary to maintain a symmetry is forbidden. It is at present a matter of taste which one adopts as a guide.

Gradually, we have uncovered evidence that the most effective and harmonious mathematical laws of Nature are those which are formulated in more than three space dimensions. The world we witness may be but a thin slice of a higher-dimensional Universe of vastly greater complexity than we had thought possible. The logic that leads to this conclusion also derives from an axiom of faith: the physicist's religious belief in the *unity* of the Universe. This belief motivates the search for a unified theory of all the forces and particles of Nature. It is unquestioned that such a theory exists, even though we could imagine two rival systems of natural laws holding sway over different regions of the Universe, and interacting in some competitive way rather like two life forms. The belief in the persistent use of symmetry, and the quest for a single unified description that utilizes the minimum possible number of components, is what dictates the development of theories about the form of the laws of Nature in the subject of elementary particle physics, where experimental evidence is often beyond the reach of current experiment. Indeed, it may well prove that the 'theory of everything', if it exists, is beyond our means and abilities to authenticate by observation. We could discover that a beautiful mathematical theory possesses unique properties that enable it to solve all the questions of principle which we have asked of it. It would possess no known faults or incompletenesses—it might even be *the* correct theory. But there is no reason why we should be able to verify it by experiment, or be able to falsify rival candidates. There is no reason why the Universe should be fashioned in a manner that allows human beings to discover its basic laws. If we are lucky we may be able to find some way of partially checking a mathematical description of Nature's most basic laws, but equally we may find that they reveal their decisive features only in environments of such extreme temperature and energy that we can never reproduce them.

Our brief encounter with the direction of inquiry into the legislation of inner space shows how far removed are the laws of Nature and the things they govern from what we see and intuit about the world. Newton persuaded generations of scientists to view the world as a gigantic clockwork mechanism, but modern particle physics reveals it to be a kaleidoscope of constantly changing patterns. But we are still left wondering how close we have managed to get to the basic level at which the laws of Nature act.

Outer space

> In the search for truth there are certain questions that are not important. Of what material is the universe constructed? Is the universe eternal? Are there limits or not to the universe? . . . If a man were to postpone his search and practice for Enlightenment

until such questions were solved, he would die before he found
the path.   Buddha

Before Einstein's formulation of the general theory of relativity there had
persisted an ancient prejudice that all the celestial motions take place relative to
some absolutely fixed background stage. It was the idea of this fixed substratum
of absolute space that Einstein's picture of dynamic space–time overthrew. The
new theory allowed the Soviet meteorologist Alexander Friedman to predict that
the entire Universe—everything that is—is in a state of dynamic change. Seven
years later, in 1929, the American astronomer Edwin Hubble confirmed this
prediction by his discovery that the light reaching us from the stars in distant
galaxies was shifted systematically towards the red end of the colour spectrum by
an amount directly proportional to the distance to the emitting galaxy. This
'redshifting', as it has become known, has a simple explanation as a 'Doppler
shift' caused by the recession of distant galaxies away from us. A familiar example
of this Doppler effect is provided by a speeding train. If we stand beside a railway
track and listen to an approaching train, then the pitch of its whistle will first rise
as the train approaches, then abruptly change and fall as the train passes and
recedes into the distance. While the train approaches, the sound waves from the
whistle reach our ears with a higher frequency than that at which they are
emitted, whereas when the train moves away the reception frequency at our ears
is lower than the rate of emission by the whistle. By measuring the change of pitch
as the train passes, its speed can be determined (this is also the principle upon
which police radar speed-traps are based).

When Hubble plotted the speed of recession necessary to explain the
redshifting of the light he observed from distant galaxies against the distance
away, indicated by their brightness, he found the systematic relationship pictured
in Figure 4.6 (although his observations only reached 50 million light years).

The notion of the expanding Universe is a subtle one which creates many
conceptual confusions. We habitually think of the expansion as if it had a centre
in space, and continually emanates outwards from that centre into space—just
like an explosion. An expanding blast-wave of this sort must have an outer
leading edge, and so, we ask, what lies beyond its edge: unfound parts of the
Universe, limbo, or what? In fact this entire image of the expanding Universe is
mistaken. The expansion found by Hubble, and interpreted by Einstein's
description of curved space–time, is an expansion *of* space rather than an
explosion *in* space. There is no accessible centre to this expansion, no edge, and
no beyond into which the Universe is expanding. Our three-dimensional
imaginations do not allow us to picture this, but we can get the basic idea by
thinking of a universe with only two space dimensions embedded in our own.
This universe is represented by the curved surface of an inflating balloon. If we
were to mark a number of equidistant points on the surface of the balloon, then as

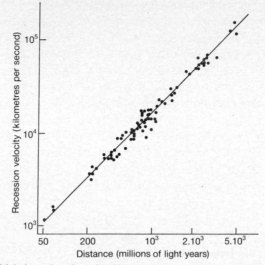

Figure 4.6 Hubble's law showing the observed increase in recession velocity of distant light sources as their distance away increases.

the balloon inflates the two-dimensional universe expands in the sense that every point on the balloon's surface recedes from every other. If we were to observe from any of these points we would see all the other points receding away from *us* as though we were the centre of the expansion, *no matter which point we observed from*. We recognize that this unusual situation exists because the centre of the expansion does not lie on the surface of the balloon. Furthermore, the surface of the balloon which constitutes the two-dimensional universe has the unusual property of being finite in extent (the surface area of a sphere is four-pi times the square of its radius) but unbounded. If we continued moving forever on the surface of the balloon we would never encounter an edge. The expansion of the universe can either continue forever, in which case the balloon continues to inflate forever (ignore the fact that, if made of real rubber, it would eventually burst!), or it can slow down and be reversed into a state of contraction, as if the balloon were eventually to be deflated.

In the movie *Annie Hall*, Woody Allen is reduced to his analyst's couch by the merest contemplation of the expanding universe. His most serious problem was the conviction that if everything is expanding then America, Brooklyn, and, indeed, he himself must also be expanding. Fortunately, this is not the case. The objects which move freely away from each other as the Universe expands are those which are not being held together by other more powerful forces of Nature. In the case of the balloon, the atoms of rubber of which the balloon's surface is composed are not being expanded. They retain a constant size because they are bound together by short-range chemical and nuclear forces which are far more powerful than the air pressure exerted on the expanding rubber. In the real Universe the entities which act as tracers for the expansion of the Universe are large clusters of galaxies. Objects of smaller size, like galaxies and stars, do not themselves expand: they are just carried along in the expansion flow.

Unique aspects of cosmology

> But the concept of a physical law is that it enables us to treat a system that is being studied as one particular case among many. The universe, however, is a unique system, and so this aspect of a physical law seems to lose all meaning when we attempt to apply it to the universe.
>
> W. H. McCrea

Cosmology is the ultimate quest for knowledge, the study of the Universe as a whole: its size, its age, its shape, its wrinkles, its origin, and its contents. This is Mankind's oldest speculation, but only in the twentieth century has it been transported from the realm of metaphysics into the domain of physics, where speculation is not unbridled but ideas must survive a confrontation with observation. Yet despite its adoption of the methodology of the more down-to-earth sciences, cosmology possesses a number of quite unique aspects that must be borne in mind as we try to unravel the laws that appear to govern the behaviour of the Universe *as a whole*. If such laws do exist they indicate something of profound significance—the existence of a logic beyond the material appearance of the Universe. With this in mind, let us examine some of the unique aspects of cosmology as a science which set it apart from all other studies of the laws of Nature.

In the terrestrial sciences one is faced with a surfeit of observational data, and the ability to accumulate even more in the future. What is often lacking is a mature and comprehensive theory which can incorporate, correlate, and explain all the available information. The sheer mass of evidence that must be aligned with a tentative theory makes it well-nigh impossible for any promising candidate to get off the drawing board. It has to be right about so many things at once. In cosmology the situation is reversed. Data is scarce, hard-won, and difficult to interpret. We cannot observe at will; we must accept the evidence which the Universe currently has on offer. Yet, in Einstein's general relativity we have a powerful instrument for providing an understanding of that which is seen.

We cannot experiment with the Universe, only observe it. Yet there are two ways in which we can keep close to our traditional method of investigating Nature by checking theoretical predictions against observational yardsticks. We can seek correlations between observed features. Our theory might predict that the most massive galaxies are the most luminous. We test this idea by looking at the relationship between the luminosity and the mass of the galaxies we observe, to discover if there is a steady trend in which the mass and luminosity increase in proportion. In practice, things are rarely so straightforward. Checks for correlations will usually exhibit some scatter of the data about a direct proportionality. There may also exist known and unknown sources of experimental error which distort the data. In the end the correlations must be evaluated statistically, and one is able to say only that a correlation exists with a certain probability. This statistical aspect is not unique to cosmology of course. It besets all repeatable experiments.

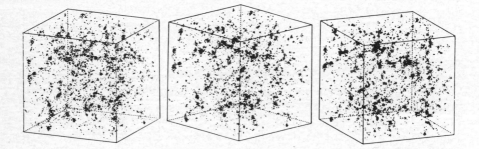

Figure 4.7 *N*-body simulations. With the help of fast computers, cosmologists are able to simulate the effects of gravity upon the motion of tens of thousands of masses. These masses are taken to represent embryonic galaxies. The computer follows the gravitational interactions between thirty-two thousand of these objects placed at random in a mathematical model of the expanding universe. The figure shows three orthogonal views of the clustering pattern that has resulted from one of these experiments performed by Marc Davis, George Efstathiou, Carlos Frenk, and Simon White. The computer stores the speed and direction of motion of each of the clustered objects as well as their position at any instant and uses this information together with the law of gravitation to compute how they change in time. The results of these simulations can be compared with observations of the clustering and motions of galaxies in the real Universe to test the correctness of the assumptions about the starting pattern and mode of clustering.

In recent years a new tool has become available: we can build computer simulations of the evolution of some aspects of the Universe. For example, one can follow the possible evolution of a simplified universe consisting of a large number of masses (each imagined to represent a galaxy) under the laws of gravitation, to compare the detailed clustering patterns which arise from particular starting conditions with what is seen in the sky today. In this way the consequences of different possible past states can be tested against observation. Some typical examples of these numerical simulations are shown in Figure 4.7. At present the scope of these simulations is limited by computational speed. The largest simulations follow the mutual gravitational interactions of about 32,000 masses—each taken to represent a galaxy. This should be compared with a total number of galaxies in the entire visible universe of about 10^{11}.

Despite the huge disparity between these two numbers, one should probably regard the existing simulations as a 'fair sample' of the Universe in the statisticians' sense—a sort of cosmic opinion poll of the effects of gravity.

It is amusing to consider the ultimate logical conclusion of bigger and better numerical simulations of the Universe. Eventually they will become so big and long-lived that they will allow the formation of stars and planets, and the evolution of conscious life-forms in the simulations. In this science-fiction scenario the simulators presumably then find themselves cast in the role of gods by the inhabitants of the simulation, some of whom will argue that all there is to their 'universe' is the content of the simulation and its laws of Nature. Occasionally, the simulators could sow a little confusion by intervening in the course of events in the simulation—altering the laws every now and then, performing the odd 'miracle'. The image of the computer scientist and the numerical simulation is an interesting analogy of the relationship between the Universe and an overseeing Deity which does not seem to have been pursued by those theologians who would be sympathetic to such a view!

(i) The Universe is Unique

The uniqueness of the Universe is, in effect, a tautology, which poses a fundamental difficulty for the application of the scientific method. We are accustomed to making repeatable observations and applying general principles to categorize the behaviour of particular events which share some crucial common elements. The impossibility of such a scheme in the study of the Universe as a whole warns us of the possibility, and even the probability, that some very special considerations are likely to play a role in elucidating its underlying structure and internal logic. These principles need have no analogous rôle to play in the understanding of particular phenomena within the Universe. If there are laws which govern the behaviour of the Universe as a whole, but which do not have any application to the behaviour of objects within the Universe, then local scientific observations are not going to reveal them.

Scientific method also relies upon repeatable observations in another way. When we encounter some physical phenomenon, we will find it to be a combination of elements dictated by the laws of Nature along with others that are accidental. Science progresses by separating these two sets. The fact that we see a red ball falling under gravity in a certain fashion might a priori be a partial consequence of redness.* The only way in which we can ascertain that it is not is

---

*This methodology of isolating the common factors is not necessarily foolproof. Let us suppose that a man decides to carry out experiments to discover how much alcohol he can consume before becoming totally inebriated. On the first night he drinks ten gin and tonics, and collapses before he can walk home. On the second night he consumes ten whisky and tonics, and again collapses before reaching home. On the third night he suffers the same fate after imbibing ten vodka and tonics. Having isolated the common factor, he concludes that in order to stay sober in future he simply needs to omit taking the tonic water.

by examining the fall of balls which differ in colour, but are identical in every other respect. In the case of the Universe we do not have this latitude and so we are hard-pressed to separate the fundamental properties of the Universe from the accidental ones.

We should also worry about the meaning we attach to the notion that a particular property of the Universe is 'unlikely' or 'improbable'. Such a use of language assumes that there exists an ensemble of equally viable alternative universes, such that if our particular Universe were situated amongst all these alternatives we would find it to be the odd-man-out in respect of possessing the particular property in question. But since we do not know whether these alternatives can exist we should approach such reasoning with caution: it is not at all clear that 'probability' any longer has a well-defined meaning.

(ii) The 'Universe' and the 'Visible Universe'

Observations of the *expansion* of the Visible Universe imply that it began about fifteen billion years ago. During this period light has had time to travel no more than fifteen billion light years, and hence events beyond this 'horizon' have not existed long enough to have been seen by us as yet. Every observer in the Universe is surrounded by a notional sphere of radius equal to about fifteen billion light years which defines his *Visible Universe*. As time passes, each observer's Visible Universe gets steadily bigger. Our own Visible Universe marks the boundary of that part of the entire Universe that can be seen or have any causal interaction with us. The distinction that must be made between the Visible Universe and *the* Universe is a vital one. Observational science knows only the Visible Universe. It is study of this finite region that leads to our laws of Nature, and only theoretical predictions about this part of the whole Universe can, even in principle, be verified or falsified by observations. Of the Universe as a whole, on the other hand, we know essentially nothing. It may be infinite in extent, in which case the finite Visible Universe will always be but an infinitesimal and possibly unrepresentative fragment of the whole. If the Universe as a whole is finite in size then we are slightly better off: at least our observations cover a finite fraction of the whole and so have a little more reason to be taken as representative.

The distinction between the Universe and the Visible Universe is often blurred in accounts of the study of cosmology. In practice, cosmologists invoke an assumption which has been elevated to the status of a 'Principle', in order to avoid constant reference to this distinction. The 'Cosmological Principle' comes in a variety of different wordings. Some versions say that the Universe is the same 'on the average' everywhere; others are more oblique, and state only the Copernican principle that there are no preferred places in the Universe. The goal of all these invocations is to underwrite an assumption that the Universe has the same gross properties everywhere. Thus, if we find the Visible Universe to contain

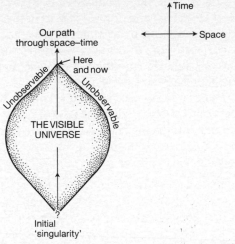

Figure 4.8 A space–time diagram showing the part of the entire Universe that is in causal contact with us. Only this region is accessible to direct observation by ourselves. The size of this 'Visible Universe' is constantly increasing whilst the Universe is expanding. The paths of the light rays which can have reached us by the present are focused into the past towards an apparent singularity of infinite density by the attractive force of gravity on light.

the same density of galaxies everywhere we look, then this assumption is supposed to allow us, with good conscience, to take the entire Universe to have the same property. Strictly speaking, this is an unverifiable assumption unless it could be shown that the structure of the Universe beyond our visible horizon has observable consequences for its local make-up. The Cosmological Principle is ostensibly just a statement about the contents of the Visible Universe, but implicitly it is an assumption we are making about the laws of Nature also. We are assuming that the laws of Nature that we observe to dictate the course of local events also hold good all over the whole Universe. This is not an assumption that we can test beyond our visible horizon but, as we shall see in a future chapter, we can offer some strong evidence that the laws of Nature we use on Earth do not differ from those which govern the structure of astronomical objects billions of light-years away.

One problem which we must take seriously with regard to the finiteness of the part of the Universe which is accessible to our observations, is the possibility that this visible portion of the Universe does not contain enough information for us to deduce the laws of Nature. There are necessarily pieces missing from our cosmic jigsaw puzzle. We do not know how numerous they are, nor how crucial a part they play in determining the overall picture. We are a long way from knowing whether the Universe is 'something' which is a particular outworking of the laws of Nature, whether it is in some sense equivalent to the laws of Nature which govern it, or whether it has no explanation in terms of laws of Nature at all. This distinction does not seem to have been adequately appreciated by those cosmologists who seek to provide a physical mechanism by which the expanding Universe can come into being out of 'nothing'. Even if successful, such a venture would still be left to explain how there came to exist laws of Nature at the moment when the Universe came into being. The ultimate cosmological conundrum appears to be the question of how and when the laws of Nature came into being, if the Universe of space and time is imagined to arise spontaneously out of nothing.

(iii) Non-local Influences

We have become accustomed to laws of Nature which are 'local'. They have the property that action between bodies does not take place instantaneously over enormous distances, but does so only locally through an underlying force field. Einstein's curved space–time is the classic picture of this sort, wherein the motion of objects is dictated by the local shape of the curved space–time in which they move. But over the largest cosmological dimensions the Universe may not always be like this. There may exist long-range influences, and even new forces of Nature, which increase in strength with distance between objects so that they are too feeble to show up on the scale of the Earth or even the solar system, but which dominate the large-scale expansion of the Universe. Einstein's general theory of relativity allows such a force to exist. It would increase in proportion to the distance between objects, but would not depend upon their identities or masses.

We have seen that quantum reality requires the existence of a non-local element which allows effects to occur instantaneously or without causes in the usual sense. It is not known whether this has any deep cosmological significance.

A curious non-local problem arises if our Universe is infinite in spatial extent. In an infinite world anything that *can* happen will happen—infinitely often in fact—if the infinite world is exhaustively random. The consequences of this reasoning are rather disconcerting. It implies that, if the Universe is infinitely large, then at this very moment there must be an infinite number of identical copies of the reader perusing identical copies of this book somewhere in the Universe. This conclusion is actually not as inescapable as it is often made to sound. The small print is that events are exhaustively random; that is, they explore all possibilities in a random fashion. Any old infinity will not do the trick. For example, all the even numbers form an infinite collection, but you will never find a single odd number in the list, because an infinite set of even numbers is not an *exhaustively* random infinity of whole numbers.

(iv) The Possible Role of Initial Conditions

When we observe the structure of the Universe, we are faced with the spectacle of individual phenomena that have evolved as a result of the Universe being both in a particular *past* state, and controlled by the laws of Nature which dictate how that state changes with the passage of time. Let us take a particular example: galaxies. We do not know why galaxies exist in all their manifold shapes. It could be that these vast islands of stars, nearly a million times denser than the ambient Universe in which they are embedded, owe their existence and their very special forms and sizes primarily to conditions at the time when the expansion of the Universe (or even the Universe itself) began. Alternatively, these features may have arisen principally as a result of the constraints and form of the laws of gravity which govern the motions of the stars and dust of which they are composed. This dichotomy between the relative influences of laws and starting conditions does not arise when we study the effects of the laws of Nature in the laboratory. In such circumstances we can control the experimental set-up in such a way that it is clear what the influence of starting conditions is upon the final result of an experiment.

In cosmology the approach towards explaining many large-scale features of the Universe has traditionally take two routes. One has attributed the present structure of the Universe primarily to special starting conditions, the other has sought to show that, no matter how the Universe began, it would inevitably take on the presently observed gross structure after about fifteen billion years of expansion. There is something to be said for both views and in doing so one must be careful not to confuse conformity with some methodological principle with truth.

Those who appeal to special initial conditions appear at first sight to be taking the lazy view, characterized by those pre-Darwinian biologists who attributed all harmony in Nature to the fact that 'it was just made that way': things are as they are because they were as they were. However, in the cosmological problem the import is a little different. Cosmologists who appeal to initial conditions do so firm in the belief that there is a wealth of unknown physics associated with events near the beginning of the Universe. If the Universe did have a beginning in time, it is a good bet that something very special indeed constrained the form of that event. For this reason, appealing to unique initial conditions to explain current aspects of the Universe—rather than to the laws of Nature—differs from the appeal to pre-existent harmony as an explanation for, say, why animals are so well camouflaged for their habitats. The drawback of this view for the working cosmologist is that any laws governing the initial conditions of the Universe are likely to be the most difficult to discover. They would require us to have a correct understanding of how quantum theory and general relativity marry together to form the laws of quantum gravity. Worse still, they could only be deduced and tested by making observations of the Universe's present structure, which is what we were originally seeking to explain. The opposing approach, which seeks to show that the present structure of the Universe is an inevitable consequence of the laws of Nature, regardless of the precise initial conditions of the Big Bang, is often termed 'chaotic cosmology'. This is because its goal is to show that if we started the Universe in an arbitrarily chaotic state it would relax inevitably into the quiescent orderly state of expansion that we observe today. If this were so, it would be advantageous from one methodological point of view, for we might be pessimistic and hold that we will never know what occurred at the beginning of the Universe, and so any explanation which tells us why the Universe has its observed structure *independently of how it began* is appealing. On the other hand, there is the disappointing consequence that if the present structure is independent of how the Universe began then no present-day observations will be able to shed light upon that most extraordinary of all problems—the origin of the Universe. And, of course, we should remember that the Universe—even the bit of it that we have seen—might not have had an origin in time; in which case the question of 'why' it exists in its present form is even less likely to admit of a cogent answer.

If the Universe's evolution in time is described by deterministic mathematical equations, so that the future is determined uniquely and completely by the past, then the present structure of the Universe will be the result of one particular set of starting conditions only.

(v) Selection Effects

Because we cannot manipulate the Universe at will, we face the dilemma of ridding our astronomical observations of the biases which afflict them. Certain types of evidence are more easily gleaned than others: intrinsically bright galaxies are more easily seen than faint ones, and hence they will tend to be over-represented in any census of the galactic population. This is just the type of 'sampling' problem which opinion-pollsters have to worry about. Suppose one were to carry out a telephone poll in a poor area of the country in order to ascertain income levels. The sampling strategy immediately eliminates all those householders who cannot afford a telephone. It has been biased. The results tell us as much about the sampling bias as they do about the intended subject of study.

For thousands of years we observed the heavens by detecting light in what is now called the 'visible waveband'. This spans from red to violet the spectrum that the human eye and ordinary optical telescopes employing lenses and mirrors are sensitive to. Light outside this very narrow band is either so energetic that it would destroy the light-sensitive rhodopsin molecules in the human retina, or it carries too little energy to excite them at all. During the last thirty years we have gradually viewed the Universe in other types of light—X-rays, radio waves, infra-red, ultraviolet—each gives a different picture of the Universe that highlights different features. If we had X-ray vision we would have evolved quite a different astrological and astronomical tradition about the appearance and meaning of the Universe in past centuries. We know that there are other ways of viewing the Universe—using detectors sensitive to neutrinos and gravitational waves—which would yield further individual pictures of what the Universe is like, but our technology has yet to advance to the stage where we can survey the Universe by these means.

In more down-to-Earth sciences, biases of this type are more easily dealt with by repeating experiments and observations with certain of the environmental conditions changed, so that the biases and their effects can be identified. In cosmology this is rarely possible. Moreover, there are far-reaching biases which are unavoidable. These are created by the fact that observers like ourselves— fragile creatures made from lightly-bound molecules—require rather special conditions of temperature and gravity to exist. Our observations are necessarily biased by the fact that the sites where carbon-based life can evolve appear to be special. We shall have a lot more to say about this problem in Chapter 7. At this stage it is important that we simply appreciate the fact that there are no pieces of 'raw data'. If we collect observations of the Universe without any understanding of the systematic biases that colour them, we are allowed to do no more than produce a catalogue of observations devoid of any reliable systematic interrelationships. No true laws of Nature could ever be deduced from them.

The goals of theory

> If the present state of the universe was exactly similar to the anterior state which has produced it, it would give birth in its turn to a similar state: the success of these states would then be eternal.
>
> Pierre Laplace

The most remarkable thing about the success of Einstein's theory of gravitation in predicting the expansion of the Universe is that it confirms the ability of Einstein's equations to provide a description of the behaviour of the Universe as a whole, rather than merely that of its constituents. No other law of Nature has been applied in such a fashion. Even if the Universe is infinite in extent it is still possible for Einstein's equations to describe its overall expansion and structure. If we can determine what the material content of the Universe is, and how this material is moving, then, at least in principle, Einstein's equations can always tell us the geometry of space and flow of time at every place. In practice we cannot determine the arrangement of all the material in the Universe in this way, and even if we could, Einstein's equations would be far too difficult for either ourselves or the smartest of our contemporary computers to solve.

The ability of Einstein's general theory of relativity to describe and predict the behaviour of the Universe as a whole produces an interesting dilemma: it describes other universes as well! It is possible to find many different solutions of Einstein's equations. Each describes a different variety of universe. Some are extremely irregular, some rotating, or even contracting, rather than expanding, others expand at different rates in every direction, while some are vibrating or even static. This profusion of possible universes shows that Einstein's theory cannot be the whole story. There is (by definition) only one extant universe, and only one solution of Einstein's equations that will describe it. So why are there so many other redundant solutions describing other hypothetical universes? Unless they are all out there populating the many worlds of Everett we are clearly lacking knowledge of a crucial winnowing principle which eliminates all but one of Einstein's universes by virtue of some deeper internal inconsistency, rather than the empirical fact of simply not agreeing with observation today. In order to do this, the winnowing principle would have to specify the starting state of the Universe uniquely. This is the belief of many modern cosmologists. It expresses the hope that in the future we shall find a new description of Nature that includes both quantum theory and general relativity unified within its compass, or superseded by a more sophisticated picture which retains their successful elements while rejecting their inessentials. Both quantum theory and relativity will, no doubt, undergo some modifications in order that they can successfully describe conditions of high density where their effects must compete in a way that we have yet to witness. Yet the idea that the unification of these two fundamental theories of Nature or any of their successors will be sufficient to determine the allowed description of the Universe uniquely still seems extremely unlikely. The real Universe is an extremely complicated structure. It has lumps and irregularities everywhere. Its *precise* structure, rather than some pretty good idealization of it as a uniform distribution of matter expanding at the same rate in every direction, is too complicated for us to determine and codify. It is an outworking of both laws and chance happenings. How can the skeletal structure of this real and vastly complicated Universe fall inevitably and uniquely out of the super quantum-relativity of the future? Even if it could, we could never successfully extract it. It would be too difficult to solve the equations of the theory for this completely accurate description of the Universe. If it is the only solution, there will be no simple idealized relatives to give us guidance. But more seriously still, if we believe that all our descriptions and laws of Nature must at some level be approximations to the true reality, then we will never find this unique and completely accurate description of reality because ultimately our equations cannot describe it.

We should stress that at present there are no problems with either quantum theory or general relativity. Both are in agreement with all the observations we have made of the world. Each is believed to be internally consistent, and each has produced strikingly accurate predictions about the structure of the Universe. Yet physicists know that they can at best be only a part of the truth. General relativity is not a quantum theory, and quantum theory does not determine the geometry of the space–time upon which it dictates the evolution of wave functions. Fortunately we have not directly encountered an environment in which both the effects of quantum uncertainty and curved space–time are manifest together in noticeable ways. Such an environment would require extremes of density and temperature trillions of times in excess of that found at the centres of the densest stars. The only site that we know of for the double-act of quantum gravity to play the major role is in the first instants of the Universe's history, as it expanded away from the inferno of the Big Bang. We cannot observe this environment directly, but we can discover whether our theories of what went on in that extreme environment lead to consequences that are in conflict with what we see in the Universe today, or even if they can shed some light on the major mysteries of the Universe's large-scale structure.

A little over ten years ago this quest looked fairly hopeless. As we looked back in time in our mind's eye towards the early stages of our expanding universe we encountered conditions of ever-increasing density, in which particles collided with higher and higher energies. It was believed that this signalled an ever-growing complexity and strength of interaction between the elementary particles of which the Universe was then composed; the farther back in time we looked, the more intractable would become the problem. But the effects of the quantum vacuum which we studied earlier in this chapter, which produce a change in the effective strengths of the different forces of Nature as the temperature of the environment increases, predict that, instead of becoming stronger and more complex in the high-density environment of the Big Bang, the forces of Nature should become *weaker* and *simpler*. This realization has made it possible to take studies of the very early history of the Universe seriously. Indeed, particle physicists now see cosmology as the only way of testing many of their most esoteric predictions about the laws of Nature at very high energies.

The legacy of the steady statesmen

> We have already learnt that geography does not matter. The steady state theory teaches us that history does not matter either.
> Herman Bondi

The Big Bang theory of the expanding universe is the central paradigm within which to co-ordinate astronomical observations, and reconstruct the past history of the universe. It has only ever had one serious rival. From 1948 until 1965 an alternative cosmological picture was advocated independently by Thomas Gold and Herman Bondi and by Fred Hoyle. The 'steady state' universe, as it became known, was a radical alternative to the traditional Big Bang picture of a universe apparently created from nothing at a finite past moment of time, and which subsequently expanded, getting ever cooler and more rarefied, into its present quiescent state. The steady state theory sought to remove all these special features from the cosmological model, and attempted to do so by taking a cue from the covariance of the laws of Nature. The conventional Big Bang cosmology (although this term was only introduced—pejoratively—by Hoyle in 1950), made heavy use of the so-called Cosmological Principle to argue that the Universe must be similar in structure in every place and in every direction, otherwise some places would be endowed with a special status that seemed un-Copernican. No observations could determine absolutely our vantage point in space. This uniformity in the structure of the Universe from place to place was imagined to apply to its physical contents—the density of matter, the average temperature, the clustering properties of galaxies, and so forth. Of course, on a very parochial level the Universe looks different in different places. The view from the surface of the Earth differs from the view from the surface of the Sun or from the centre of our Galaxy. But the Cosmological Principle is to be applied only over very large dimensions exceeding the dimensions of galaxies. It claims that the larger the dimensions over which one surveys the Universe, so the smaller the averaged non-uniformities should become. The steady state theory of the Universe sought to extend the Cosmological Principle from the realm of space into the domain of space and time. The 'Perfect Cosmological Principle' claimed that the Universe was not only similar from place to place but also from time to time: no astronomical observations could absolutely characterize the cosmic epoch at which we live. This implies that the Universe must be in a steady state, which looks the same at whatever moment one views it. The motivation for such a philosophy came partly from the fact that the laws of Nature were held to be the same everywhere in the Universe and at all cosmic epochs. This was held to be the basis of rational scientific inquiry. The Perfect Cosmological Principle dictates that the structure of the Universe should possess the same symmetry as the laws themselves. From what has already been said in this chapter about the relationship between symmetrical laws and their asymmetrical outworkings, one can see that such a motivation would be flawed. If the structure of the Universe is a manifestation of the laws of Nature, it need not display all of the symmetries possessed by those laws. How is it possible for the Universe to be in a state wherein its gross structure is the same at all times? The simplest option is a static universe which remains the same yesterday, today, and forever. Yet this is immediately in conflict with the observed expansion of the Universe. The Visible Universe is not static. The steady statesmen sought to find a model of the Universe which expanded but still presented the same steady aspect at all times, rather like a steadily flowing river. This does not seem possible. For if the Universe is expanding, then any marked volume of it will contain a steadily falling density of matter as time passes. If we use the average density of matter as a clock, the past differs radically from the present: the past has high density, the present has low density.

The only way of maintaining a steady aspect which does not allow any physical property of the Universe to be used to distinguish the past from the future is if there is a continual steady creation of matter everywhere at precisely the rate required to offset the dilution brought about by the expansion. This creation rate turns out to be far too small to be detected directly by observations. The cosmological model that results has the same expansion rate at all times—otherwise a measurement of the expansion rate would distinguish one moment of cosmic time from another. It can have no beginning and no end—otherwise these cataclysmic moments would constitute further 'special' times. The contrast with the Big Bang model of the expanding universe could not be greater.

The continuous creation of matter proposed in the steady state theory of the Universe was a radical idea. It required a new law of Nature here and now. Whereas the Big Bang theory envisaged universal creation (along with the creation of the Laws of Nature themselves) at some particular past time, the steady state theory required that creation occur at all times governed by eternally existing laws. Unfortunately, no explanation for the process of continuous creation could be found from quantum physics. Another awkward problem, which Hoyle later admitted was the principal reason for his subsequent disenchantment with the continuous creation idea, was the observed asymmetry between matter and antimatter in the Universe today. Hoyle believed that the continuous creation process would have to produce particles and their antiparticles at equal rates, whereas the observed Universe displays no evidence for astronomical sources of antimatter. We have found no antiplanets, no antistars or antigalaxies, and the only antiparticles found in cosmic rays have precisely the characteristics of particles formed in the debris from collisions between primary particles of matter. Eventually the steady state model fell into conflict with astronomical observations. It was confirmed that the Universe did change its aspect with the course of time. The discovery of quasi-stellar objects preferentially at a particular distance from us is equivalent to their having appeared preferentially during a particular period in cosmic history, since the finite time that light takes to travel from distant sources to our telescopes means that objects far away emitted their light at an earlier cosmic epoch than nearby ones. The final blow to the steady state picture of the Universe was struck in 1965, when the 'echo' of the Big Bang was discovered. In 1948 two young American researchers, Ralph Alpher and Robert Herman, had predicted that if the Big Bang picture of a hot and dense past was correct, then there ought to remain some fossil evidence of that fiery past in the form of radiation fall-out cooled by the expansion of the Universe to a temperature only about five degrees above absolute zero. In 1965 Arno Penzias and Robert Wilson happened upon this radiation field whilst calibrating a sophisticated radio receiver designed for satellite tracking. The radiation had a temperature three degrees above absolute zero—almost exactly as predicted by Alpher and Herman—and subsequent observations have revealed that its spectrum carries the distinctive Planckian signature of heat radiation. There is no way of producing radiation with these characteristics locally in the Universe. Subsequently, the Big Bang theory received further confirmation from its successful prediction of the cosmic abundances of helium, deuterium, and lithium, all of which would be produced by nuclear reactions during the first three minutes of the expansion after the Big

Bang. Since the steady state theory never possessed a past state of high temperature and density—the steady state universe is always the same as it is today—it cannot naturally account for the existence of the remnant radiation, and does not predict the abundances of the light nuclei in the way that the Big Bang theory so neatly does.

The steady state theory died in the mid-1960s, despite the valiant attempts of some of its perpetrators to demonstrate that reports of its death were greatly exaggerated. We have devoted so much space to it not simply because the interested non-astronomer so often thinks that the old Big Bang *versus* Steady State dispute is still going on, but because the steady statesmen endowed us with an important legacy that leads to some of the most profound problems contemplated by modern cosmologists, and these problems have an important bearing upon our expectations concerning the laws of Nature as they relate to the Universe as a whole.

Chaotic cosmology

> God has put a secret art into the forces of Nature so as to enable
> it to fashion itself out of chaos into a perfect world system.
>
> Immanuel Kant

Ever since Hubble's first observations of the expansion of the Universe, it has been known that the expansion possesses surprising properties. It proceeds at virtually the same rate in every direction of the sky. Since the discovery of the relic radiation, it has been possible to determine the level of this directional fidelity, or 'isotropy' as cosmologists call it, very precisely by searching for differences in the temperature of the radiation between one direction and another over the sky. The differences are staggeringly small—smaller than one part in ten thousand. This tells us that over the largest visible dimensions the Universe must have the same density and expansion rate from place to place. Any significant non-uniformities from one place to another would also create directional differences in the temperature of the radiation as it travelled from the Big Bang through gravitational fields in different parts of the Universe. This observational evidence for isotropy and uniformity is support for the truth of the Cosmological Principle. But we still lack an explanation of why Nature should adhere so closely to the Cosmological Principle. On the grounds of probability alone one might have thought that there are simply so many more ways of having a non-uniform Universe than a uniform one that we should have found the Universe to be highly disordered and irregular. Furthermore, our experience of the growth of entropy and disorder with the passage of time in accordance with the Second Law of thermodynamics might have led us to suspect that, even if the Universe began in an orderly fashion, there have been fifteen billion years in which disorder could make its inexorable increase felt. So why is the Universe still so regular?

The problem of why the Universe exhibits a remarkably isotropic and uniform structure in the face of so many irregular alternatives was first taken seriously by the steady state cosmologists. In 1963 Hoyle and his student Jayant Narlikar proposed that this uniformity could be understood in the context of the steady state universe without any difficulty. The steady state universe has the property that, if it is disturbed slightly from its steady state, then it tends to return to it. We say that it is stable. Like a marble rolling inside a bowl, it has the property that all disturbances tend to create opposing effects that restore it to the lowest point. By this token it was claimed that once in a uniformly expanding state the steady state universe would stay there, held by the restoring force supplied by the continuous creation of matter, whereas the Big Bang model would gradually drift away from it. Thus it was argued that the steady state theory could explain the fact that the Universe expands isotropically, without appealing to initial conditions at some moment of creation in the past which are beyond the scientific method to ascertain.

Following the demise of the steady state theory, and the discovery of the remarkable isotropy of the relic radiation, the Big Bang cosmologists took up the basic philosophy of Hoyle and Narlikar: that one should attempt to explain the way the Universe is regardless of how it began, by showing that no matter what the state in which it started out, or what events occurred during its subsequent history, it would inevitably tend towards the observed state of isotropy and uniformity after many billions of years of expansion. This philosophy was dubbed 'chaotic cosmology' by Charles Misner, its principal propagandist, because its goal was the demonstration that, no matter how chaotic the state of the Universe when its expansion began in the dim and distant past, there would always arise frictional processes which would smooth out the irregularities and anisotropies in the course of the expansion history. If true, such a scheme has some attractive features. It promises an explanation for the present structure of the Universe independent of the unknown and perhaps unknowable initial state of the Universe. It is, in effect, a proof of the Cosmological Principle. The burden of responsibility for explaining the present state of the Universe would be shifted entirely onto the laws of Nature, leaving the initial conditions to play little or no rôle at all.

Subsequently, the weaknesses in this attractive scheme were brought to light. Most fundamentally one can see that, if the evolution of the Universe is described by a system of deterministic laws of the sort provided by general relativity and assumed by the chaotic cosmologists, then it is not possible for the present state of the Universe to have arisen from all possible starting states. Deterministic laws mean that any starting state will evolve into *some* particular state after fifteen billion years of expansion; but they also mean that any fifteen billion year-old state that you care to dream up must have evolved from *some* initial state. To confound the chaotic cosmologists one simply needs to write down a cosmological model which is a solution of Einstein's mathematical equations, but which does not describe the Universe we live in, and then follow it backwards to determine the starting state that gives rise to it. As a result, we have an example of a starting state which cannot evolve into the present state by the present time. It is not possible to eradicate totally the influence of the initial conditions upon future states of a deterministic system.

The first fall-back position one could move towards in order to retain the philosophical appeal of the chaotic cosmology programme is one in which one attempted to show that, in some sense that needs to be defined carefully, the starting states which lead to universes as uniform as ours after fifteen billion years of expansion are *more probable* than those that do not. In order to be compelling, such a demonstration would need to show the balance of odds to be something like 100 to 1 or better, rather than, say, 6 to 4. We want an overwhelming probability. We require things to be as in the criminal courts, 'beyond reasonable doubt', rather than merely 'on the balance of probabilities' as suffices in the civil courts. Ideally, one would like a situation along the following lines. Represent all the possible starting states for the Universe by the points on this page. If we colour red those that lead to a Visible Universe like our own by the present time, and blue those that do not, then we would like to see only isolated points of blue in a sea of red, so that the slightest deviation from a blue point would always land you on a red point. By contrast, a small deviation from any red point would leave you on another red point. During the 1970s investigations of this issue invariably came up with the answer that the collection of starting states that give rise to universes totally *unlike* our own are overwhelmingly the more probable. The page is a sea of *blue* in which there sit a few isolated points of red. Furthermore, the conjuring up of elaborate frictional smoothing processes arising in the first moments of the Universe's expansion was a hopeless cause. All these physical processes involve means of equalizing disparities in temperature and density between different places by transporting radiation or particles. Such transport cannot occur faster than the speed of light, and so after the Universe has aged a time T the smoothing can only have been effective over regions of size equal to T times the speed of light. Unfortunately, at the times when the processes are strong enough to actually smooth things out, the maximum size of these regions is tiny— smaller than the solar system in fact! They can provide no explanation for the uniformity of the Universe over its largest observed dimensions today.

Interest faded in this philosophy, and attention focused upon how the regularity of the universe could have been built into it from the very beginning— the opposite philosophy to that of the chaotic cosmologists—until 1980. In that year Alan Guth, an American particle physicist, changed the direction of the cosmological speculation by showing how the balance of probabilities could be swung overwhelmingly in support of the chaotic cosmology philosophy.

Inflation

> Large streams from little fountains flow,
> Tall oaks from little acorns grow.
> David Everett

Guth's original theory, entitled '*The Inflationary Universe*' (an amusing reflection of the times in which we live!), has since undergone a number of important revisions, but the basic idea remains the same. The early investigations of the chaotic cosmologists in the 1970s took place before the revolution in particle physics that led to the 'gauge age'. During those earlier days it was believed that all forms of matter exert gravitational attraction. By the 1980s this faith was totally eroded. Physicists then came to believe that there exist forms of matter at very high temperature and density which, in effect, *anti-gravitate*. If matter fields of this special type arise for a very brief period of the universe's life in the first instants after the expansion begins (no more than the period from 10^{-35} to 10^{-33} seconds is needed!) then a number of remarkable consequences follow which appear to explain the concurrence of several of the Universe's large-scale features.

The prime consequence of the influence of the special anti-gravitating matter upon the Universe is that it causes the expansion to *accelerate* when it would otherwise have been decelerating following its initiation. This means that the part of the early Universe which expands by the present time to constitute our Visible Universe will be much smaller than it needs to be in the standard decelerating picture of the expanding universe. In fact, that region can be small enough to lie within the smoothing domain of frictional forces at very early times—some 10^{-35} seconds after the expansion starts. Moreover, when the Universe enters this accelerating phase of expansion, all directional disparities in its rate of expansion are rapidly rendered negligible, and the expansion of the Universe is driven closer and closer to the critical divide between the ever-expanding and recollapsing universes, so offering an explanation of what had long been a mysterious property of the Universe's present state of expansion.

The resolution of the Universe's uniformity that inflation offers is a fascinating one because, unlike the chaotic cosmology programme, there is no attempt to *dissipate* any primordial chaos that might have existed when the universe began (or have been inherited from the infinite past). The entire Visible Universe is imagined to lie within the presently expanded domain of a microscopic region which it is quite reasonable to believe had been smoothed by physical processes going on within it in the distant past. The smoothness of the Visible Universe is just a reflection of that microscopic smoothness. If the accelerated period of inflationary expansion had not occurred, then those smoothed microscopic regions would only have expanded to an insignificant size by the present, and no explanation of the smoothness of the entire Visible Universe would be possible (Figure 4.9).

It is fascinating to realize that if the Universe as a whole had started in a fairly random fashion, then although any microscopic region would be kept smooth, it might be smoothed to quite a different density from that of all its neighbours. All of these causally disjoint regions could inflate, perhaps by different amounts, to produce vast regions within our present Universe, but which lie beyond the horizon of our Visible Universe. The inflationary universe picture suggests that the Cosmological Principle is not true of the whole Universe, only that it is locally true of the visible part of it. Inflation does not explain the uniformity of the Visible Universe by eradicating primordial chaos, but by sweeping its effects out of sight beyond the boundary of the visible part of the Universe.

In retrospect one can see that some of the nice features of the inflationary universe scenario were present in the steady state universe, which always possessed *accelerating* expansion. The period of inflationary expansion suggested first by Guth was essentially identical to any segment out of the history of the steady state universe, and the effect of the continuous creation was to mimic one of the curious anti-gravitating matter fields required to effect the period of accelerated expansion in the very early stages of the Big Bang universe. Indeed, Hoyle's version of the steady state theory had explicitly introduced a matter field of this sort which he called the 'C[reation]-field'.

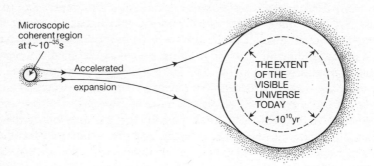

Figure 4.9 The inflationary universe. A schematic representation of the inflationary universe picture wherein the entire Visible Universe evolves from the accelerated expansion of a single causally coherent region of microscopic extent. Without the phenomenon of accelerated expansion these smooth microscopic regions would have evolved into regions far smaller than the size of the observable universe today.

There is one further conundrum. Although the observed universe is remarkably uniform over its largest dimensions, it is not perfectly uniform. We would not be here if it was. There exists a well-defined population of material aggregates in the Universe—planets, stars, galaxies, and clusters. Why do they exist?

One had always been faced with a choice in cosmology: either one could assume the existence of a smooth expanding substratum, and seek to explain the existence of the small irregularities that exist upon it—the 'frills and furbelows' of the Universe as Herbert Robertson once called them—or, like the chaotic cosmologists, one could take the view that it is the smooth underlying substratum that is most in need of an explanation. One of the principal bugaboos of the steady state theory was that the *perpetual* state of accelerated expansion led to the total eradication of all irregularities in the density of the Universe. No galaxies could survive. The accelerating expansion dispersed matter more efficiently than any force of gravitational attraction could condense it.

Because the inflationary phase of accelerated expansion in the Big Bang model lasts only for a short period it does not lead to the obliteration of all irregularities. Better still, it predicts that there ought to exist a very special distribution of small irregularities, because each microscopic region that undergoes inflation must possess a minimum level of quantum graininess from place to place because of the intrinsic quantum uncertainty that prevents the state of matter ever being fixed and definite. These quantum fluctuations are amplified by the inflationary expansion into a special form which obeys the Perfect Cosmological Principle, and could well give rise to all the galaxies and clusters of galaxies that we now witness.

There is one weak point in the inflationary explanation of the uniformity of the Universe, where the 'axiom of faith' is unconsciously invoked. If the smoothness and regularity of the Visible Universe is to arise as the inflated image of some microscopic primordial region, then we must maintain that the structure of space stays smooth as one probes to smaller and smaller dimensions. We do not know whether this is true. The smallest length-scale which can be examined by direct observation is only about 10^{-15} centimetres, whereas inflation is concerned with what the structure of space looks like on scales smaller than 10^{-30} centimetres. If space is turbulent like a foam on or below these length-scales, then inflation fails as an explanation of the current structure of the observed Universe. The present form of the Universe would be just the expanded image of this microscopic chaos. Picturesque examples of how there can exist irreducible structures on the arbitrarily small length-scales are provided by the structures which Mandelbrot has called 'fractals' (see Figure 4.10 and also p. 289 for further discussion): Space, or even space–time, might very well exhibit such a bottomless pit of complexity in the extreme environment of the Big Bang.

The inflationary 'paradigm'

> Truth is not that which is demonstrable, but that which is ineluctable.
>
> Antoine de Saint-Exupéry

The logic behind the inflationary explanation of the Visible Universe's structure

Figure 4.10 The Mandelbrot Fractal Set. Within this pattern is stored such information as the collection of possible frequencies that any beating system (such as the human heart) will pass through as it passes out of control into an arhythmic state. The intricate, endless complexity is generated in the horizontal (x) and vertical (y) plane by repeatedly iterating the algorithm $(x, y) \to (x^2 - y^2 + A, 2xy + B)$ where A and B are some fixed numbers. For most choices of A and B the iterates fly off to very large values off the picture if the starting values of x and y are both zero, but for others they remain within a region whose boundary exhibits never-ending structure when viewed with ever-increasing magnification. The dark region within this boundary is the Mandelbrot set. [Reproduction of Plate 189 from *The fractal geometry of nature* by Benoit B. Mandelbrot. Copyright © 1983. Reprinted with the permission of W. H. Freeman and Co.]

is to show that however the Universe began—whether in chaos or in order—the inevitable effect of the laws of gravitation is to deliver it to us in the form we find today.

Unfortunately, it is likely to be impossible for us to ascertain whether inflation did occur. How could we tell whether the special features that inflation endows upon the observable Universe were so endowed or whether they are the reflection of very special initial conditions? One might feel that methodologically the latter is a less appealing 'explanation', but this anthropocentric prejudice towards explanations that are convenient for our beliefs about the way to do science is no argument for or against their *truth*. Moreover, all the individual consequences of the inflationary universe picture had been suggested as the consequence of other quite different cosmological pictures. The principal novelty of the inflationary model is that it manages to obtain them all simultaneously from one cause.

Although the prospects look bleak with regard to ever knowing whether inflation is the correct explanation of the Visible Universe's large-scale structure, it is possible to rule it out by determining whether the present expansion of the Universe is as close to the critical divide separating indefinite expansion from future re-collapse as it predicts, and to ascertain whether the irregularities in the Universe follow the particular pattern that inflation creates.

The wielding of laws of Nature in the manner either of the chaotic cosmologists or of the 'inflaters' to explain the universe in a way that puts no weight upon its structure at the beginning, can be seen as the natural inheritance of a style of reasoning that began in earnest with Wallace and Darwin. The pre-Darwinian explanation for the detailed structure of the living world was an appeal to special initial conditions: things were made that way. The result of invoking evolution by natural selection is to downgrade the influence of special starting states which presuppose some element of Divine design, and place all the responsibility for what is observed upon the physical processes and the laws that govern their development in time. However, there are important differences of detail. The cosmological scenario in which appeal is made to general physical processes to render the future state of the universe inevitable has as its goal the *predictability* of that future state. Natural selection is quite the opposite. Given a soup of chemicals in a complex environment, its precise future evolution cannot be predicted because the system possesses a sensitive dependence upon the precise nature of the starting state which we shall discuss in Chapter 5. Inflation has the opposite property that the final result is insensitive to the starting state. A good analogy to these two approaches to the role of starting conditions in cosmology is the 'IQ debate' between educationalists over the relative roles of 'nature and nurture'. There, the dichotomy is between those who believe that intelligence is primarily genetically inherited (that is, dependent upon 'initial conditions') and is insignificantly influenced by environment, and those who maintain that it is principally conditioned by environmental factors (that is, by the 'laws' of environmental development).

The future

> The universe is simmering down, like a giant stew left to cook for
> four billion years. Sooner or later we won't be able to tell the
> carrots from the onions.   Arthur Bloch

Even if the inflationary universe picture is successful in its bold attempt to explain many of the gross structural features of the Visible Universe in a way that makes little appeal to how the Universe began, it leaves many fundamental cosmological questions unanswered. Although it predicts that the Universe should be observed to expand tantalizingly close to the critical divide between perpetual expansion and eventual re-contraction, it does not tell us on which side of this cosmic Rubicon we lie. This might be thought somewhat unsatisfactory. For would one not expect the most fundamental principle, as yet unknown, which dictates on which side of the divide we lie, also to have something to tell us about how close to it we should lie?

The divide between the universes doomed to expand forever and those doomed to re-collapse to a Big Crunch has a far wider significance than the creation of cosmic agoraphobia or claustrophobia. If we believe that the Cosmological Principle holds good over the whole Universe, then this divide may separate those universes which are infinite in volume from those which are merely finite. We say 'may' because this is a question that we cannot decide by observation. The universes which expand forever are usually cited as necessarily possessing infinite volume. Most elementary cosmology texts tell one this. However, this deduction is based upon the assumption that the *topology* of space is simple in these ever-expanding universes; or more specifically, that it lies flat like a sheet extending to infinity in all directions. However, it could be that the sheet of space has been wrapped up. If rolled up in this way in only one of the three dimensions of space this would correspond to a sheet being glued along two edges to form a cylinder. If the two circular ends of the cylinder are now glued together then it becomes a toroidal ring, like a doughnut. It is possible for space in the ever-expanding cosmologies to be identified in this manner in all three directions of space. In the event of this being the case, the spatial volume of the Universe is *finite* even though it expands forever.

At present we do not know what specifies the topology of space. Topology deals with those properties of space that remain the same when its shape is deformed in ways that do not tear it. It is quite distinct from geometry. A flat sheet of paper can be bent to form a crescent. This operation leaves its topology unaltered, but changes its curvature and geometry—the shortest distance between two points on the paper changes from being a straight line to a curve. If we were to tear a hole out of the paper we would alter its topology. We recall that Einstein's picture of curved space–time shows how the presence of mass and energy determines the geometry of space and the flow of time. But they do not determine the topology. Usually, cosmologists assume that the topology of the Universe is simple rather than one of these unusual identified (by joining together, as when a sheet of paper has its ends 'identified' to form a cylinder) or punctured forms, but we do not *know* that it is simple, nor do we know of any reason why we should expect it to be simple. Indeed, if the Creator were simply choosing topological spaces out of a cosmic hat, one would find that almost all were of the exotic identified type. It is the simple ones that are special. Furthermore, those particle physics theories which predict the existence of additional spatial dimensions in the Universe require that the extra dimensions have one of these unusual topologies in order to trammel them up to an unobservably microscopic extent. Why should the topology of the remaining directions be qualitatively different?

If our three-dimensional universe possesses one of these identified topologies, then this may in some way be dictated by laws of Nature. The dimensions over which the identifications will have been made in each of the three dimensions of space that we see are new fundamental constants of Nature. What determines their values? If there is indeed a unified 'theory of everything', then it would have to relate these lengths to the other constants of Nature. Moreover, the topology of space influences the part that can be played by symmetry in Nature.

Thus we see that, even if we assume that the entire Universe is well-represented by the part of it that is visible to us, we will not be able to decide whether space is really infinite. The boundary could always lie farther away than light has had time to reach since the expansion began. The rate of expansion alone will not reveal the answer to us, because of the possibility of a complicated spatial topology. If the Universe is quite different beyond our horizon from how it appears within it, as the inflationary idea suggests, then things are quite hopeless. Moreover, because inflation predicts that we should be within one part in ten thousand of the critical divide, fluctuations in the density around the critical value can tip the balance in one direction or the other. For the sake of illustration, suppose that today we could add up all the matter in the Visible Universe. We might find it to be just one part in ten thousand above the critical value, so indicating future re-collapse. But we might then return tomorrow when the Visible Universe is one light-day bigger, and re-total the contents only to find that the new density is now just below the divide by one part in ten thousand, because the new part of the Universe that has become visible was rather less dense than average. Our previous day's conclusion about indefinite future expansion would have to be revised. No precise answer to the question 'what is the density of the Universe?' can be forthcoming in a Universe like this. We could be living in a 'bubble' that mimics a collapsing universe within an infinite ever-expanding Universe; or we could be living in a rarefied bubble apparently going to expand forever, but yet imprisoned within an over-arching Universe that is going to collapse in upon itself. The fact that we habitually assume the truth of the Cosmological Principle is the only reason why we can contemplate making statements about the Universe, but the fact that we cannot demonstrate its truth, even if it is true, means that we must face the conclusion that we are unable to decide whether the Universe of space is infinite.

The question of the physical consequences of the long-term evolution of the Universe has been one of recurrent fascination and fundamental significance ever since thermodynamicists realized the existence of the famous Second Law of thermodynamics. In the pre-war era there was a morbid philosophical interest in the pessimistic forecast of what lay in store for the denizens of a universe set to expand for all future time (the inhabitants of the re-collapsing universes having been obliterated in the Big Crunch at some finite time in the future). In Britain both Sir James Jeans and Sir Arthur Eddington told the story of the 'Heat Death' of the Universe wrought by the inexorable increase of entropy and the approach towards a state of thermodynamic equilibrium in the far future. It was argued that life must eventually die out. The infinite future was bleak from the point of view of mind and meaning. This doctrine created considerable interest amongst philosophers and theologians of all shades of opinion. Some, like Bertrand Russell, saw the Heat Death as a fatalistic affirmation that

> all the labours of the ages, all the devotion, all the inspiration, all the noonday brightness of human genius, are destined to extinction in the vast death of the solar system, and the whole temple of Man's achievement must inevitably be buried beneath the debris of a universe in ruins—all these things, if not quite beyond dispute, are yet so nearly certain that no philosophy which rejects them can hope to stand. Only within the scaffolding of these truths, only on the firm foundation of unyielding despair, can the soul's habitation henceforth be safely built.

while others, like Edmund Whittaker, found in it confirmation of a deeper, immaterial reality:

> The knowledge that the world has been created in time, and will ultimately die, is of primary importance for metaphysics and theology: for it implies that God is not Nature, and Nature is not God; and thus we reject every form of pantheism, the philosophy which identifies the Creator with creation and pictures him as coming into being in the self-unfolding or evolution of the material universe . . . The certainty that the human race, and all life on this planet, must ultimately be extinguished is fatal to many widely held conceptions of the meaning and purpose of the universe, particularly those whose central idea is progress, and which place their hope in an ascent of man.

It is of considerable interest to recognize the extent to which the form of the laws of Nature have affected general philosophical speculation on the meaning and future of life in the Universe. Although eschatology is traditionally a religious matter, and the fatalism of the Second Law looks so far to the future that it could never worry a single living soul, the concept of the Heat Death of the Universe succeeded in deeply influencing an entire generation of thinkers—and still does so today.

Much has changed since Eddington and Jeans propagandized the Heat Death. We have discovered much more of the 'furniture of the earth'. We have recognized the possibility of creating artificial forms of life based upon information-processing hardware that need not be tied to the matter of flesh and blood of which we are composed. While we can be sure that our own species cannot survive into the indefinite future in its present biochemical form, we cannot rule out the possibility that it might transfer its essence and information content to a different medium.

At the purely physical level we can view the phenomenon which we call 'life' to be a form of software. This software could be run on a variety of different forms of 'hardware'. It so happens that in our particular embodiment this hardware consists most basically of DNA molecules, along with other complex biochemicals. All the genetic information stored in this 'hardware' constitutes the software program which we call 'human life'. Of course, some parts of the life program are only present to control pieces of hardware (the movement of arms and legs for example), but the information necessary to run the program could be stored in non-chemical forms. If we were clever enough we could store that information in another medium—magnetic tape, compact disc, paper, or whatever—and then run the 'life' program on a different type of 'hardware'—perhaps some soup of elementary particles, or a robotic form. To some extent transplant surgery is already on the way towards replacing our hardware by facsimiles. What would one say of an individual who had been replaced, cell by cell, by artificial components; at what point would one say that he or she ceases to be a person? In practice we ignore physical disabilities and artificial limbs as inessential to the human personality; would we do the same when things were taken further? There seems no *logical* reason not to. The question of deciding whether 'life' in this most basic and abstract sense of information-processing or computation—but not necessarily on the primitive devices that we term 'computers'—can survive forever in an ever-expanding universe amounts to determining whether, on the basis of what we currently know about physics, there is any fundamental reason why an infinite number of computations could not be carried out by a computer in the future. Surprisingly, it has been found that such computation is possible. This does not mean that it will happen, of course. We might be the only form of life in the Universe, and be annihilated by a nuclear war next week. But the Heat Death does not prevent the processing of an infinite number of 'bits' of information in the future. This is a way of saying that an infinite number of thoughts can be thought even if the Universe exists for only a finite proper time.

This scenario of self-reproducing robots perpetuating 'life' into the infinite future is not to everyone's taste (one feels that such a life is not *really* worth living!). As with the original Heat Death picture, there is a division of opinion as to its appeal and significance for wider philosophical and theological issues. There are those who recoil in horror at the idea that the image of Man must one day be changed, while the Christian apologist and expert on artificial intelligence, the late Donald MacKay, regarded the possibility of information transfer to different forms of hardware as an image of bodily resurrection. He writes of the fallacy that the death of the body necessarily signals the end of life:

> I am not suggesting that human beings have no more individuality than a mathematical equation, [but] the ending of a particular embodiment is no barrier in principle to the re-embodiment of the very same original. Conversely, it shows the fallacy of confusing the final destruction of an embodiment with final destruction of what is embodied. I have been suggesting that in this life our bodies physically express (represent from an observer's standpoint) our conscious agency—our

thinking and doing and suffering and all the rest of it. I am now arguing that if this is the kind of relationship that holds between us and our bodies, then there is no more obstacle in principle to [re-]embodiment—'resurrection' is the biblical name for it—than there is in the case of the man whose computer has been destroyed, but who wants the same computation to continue in a new embodiment.

Creation out of nothing?

You can't get owt for nowt. Anon

We have seen that inflation tells us nothing about the ultimate fate of the Universe. It has almost as little to say about the beginning of the Universe. This is a consequence of its desire to present the Universe's present structure as a logical consequence of laws of Nature independent of how it began. If no trace of initial conditions survive the inflationary phase to remain accessible to present-day observations, none of our observations of the Universe can help us answer the question 'did it have a beginning?'

Cosmology has traditionally given support to the idea that the Universe was created out of nothing at a particular past time. However, it is worth remarking that many Christian theologians, both Catholic and Protestant, do not regard the idea of *creatio ex nihilo* as either stated or implied by the biblical text. They would claim that creation is described as having occurred out of darkness and chaos. Supporters of this hermeneutic suggest that the idea of *creatio ex nihilo* arose historically as a counter to particular Gnostic heresies. However, it is undeniable that the Western Christian tradition has assumed that *creatio ex nihilo* is implicit in the Old Testament writings.

The Big Bang theory leads naturally to the picture of a Universe expanding from nothing, so much so that Steven Weinberg once remarked that for him the principal appeal of the steady state picture was that in removing the possibility of a beginning in time, it offered the least possible resemblance to the traditional religious picture of the creation of the universe: 'The steady state theory is philosophically the most attractive theory because it *least* resembles the account given in Genesis.' This reaction was no doubt partially provoked by the appeal made to the Big Bang picture by those cosmologists of a Christian persuasion. Milne and Whittaker were the most famous apologists of this ilk, and Whittaker's statements about the accord between the Christian tradition and the picture of the expanding universe were used by Pope Pius XII in his 1951 citation of scientific evidence for the Catholic world view. This methodology lives on in the writings of scientific historians like Stanley Jaki, whose passionate opposition to some cosmological model universes springs from his ideas about their supposed philosophical consequences in the past when they emerged in primitive form at different times within diverse cultures.

Universes without a beginning in time cause problems for our limited imaginations. In some circles the implicit beginning of the Universe implied by the naïve Big Bang model is still used as a logical argument for the existence of God. For, it is claimed, everything must have a cause, and so there must be a Cause of the Universe that is in essence 'other' than the Universe. However, the logic of this particular argument is not very compelling. Anyone who can live with the concept of the Deity as an uncaused cause can surely live with the Universe itself as the uncaused cause. Moreover, there is a sleight of hand in the argument. 'Everything we encounter has a cause, so the Universe must have a cause', the argument goes; but the Universe is not a 'thing' in the same sense. It is a collection of things. Every person has a mother, but this does not mean that we can conclude that every society or every nation has a mother.

If we grant the truth of the Big Bang picture then we are faced with the question of whether the apparent beginning of the expansion of the Universe fifteen billion years ago implied by the current rate of expansion signals the beginning of just a period of expansion in the history of the Universe, or whether it signals the beginning of the Universe as a whole in some sense. The latter, more dramatic, option has far wider philosophical ramifications, because it requires the framework of space and time, as well as the laws of Nature, to come into existence simultaneously. It is this aspect that underpinned Whittaker's deductions of a Creator from cosmology:

> when the development of the system of the world is traced backwards by the light of laws of nature, we arrive finally at a moment when that development begins. This is the ultimate point of physical science, the farthest glimpse that we can obtain of the material universe by our natural faculties. There is no ground for supposing that matter . . . existed before this in an inert condition, and was in some way galvanized into activity at a certain instant: for what could have determined this instant rather than all the other instants of past eternity? It is simpler to postulate a creation *ex nihilo*, an operation of Divine Will to constitute Nature from nothingness.

It is interesting to detect in these words the ancient 'why not sooner?' argument against the Universe having had a beginning at a particular moment in time. The argument is first associated with Parmenides and Aristotle, but it is most commonly linked with Augustine who, in response to the question 'what was God doing before the creation of the world?', answered that there was no time before Creation, and hence the question was not cogent.

During the early 1960s there was much confusion as to the reality of the apparent beginning to the Universe implied by winding the expansion history of the Universe backwards down to conditions of zero size and infinite density. The cosmological solutions of Einstein's equations which model the present state of the expanding universe most successfully are those in which the expansion proceeds at exactly the same rate in every direction. This is only an approximation of reality, but none the less currently a very good one indeed. But in the past things might not have been so symmetrical. Could it not be that the retrodiction of a past 'Bang' in which the entire Universe was squeezed to a point arises entirely as a consequence of following the expansion backwards in a perfectly symmetrical fashion? If one disturbed the symmetry slightly, then the disparities would grow as one went backwards, and the Bang might be de-focused and have never really occurred. Or could not some unknown force of Nature emerge to change the result of our naïve extrapolation back to a beginning? Or more subtly, maybe we have simply chosen a defective way of mapping the expansion of the Universe, and it is only this *description*, rather than the Universe itself, that becomes degenerate at the apparent Big Bang? Something like this occurs on a geographer's globe as we approach the North and South Poles. The meridians intersect and pile up to create a 'co-ordinate singularity' at the Poles, but nothing odd happens on the Earth's real surface. Another set of map co-ordinates could be introduced near the Poles to overcome this pathology of our mapping description.

During the mid-1960s Roger Penrose, an English mathematician, pioneered a mathematical approach which enabled the concept of a 'beginning' to the Universe to be made precise, thereby avoiding any ambiguity regarding phoney beginnings created merely by the breakdown of our way of mapping the evolution of the Universe mathematically. The conditions under which it can be concluded that the Universe must possess a beginning in the past can then be spelt out precisely. In collaboration with Stephen Hawking, Penrose demonstrated that *if* (i) gravity is attractive and acts upon everything, (ii) the Universe is expanding today and contains a sufficient quantity of matter, and (iii) time-travel is impossible, *then* general relativity requires space and time to come to an end somewhere in our past. The Universe of space and time must have an 'edge'.

A number of comments need to be made about this result. First, it is a *theorem* not a *theory*. That is, if the premises hold then the conclusion follows by logic alone. If any of the assumptions are found to fail then the conclusion need not follow, although of course it may. Conditions (i)–(iii) are thus *sufficient* but not *necessary* to establish the existence of a path through space and time which must come to a full stop in the past. This full stop we call a '*singularity*'. It would represent the beginning of time for an observer who happened to travel backwards along it. Before we examine the validity of the assumptions (i)–(iii) used in the theorem, let us examine some unexpected features of the singularity that it predicts. First, notice what the theorem does *not* say about the initial singularity. It does not require every point in the Universe to experience it; nor does it require all the points that do to do so at the same time; nor does it say anything about infinite densities or the other accoutrements of the popular Big Bang. There have been subsequent attempts to prove, under conditions similar to (i)–(iii), that the general singularity is all-encompassing, and caused by the fact that physical quantities like the density are becoming infinite, but these have not yet been completely successful. These questions remain open.

In Chapter 6 we shall have more to say about the meaning of singularities in space and time. Here, we shall confine our attention to discussing the conditions which must be satisfied if their occurrence is to be compulsory. The most striking of these is the stipulation that time-travel must be impossible. It is easy to see why we need to make this stipulation. If all the possible paths through time were closed loops then none of them could come to an end at a singularity. This possibility must be excluded by hypothesis because Einstein's law of gravitation actually permits time-travel. In 1950 Kurt Gödel, the famous creator of Gödel's Incompleteness theorem, found a solution of Einstein's equations which described a spinning universe in which time-travel could occur. This particular universe possessed other properties which ensured that it could not describe our own, but none the less we do not know whether the solution of Einstein's equations which most resembles our own Universe also includes the possibility of time-travel. If Einstein's theory is the law of gravitation, then time-travel appears to be possible in some situations. We do not know whether the realization of these situations is compatible with the other laws of Nature. If we are unconcerned about the influence of time-travel upon the Hawking–Penrose theorem, then the most worrying aspect would seem to be the possibility of a factual contradiction of the 'what if I killed by own grandmother?' variety. But if the recurrence time of each time-path were immensely longer than the present age of the expanding universe then we would have yet to experience these strange acausal effects. In fact, time-travel seems to be a property of very few of Einstein's universes.

Of the remaining assumptions that go into the singularity theorem (ii) is seen to be satisfied in our experience, but (i) is the crucial one. Remarkably, it turns out that the condition needed to produce inflationary expansion in the very early universe is precisely the *opposite* of (i).

In the mid-1960s when the singularity theorems were first composed, the condition (i) appeared to be eminently reasonable. Today, the particle physicists have completely altered our view about this. There is no reason to expect that all the forms of matter that may appear at the very high energies experienced on approach to the singularity will exhibit effective gravitational attraction. This means that the existence of a beginning cannot be regarded as mandatory. Nevertheless, as we said earlier, a violation of condition (i) during some period of inflation does not mean that a singularity cannot occur, only that it need not. There exist many cosmological models which could describe the early stages of our Universe, and which possess an initial singularity and yet also experience inflation. Yet the plethora of possibilities means that the singularity theorems of Hawking and Penrose do not allow us to conclude that there was a beginning in time, even if we assume that the laws of Nature remain unaltered all the way back to the earliest possible moments.

Attempts have been made in recent years to produce an improved version of Einstein's theory of gravitation which incorporates the essentials of quantum theory, and thereby treats gravity as a quantum field. This is one of the frontiers of research in modern theoretical physics. There is no particularly compelling candidate for a theory of quantum gravity as yet. Many believe that superstring theories will provide the front runner. Without indulging in the formidable technicalities of quantizing Einstein's general relativity, one can see that the very idea of marrying general relativity to quantum theory is problematic. The central message of general relativity is that the distribution of matter in the Universe determines the geometry of space and the rate of flow of time, but the inherent uncertainty of the quantum theory tells us that we can never know precisely where the matter is. How then do we know what is the structure of the space and time within which we wish to place the masses? The new theory of quantum gravitation may, or may not, predict a beginning to the Universe.

If a singularity really did occur everywhere in the Universe, then there does not exist anything before the moment of that singularity. No meaning can even be attached to the idea of 'before'. There is no time prior to the singularity, and the laws of Nature break down there if we extrapolate them backwards in time. Alternatively, we might say that they come into being spontaneously at the singularity (in Chapter 6 we shall discuss this feature in more detail). Thus, in a description of the history of the Universe which inevitably predicts the existence of an all-encompassing singularity at a finite time in the past, the Universe must be regarded as having been created at that moment *ex nihilo*. The theory provides no explanation as to *why* this happened; it merely tells us that, under the assumptions (i)–(iii) listed above, the Universe and its laws cannot always have existed: it must have come into being spontaneously at some moment in the past. Likewise, if it contains sufficient material, the universal expansion will one day reverse into contraction, and collapse to a final singularity at which space and time and the laws of Nature will go out of existence. If conditions (i)–(iii) do not hold, then the Universe can still collapse if it contains sufficient material, but we do not know what its final state might be. It could conceivably bounce back into a new phase of expansion, and then continue in cyclic recurrence forever.

In recent years cosmologists have begun to discuss the spontaneous creation of the Universe as a problem in physics. Those who do this assume that a future synthesis of quantum theory and relativity which reveals how gravity behaves when matter is enormously compressed will evade the prediction of a real singularity of the type required by the singularity theorems. Although the assumptions of the singularity theorems are not expected to hold near the singularity, we do not know whether to expect a singularity or not as yet. But even in the absence of this singularity to denote the beginning of the Universe, it has been speculated that the application of quantum theory to the whole Universe may allow physical content to be given to the concept of 'creation of the Universe out of nothing'. The goal of this research is to show that the creation of an expanding universe is inevitable. The reason there is something rather than nothing is that 'nothing' is unstable.

There are two types of picture one can paint of this process. We recall how our earlier discussion of the quantum vacuum introduced the notion of a virtual fluctuation, that is, a spontaneous appearance of energy E for a time T, such that the product E times T is less than Planck's constant of action h, and hence unobservable. One way of viewing this is to say that Nature allows energy conservation to be violated locally, so long as it is not observable. Now, suppose the whole expanding Universe were a vacuum fluctuation with total energy very close to zero, this would allow it to persist for a very long time, with T perhaps equal to tens of billions of years. Curiously, if the Universe is of the closed sort that lives for only a finite time, then the only sensible way to define its 'energy' gives it a value of zero. In that case it could appear 'out of nothing' without violating the conservation of energy at all. Although one has the intuitive feeling that creating something out of nothing must violate some sort of conservation law, it may well be that all the conserved quantities—like electric charge, spin, and total energy—are zero (or even non-existent concepts) for the Universe as a whole, and therefore their local conservation actually presents no impediment to the spontaneous appearance of the Universe 'out of nothing'.

At a more sophisticated level than these heuristic arguments, there exist some solutions of the embryonic theory that is used to study how a quantum Universe would behave which are mathematically identical to a situation in which an entire Universe has appeared from nothing at all. This indicates that a quantum 'nothing' is unstable, and will eventually turn into something else. Before one gets carried away with all these dramatic ideas, it is appropriate to be more critical about the use of language. Suppose that there existed a robust physical theory of quantum cosmology which allowed the expanding Universe to appear spontaneously as an event in time, but before which there had been no expanding universe and no matter. One must still grant the existence of a considerable body of pre-existing laws of Nature in order to get away with this trick. In no sense in which the word 'nothing' is commonly used is one really demonstrating 'creation out of nothing'. One assumes the pre-existence of quantum laws, quantum fields, time, space, and presumably mathematical logic. At present there is no real sense in which they are being done away with. What quantum cosmology tries to exploit is the fact that in quantum mechanics there is a real sense in which there are effects without causes. We saw this even in the simple two-slit experiment described in the last chapter. Energy is deposited on the target screen, but in no meaningful sense can one attribute that event to a specific cause. One cannot 'find' the emitted neutron that was responsible. One cannot uncover which of the two slits it went through.

This last aspect of ordinary quantum mechanics is bound up with the fact that the observer and the observed cannot be separated in quantum reality as they are in its traditional classical counterpart. But when we come to apply quantum theory to the Universe as a whole, new and peculiar problems emerge. First, quantum theory does not allow us to make a definite prediction of what a particular experimental measurement will yield—only of the probability of getting a certain result when a measurement is carried out. However, in cosmology the entire notion of probability is a strange one, because there is only one Universe. Einstein's rejection of the probabilistic interpretation of quantum theory as the final word would only have been reinforced by his contemplation of what it would mean for his other creation—relativistic cosmology. Stranger still, if Bohr's doctrine is adhered to, then no meaning can be attached to a quantum Universe until it is observed. But who observes it? Who transforms the wave-function of the whole Universe, with its latent information regarding the information content of all the possible states the Universe could be found to reside in, into a definite state by the act of observation? The theologians have not been very eager to ascribe to God the role of Ultimate Observer who brings the entire quantum Universe into being, but such a picture is logically consistent with the mathematics. To escape this step cosmologists have been forced to invoke Everett's 'Many Worlds' interpretation of quantum theory in order to make any sense of quantum cosmology. We recall that this picture does not require the wave function to undergo that miraculous and unexplained 'collapse' into a definite state in order to give meaning to things. The wave function of the Universe never collapses. The Universe as a whole need never be observed by anyone in the Many Worlds picture. The price one must pay for this simplification is that one must grant the multiplication of worlds which are produced by every physical interaction that ever occurs. It is for this reason that almost all the strong supporters of the Everett Interpretation of quantum reality are cosmologists. It is the only interpretation of quantum reality which even allows the subject of quantum cosmology to exist as a scientific study.

Philosophers have made half-hearted attempts to introduce the concept of creation out of nothing into logical deductions about the physical world. Their reasons for so doing are often the desire to undermine traditional arguments for the existence of God derived from natural law. These arguments possess all manner of problems, and indeed, because they are logical arguments, they can never do more than offer a choice. They only prove something if the initial assumptions on which the argument is predicated are believed. One always has a choice between believing or disbelieving the assumptions, and, along with them, the conclusion of the argument. Arguments that look to the fact of the existence and origin of the Universe as a confirmation of the existence of God are not terribly compelling to scientists. An interesting common objection to the view that God is the author of the laws of Nature is to ask why God created the particular universe and set of natural laws that He did. If there was no reason at all for the choice, then we have found something that is not subject to the laws of Nature. If there were constraints which imposed the necessity of the actual choice, then God is subject to some higher law of Nature. However, the chaotic cosmology philosophy, like the inflationary universe picture, implicitly recognizes the fact that the Universe could have been initially structured in any number of different ways, all of which result in the same universe at later times. Thus the 'either/or' of the counter-argument is somewhat undermined. Uniqueness can be arranged to emerge from random choices even to the extent that our limited imaginations can conceive.

Cosmology and the law

A judge is a law student who marks his own examination papers.

H. L. Mencken

The previous sections display a striking historical fact: the creation of the Universe—once a question of concern only to theology and metaphysics—is now discussed in scientific papers and regarded as amenable to *scientific* enquiry. The prestigious *Physics Letters* has recently published a paper by a respected author under the title of 'Creation of the Universe From Nothing'. A significant number of the world's leading theoretical cosmologists are now interested in the problems that can be associated with that title. Amusingly, it has not been widely recognized that those leading American physicists now engaged in such work, and who propagate their results through classes and seminars in state universities, are actually in violation of American law!

On 7 December 1981 there began, in the American state of Arkansas, a theatrical court case brought by the American Civil Liberties Union as a challenge to the 'Balanced Treatment for Creation Science and Evolution Science Act' which had slipped through the Arkansas legislature in sinister circumstances, to be signed into law by Governor White. One month later, following 'expert' testimony by all manner of scientific and religious celebrities, Judge Overton delivered a judgement against the 'creationists' desire to have their disguised religious beliefs taught as science on an equal-time basis in public schools. Although I believe this decision to have been right, it is most instructive to study the muddled nature of Judge Overton's lengthy statement as to the reasoning behind his decision. On the 'sudden creation of the universe, energy, and life from nothing' he rules:

> 'creation out of nothing' is a concept unique to Western religions. In traditional Western religious thought, the conception of a creator of the world is a conception of God. Indeed, creation of the world 'out of nothing' is the ultimate religious statement because God is the only actor . . . The only one who has this power is God . . . The idea of sudden creation from nothing, or *creatio ex nihilo*, is an inherently religious concept.

Regardless of the merits of the case, and disregarding the remarks about 'creation out of nothing' being unique to Western religions, the latter part of the quoted extract in effect denies 'creation out of nothing' as a scientific study. It defines it as religious, and classes it as a 'creation science' along with the ragbag of anti-evolutionary ideas promulgated by the Creation Research Society and their sympathizers. Moreover, Overton concludes that all such 'creation science' 'has no scientific merit or educational value as science'. Finally, since inherently religious ideas cannot be taught in public educational institutions, the Judge's inexorable logic would appear to throw the baby out with the bath-water, and forbid the teaching of modern Big Bang and quantum cosmology in state universities in the USA.

The nature of time

Had we but world enough, and time . . .

Andrew Marvell

In our discussion of the latest speculations concerning the beginning of the Universe in time we have used the concept of 'time' in a happy-go-lucky fashion. We have assumed that there is no subtlety associated with the concept of time, no way in which its meaning is tied to the evolution of the Universe it is clocking. We have assumed implictly that if we measure time in the way that Einstein taught us—with a clock that is not in relative motion with respect to us or feeling a different gravitational field—then its meaning is unambiguous. This measure of time is Einstein's *proper time*. When we said that the singularity theorem of Hawking and Penrose predicts a beginning to the Universe a finite time ago, we meant that this would be measured as a finite number of 'ticks' of a clock carried by an 'observer' who was taken back to the singularity. Likewise, if we were to fall into the final Big Crunch of a collapsing universe, then we should cease to exist after a finite number of ticks of a co-travelling clock. When we have spoken of the 'age' of the Universe, we have meant the period of time measured by a hypothetical observer who expands along with the Universe from the initial singularity to the present. But there is nothing in Einstein's theory of special relativity that requires proper time to apply to the Universe as a whole, where the force of gravity enters in an essential role. Furthermore, whereas Einstein taught us that in the world of special relativity (where gravity is ignored) there is no absolute standard of time that is better than any other, and by use of which everyone in the Universe can agree on the time unambiguously, this is no longer true in general relativity. In relativistic cosmology there exist absolute times. For example, in an expanding Universe of uniform density, anyone in the Universe could use the local measure of that density to determine the absolute time since the beginning of the expansion. The density clock provides a universal and absolute measure of cosmic time that would ensure identical time-keeping by horologists all over the Universe who could never confer in any way. Many other cosmic times are also conceivable. The question is: 'which, if any, is the fundamental cosmic time?'

Suppose we were to travel backwards to the Big Bang through a time-reverse of the expansion history of the Universe. At first, time could be kept by using a wrist-watch. But eventually conditions would become too hot for plastics, metals, and even atoms to exist (and us as well, but it is the story-teller's license to ignore that little detail). Time would have to be measured by using the vibrations of an atomic nucleus. But eventually nuclei would be crushed into their constituent elementary particles. Every material artefact that could be arranged using particles of matter would eventually be disrupted as the singularity was approached. Everything would be crushed to zero size. What meaning can we associate with 'time' in an environment like this, where there are no material 'clocks' to measure it with? There appears to be one natural choice: the curvature of the geometry of space and time. It extends everywhere. It exists for as long as the Universe does, and it changes inexorably as the Universe evolves in time. A 'clock' that measures the change in the Universe's geometry is not tied to any man-made artefact. It is observer-independent, yet observers can use it to tell the time universally. This all sounds well and good, but does it have any fundamental significance?

As we back-track to the initial singularity, the curvature and the density probably become infinitely great, and our 'curvature clock will measure an infinite interval of curvature time during a corresponding finite interval of conventional proper time. Now imagine that an information-processing device (a sort of gravitational Maxwell's demon) exists that senses curvature time. That is, its subjective or psychological time would be experienced to flow in unison with curvature time. This creature would take an infinite period of its subjective time to actually reach the singularity. Similarly, in a closed, re-collapsing Universe this creature would take an infinite period of curvature time to reach the end of the Universe. It would live forever in its own subjective time. The tables are turned in the far future of an ever-expanding universe. Here, the observer living by proper time has an infinite future, but, according to curvature time, he has only a finite future if the Universe tends towards a state of global uniformity. His subjective time will slow down in unison with the deceleration of the Universe.

The essence of the curvature time idea is that time only has meaning when things happen. It measures the rate at which they happen. In practice we are conscious of many different time arrows. We have a subjective intuition of the passage of time which enables us to discriminate the past from the future in our actions. In thermodynamics we find that the future is defined by the direction of entropy-increase. There are many other arrows of time which we could use.

If we are to relegate the beginning of the Universe to a past infinity we need to display some evidence that there is a real curvature clock whose 'ticks' correspond to real, physically distinguishable, events. Otherwise we run the risk of merely recreating one of Zeno's paradoxes of the infinite, in which we say that we cannot reach the singularity because we can always divide the time required to reach it in half. However, this fallacy we can avoid. The most general type of universe that has been extracted from Einstein's general theory of gravitation behaves in a most extraordinary fashion as we trace it back to the singularity in past time. The singularity is reached in a finite proper time, but *en route* an infinite number of oscillations occur, juggling the rates of expansion through different directions and values in a chaotic fashion. Each of these oscillations is a physically real and distinct event. If we could exist in the early stages of such a universe we could observe them occurring. They are caused by the curvature of the universe being different in all directions. Thus, an infinite number of physically distinct sequential events occur in a finite interval of proper time. If we use the universe's oscillations as a curvature clock, then it takes an infinite past time to reach the singularity. We shall have more to say of this state of affairs in Chapter 6.

We see that even in non-quantum physics the meaning of time creates fundamental cosmological conundrums. If we try to create a quantum cosmology, then it is very difficult to see how the concept of time can be introduced at all. Time has a strange status in quantum theory that leads to its disappearance from the mathematical description of a quantum cosmology. The problem of time in quantum cosmology remains at present a mystery.

Where have all the dimensions gone?

> Little laws have bigger laws
> From which they're forced to follow,
> But a bigger law's a bigger guess
> That's harder still to swallow. Anon.

In our exploration of the direction of research into the inner space of elementary particles we saw that there are powerful reasons for entertaining the idea that there exist many more dimensions of space than the three in which we have our being. The additional dimensions pay their way by allowing a unification of fundamental laws of Nature to occur at very high energies. When we come to explore the cosmological consequences of this idea we are confronted with a puzzle. In order to achieve a self-consistent unification of the forces of Nature, and yet explain the manifest tri-dimensionality of the Universe at present and during all the periods of its history about which we have direct or indirect evidence, it is necessary for the extra space dimensions to be infinitesimally small in extent during those periods. If the Universe commences expansion in a fully ten- or twenty-six-dimensional fashion, then events must have occurred which resulted in all but three of the dimensions of space being flattened out to an imperceptible level. We do not yet understand how this could happen, nor why it should happen to all but *three* (rather than five or seven, say) of the dimensions. At first it was viewed as being a challenge to explain the containment of the extra dimensions to microscopic size, but now it is appreciated that the tiny extent of these extra dimensions is naturally consistent with the force of gravity. The major mystery is how three of them could escape and expand to a size of fifteen billion light years—about 10^{60} times bigger than the others. It is as if the process of 'inflation' has managed to act upon three of the directions of expansion, but not upon the others. Maybe this is a clue to what did occur. We do not know. There could be a law of Nature which dictates the number of dimensions that inflate to huge extent; or this crucial number could be simply the result of 'chance'.

5

Why are the laws of Nature mathematical?

> Without a parable modern physics speaks not to the multitudes.
>
> C. S. Lewis

A puzzle

> Philosophy is written in that great book which ever lies before
> our eyes, I mean the universe, but we cannot understand it if we
> do not first learn the language and grasp the symbols in which it
> is written. This book is written in the mathematical language,
> and the symbols are triangles, circles, and other geometrical
> figures, without whose help it is humanly impossible to
> comprehend a single word of it.
>
> Galileo

The words of Galileo, quoted above and written in 1623, recognize a profound fact of life: the laws of Nature are mathematical in character. They need not have been expressed in the language of mathematics, but somehow the nuances of that language are infinitely adaptable to the facts of experience. We have found natural phemomena to be most deeply and most completely described by mathematical formulae. Why should this be? Is it telling us something important about the intrinsic logic governing the Universe—that 'God is a mathematician'—or is it just the reflection of our own minds in the medium we have chosen to express Nature's regularities? Then again, is it conceivable that the Universe's structure possesses some intrinsically mathematical aspects, but these form only a small fraction of its total information content, albeit the only part we have been able to discover reliably? Is reality mathematical, or is mathematics simply the nearest to reality that we can realistically get? In order to explore the content of these options we shall have to examine the different possible views of mathematics that are on offer. In some respects they are analogues of the philosophical positions outlined in the first chapter, but they possess other, unexpected ingredients.

We have grown so used to the extraordinary fact that we can reliably predict the course of natural phenomena by means of little squiggles on a sheet of paper that we may have ceased to be impressed by it. But we can see that something like mathematics—a language that possesses a form of built-in logical structure which constrains its form and direction—is necessary, although not sufficient, for the representation of laws of Nature. Mathematics allows statements to be made unambiguously, in a manner that is value-free and culturally independent. Only when hypotheses can be stated precisely in this way can they be compared and tested experimentally or observationally. Science has to use such a language if it is to progress in a meaningful way. Mathematical language permits irrelevancies to be excluded from the analysis of particular problems in an unambiguous fashion. It also allows us to avoid unnecessary thinking, by building certain necessary aspects of logical consistency into the mathematical formalism itself *ab initio* so we are assured that they will hold whenever we employ that formalism. Ordinary language does not contain these advantages. The grammatical correctness of an English sentence does not tell us anything at all about the truth or self-consistency of what it states. Some of the most powerful communicators ignore basic linguistic rules when they write, and thereby create a personal literary style. Mathematics allows no such artistic licence. Respect must be paid to its internal logic or it serves no useful purpose at all. There is freedom to dictate what this logic will be in some cases, but once decided upon, to break its rules is to lose the game.

Mathematics is a way of representing and explaining the Universe in a symbolic way. In this respect it has a superficial resemblance to many mystical religions, but it differs from, say Buddhism, in that its symbols and their meanings are unambiguous. They do not depend on the hearer's interpretation of their meaning, nor upon personal associations that he might care to draw with other concepts. They are predicated upon a set of inviolate interrelationships.

In everyday life we often require particular concepts or descriptions to be unambiguous, but we usually avoid recourse to mathematical symbols by either defining new words or, more commonly, by giving ordinary words additional technical meanings (for instance, in earlier chapters we have done this with common terms like inflation, force, work, information, nucleus, and hardware). However, when this is done in science the technical terms are defined mathematically, and the words are used subsequently as a shorthand representation of the precise mathematical concept. This has an interesting sociological consequence that no doubt inspired past advocates of the 'Two Cultures' view of modern societies. Whereas the average mathematician or scientist usually reads novels, watches films and plays, or listens to music, the professional writer, linguist, actor, or musician has in general cultivated comparatively little knowledge of or interest in science. The principal reason is often the barrier created by the special language—mathematics—in which the 'poetry' of science and mathematics is expressed, rather than any pathological lack of interest in technical matters. The arts require the human senses for their appreciation, and partial enjoyment of them is possible without specialized knowledge of a new form of language. The popularization of science means, in practice, the re-expression of mathematical ideas in terms of everyday language, and judicious uses of analogy to represent abstract concepts and their interrelationships in terms of familar non-mathematical ones.

What is mathematics?

> Mathematicians are a species of Frenchman: if you say something to them they translate it into their own language and presto! it is something entirely different.
>
> Goethe

It is not easy to say what mathematics is, but 'I know it when I see it' is the most likely response of anyone to whom this question is put. The most striking thing about mathematics is that it is very different to science, and this compounds the problem of why it should be found so useful in describing and predicting how the Universe works. Whereas science is like a long text that is constantly being redrafted, updated, and edited, mathematics is entirely cumulative. Contemporary science is going to be proven wrong, but mathematics is not. The scientists of the past were well justified in holding naïve and erroneous views about physical phenomena in the context of the civilizations in which they lived, but there can never be any justification for establishing erroneous mathematical results. The mechanics of Aristotle is wrong, but the geometry of Euclid is, was, and always will be correct. Right and wrong mean different things in science and mathematics. In the former, 'right' means correspondence with reality; in mathematics it means logical consistency.

Before we can draw any conclusions from our good fortune in finding our experience of reality to be so well described by mathematics, we need to have some understanding of what mathematicians think mathematics is, or at least what they think it could be. We have learnt some mathematics; we can write down some mathematical formulae which are 'true': what is it that makes them so?

There are essentially four interpretations of mathematics, and the view to which one subscribes will determine to a large extent the assessment one makes of the remarkable effectiveness of mathematics in describing Nature. Conversely, the natural applicability of mathematics to the world of experience might enter as a key witness in deciding the merits of the competing interpretations of mathematics. Each interpretation is a possible view of what is meant by the statement that some mathematical statement is 'true'. We shall call the four options *Platonism*, *Conceptualism*, *Formalism*, and *Intuitionism*. Their *curricula vitae* read like this:

Platonism: This is the view that mathematicians discover mathematics rather than invent it. All the concepts they arrive at and find useful, like groups and sets, triangles and points, infinities, and even numbers, really exist 'out there' independently of you and me. There would exist mathematical quantities even if there were no mathematicians; they are not creations of the human mind but manifestations of the intrinsic character of reality. 'Pi' really is in the sky. These mathematical objects do not exist in the space and time we experience. They are abstract entities, and mathematical truth means correspondence between the properties of these abstract objects and our systems of symbols.

The reason for the Platonic association with this notion should be clear from our discussion of Plato's Ideal Forms in Chapter 2. Mathematical ideas like the number 'seven' are regarded as immaterial and immutable ideas that really exist in some abstract realm, whereas our observations are of specific secondary realizations like seven dwarfs, seven brides, or seven brothers.

On this view we could use mathematical entities as a language with which to communicate with alien beings from other worlds, and be confident that they would have discovered many of the same mathematical structures that we have. They would not, almost certainly, be using the same symbols or language as ourselves to represent those mathematical entities, but none the less they would be expected to have representations of the same idea, just as surely as the words 'seven', 'sieben', and 'sept' code the same information to English, German, and French speakers. Of no other part of our human experience can this be said. Our art and ethics, our forms of government and styles of literature would probably be unintelligible to alien beings because they do not describe something independent of our minds. Platonists would be confident of using mathematics as a universal language because they hold it to be a description of something ethereal and absolute. The reason why mathematics is so accurate in capturing the workings of Nature is, therefore, attributed to the simple but inexplicable fact that Nature really *is* mathematical, and is in fact the basic source of our mathematics. Despite being idealists, mathematical Platonists often regard themselves as realists simply because they hold the most straightforward interpretation of mathematics. They regard mathematics as an absolute truth— 'God is a mathematician'—whatever your definition of God.

This has given rise to a mystical offshoot dubbed 'neo-Platonism', one of whose Soviet proponents has likened mathematics to the composition of a cosmic symphony by independent contributors, each moving it towards some grand final synthesis. This goal cannot, he claims, be something so mundane as a description of the world or the application of mathematics to the solution of practical problems. It has an other-worldly quality because completely independent mathematical discoveries by different mathematicians working in different cultures so often turn out to be identical. But few professional mathematicians would seek to draw such cosmic conclusions. Most work as though Platonism were true, and seek to 'discover' new mathematical structures that are interesting because they are rich in internal properties and possess unexpected connections with other, superficially unrelated, branches of mathematics. Talk of 'existence' proofs of particular mathematical objects is a reflection of this subconscious attitude. However, if pressed to defend this common-sense approach, few pure mathematicians would seek to do so. On the other hand, the applied mathematician and other 'consumers' of mathematics (physicists and economists, for instance) provided 'ready-packaged' for them by others generally regard mathematics as a useful 'black-box': a tool for obtaining answers to specific problems whose effectiveness is promised by the innumerable successes in the past.

Conceptualism: This is the complete antithesis of Platonism, and is popular with sociologists rather than mathematicians and scientists, most of whom would instinctively reject it. It maintains that we create an array of mathematical structures, symmetries, and patterns, and then force the world into this mould because we find it so compelling. Ultimately, the choice of what mathematics we construct is culturally derived. We invent mathematics, we do not discover it. Mathematics is what mathematicians do. The suspicion that there is some truth in this view has led to a gradual and often unnoticed shift in the way applied mathematicians describe what they are doing. Whereas the classical mathematicians of yesteryear would write treatises or give lecture courses on 'the mathematical theory of . . .', today there is a growing emphasis upon the less grandiose term 'mathematical modelling'. This illustrates the fact that the conceptualist does not regard the Universe as being intrinsically mathematical: God is not a mathematician, but what He does can be fairly accurately described by mathematical 'models'. Mathematics is entirely a product of the human mind. Conceptualists would not expect to be able to communicate with the Andromedans by using our mathematical concepts.

This viewpoint might have significant consequences for the physicist. It could, for instance, be argued that our so-called constants of Nature, (quantities like Newton's gravitational constant G), which arise as theoretically undetermined constants of proportionality in our mathematical equations, are solely artefacts of the particular mathematical representation we have chosen to use for the gravitational force. In that sense 'G' is seen as a cultural creation. It reflects our inclination to express natural phenomena in a particular way. It also reflects the views of Kant regarding the innate categories of thought whereby all our experience is ordered by our minds. Whether or not the 'thing in itself' is intrinsically mathematical (as the Platonist believes), we can experience it in no other way than the mathematical. Thus, our minds imprint mathematical ideas upon experience. It also displays something of the opposing belief that cultural elements completely condition our mathematical experience of Nature, because Nature has impressed mathematics onto our minds during the course of our evolutionary adaptation. We honed our ability to formulate and manipulate abstract symbols most effectively where they were based upon things that actually exist in the real world. This latter viewpoint has far-reaching consequences, for it maintains that mathematics has been shaped by our experience. If this is so, then we should not expect it to hold good when confronted with newly discovered phenomena outside of everyday human experience. It could be said that many aspects of the world of physics, be they to do with astronomy or particle physics, turn out to be unfamiliar in scale, but not in the type of mathematical reasoning involved, so we should not necessarily be impressed by the fact that our human, evolved mathematical sense provides a good description. However, there are places where no past evolutionary history can have produced the requisite concepts; the peculiar aspects of quantum reality which we introduced in Chapter 3 may be signalling to us that it is not just our physical theory that needs improving if we are to deal with the microworld. Indeed, mathematics may not even be the language that most naturally describes what is happening. There have been investigations into the possibility that classical logic (where statements are either true or false) does not apply at the quantum level, but is replaced by a three-valued 'quantum logic' which allows the additional status of 'undecided' to be associated with a statement; thus a statement that is not true need not be false. In this way a different answer can be given to the question 'what slit does the particle go through in the two-slit experiment?' However, the solution to the problems of quantum reality may be far more radical. It may require some other type of description: a new language that does for logic what logic does for mathematics.

Conceptualism is a form of anti-realism directed against theories about, rather than the existence of, a bedrock of underlying reality.

One can certainly detect specific cultural biases in the way mathematics has been and is being done. The British style eschews general formalism for the sake of elegance alone, is biased towards applications, and motivated by the desire to solve practical problems. The French, by contrast, are attracted by formalism and abstraction, as epitomized by the encyclopaedic project of the Bourbaki group. Are these national styles a harmless irrelevance, or do they indicate a deep-seated subjectivity that colours the development of all human mathematics, and thereby dictates the possible laws and explanations that are available to scientists in their representations of Nature?

Formalism: The next of the 'isms' is one that grew up primarily near the turn of the century. Logicians had uncovered a number of embarrassing logical paradoxes, and there had begun to appear mathematical proofs which established the existence of particular objects but offered no way of constructing them explicitly in a finite number of steps (in this respect they might be compared with certain legitimate configurations which can be laid down in John Conway's board game 'Life', but which cannot be reached from some starting state in a finite number of moves). These logical entities arose as a consequence of properties of infinite collections of objects, when it was assumed that infinite collections of things were governed by the same common-sense logic that applies to finite collections (like the assumption that the collection either possesses a certain property or it does not). Some mathematicians were nervous about such a step of faith, because infinite sets were not physically realizable. In response to these doubts David Hilbert proposed a programme to eradicate such ambiguities. The underlying philosophy of Hilbert's formalist programme was to define mathematics as nothing more than the manipulation of symbols according to specified rules. The resultant paper edifice has no special meaning at all. It should, if the manipulations were performed correctly, result in a vast collection of tautological statements: an embroidery of logical connections. Sometimes the word 'model' is used in this approach, but with a different meaning to that intended by the conceptualists: here a 'model' for a set of axioms is some collection of conceptual entities that satisfy them, and hence to which the resulting tautologies (which we call theorems) derivable from the axioms apply. The focus of attention is upon the relations between entities and the rules governing them, rather than the question of whether the objects being manipulated have any intrinsic meaning. The connection between the world of Nature and the structure of mathematics is totally irrelevant to the formalists. One of the aims of this approach was to avoid having to worry about the *meaning* of non-intuitive objects like infinite sets. Attention was focused upon relationships between concepts rather than on the concepts themselves. The only goal of mathematical investigation was to show particular sets of axioms to be self-consistent, and hence acceptable as starting-points for the logical network of symbols.

From what we have said one might consider Euclid to be an archetypal formalist. In retrospect this appears so, but we have to remember that he abstracted his axioms from observation of the real world. All his theorems were visualizable by drawing points and lines in the sand and measuring angles. Modern mathematics does not require its sets of axioms to possess visualizable or self-evident properties. It is sufficient simply that they be self-consistent. This view of mathematics—that it is a logical game, like chess, with lists of pieces and rules—is opposed to the Platonic picture, for it regards the mathematical rules and axioms as entirely our own creations. They have no independent meaning except through their interrelationships. We determine these relationships by setting the rules of the game. Formulae exist: mathematical objects do not. For the formalist, the utility of mathematics in describing Nature is a curiosity which had nothing to do with mathematics. For the formalist, a mathematical theory is only intelligible if it is meaningless.

Intuitionism: This interpretation was also a reaction to the use of non-intuitive concepts in mathematical proofs. To avoid founding whole areas of mathematics upon the assumption that infinite sets share the 'obvious' properties possessed by finite ones, it was proposed that only quantities that can be constructed from the natural numbers 1, 2, 3, . . . in a finite number of logical steps should be regarded as proven true (prior to Cantor's work on infinite sets mathematicians had not made use of an actual infinity, but only exploited the existence of quantities that could be made arbitrarily large or small—this idea forms the essence of the rigorous definition of a 'limit' introduced in the nineteenth century by Cauchy and Weierstrauss). Each step must unambiguously specify the next logical step to be taken. For this reason it is also referred to as *constructivism*. The name 'intuitionism' reflects the idea that only the simplest intuitive ideas could be used. Anything outside our experience must be constructed from the simplest ingredients by a sequence of intuitively familiar steps. This approach is analogous to the operationalist stance we introduced in the first chapter; whereas the operationalist restricts attention to measurable quantities in order to avoid introducing 'obvious' concepts like simultaneity which may turn out to be experimentally meaningless, the intuitionist retains the 'obvious' to avoid arriving at the meaningless.

There is a parallel between the goal of the intuitionists and that of some interpreters of the quantum theory which we introduced in the last chapter. Both Bohr and the intuitionists were trying to introduce new views of physical and mathematical quantities that divorced them from objective reality. A quantum measurement reflects one's state of knowledge about physical reality. A mathematical formula, according to the intuitionists, describes only the set of computations that has been carried out to arrive at it. It is not a representation of any reality existing independently of the act of computation.

Intuitionists exclude all arguments that prove the existence of something without providing the recipe for constructing it. Of course, if we can prove that there exists *some* special member of a *finite* set of objects then, although this result is not acceptable to the intuitionist, it can be made so by working through the finite collection of objects (perhaps by searching through the mathematical possibilities using a fast computer, as in the recent proof of the famous 'four-colour theorem') so as to isolate the special member explicitly. This is a legitimate constructive procedure. However, if we had proved only the existence of a special member of an *infinite* set, then this result would be judged unacceptable because it could not be checked constructively in a finite number of steps using the computer. It is very interesting that the famous singularity theorems of Hawking and Penrose (see the previous chapter), which establish the existence of a beginning to space and time in the Universe's past if a number of observationally checkable assumptions are met, do *not* meet the intuitionists' requirements. They predict the existence of some path (or paths) through space and time which inevitably comes to an end a finite time ago, but do not construct it explicitly. The best we could do is to find explicit solutions of Einstein's equations of general relativity that possess initial singularities. However, the only such solutions that we have been clever enough to find are rather special cases.

The early enthusiasts for the constructivist approach, like Kronecker and Brouwer, proposed rebuilding the whole of mathematics constructively, avoiding the use of non-intuitive entities like infinite sets. Not surprisingly, this overly positivist proposal did not meet with great enthusiasm. It would have decimated mathematics (remember that depressing feeling when it was suggested that your high-school essay or scientific paper be completely rewritten?!). Hilbert believed that Brouwer's programme would be a disaster, even if it succeeded. He claimed that after the constructivists had finished with mathematics, 'compared with the immense expanse of modern mathematics, what would the wretched remnant mean, a few isolated results, incomplete and unrelated, that the intuitionists have obtained'.

Brouwer's dogmatic approach actually created quite a rumpus within the world of mathematics. He was one of the editors of the German journal *Mathematische Annalen*, the leading mathematics journal of the day, and declared war on mathematicians who did not accept his constructivist philosophy by rejecting all papers submitted to the journal that employed non-constructive notions like infinite sets or his favourite *bête noire*, the Aristotelian law of the excluded middle, which states that something is either true or false. This created a crisis which the other members of the editorial board resolved by resigning and then re-electing a new editorial board—only Brouwer found that he was not on it! The Dutch government viewed this as an insult to their distinguished countryman, and responded by creating a rival journal with Brouwer as editor.

In practice, the intuitionists did not regard all mathematical statements as either true or false. They stipulated a third category: *undecidable*. This three-fold logic is reminiscent of Scottish courts of law, where the verdict of 'not proven' may be returned, whereas English courts require a verdict of either 'guilty' or 'not guilty'. The 'undecided' state of limbo was the fate of statements that could not be demonstrated true or false in a finite number of constructive logical steps.

The intuitionist does not admit as valid arguments any that begin with statements like 'in the infinite decimal expansion of pi there either exists a run of one hundred consecutive odd numbers or there does not'. Such properties of pi are in the undecidable limbo. This restricted logic also outlaws a classic method of proof called *reductio ad absurdum*, because it no longer follows that the negation of the negation of some statement S implies that S is true. The intuitionists are therefore working with a system of logic that lacks one of the most useful procedures for generating new true statements. They are fighting, as it were, with one hand tied behind their backs. Of course, anything true in this reduced system of logic will also be true in ordinary logic, but not vice versa. The intuitionists have a much more demanding criterion of mathematical truth.

Few mathematicians are intuitionists, but recently there has been a revival of interest in this approach. Like formalism, intuitionism is decidedly non-Platonic. It regards mathematics as invented, not discovered. Moreover, it is constructed by *human* manipulations from a set of basic, intuitively obvious notions—but intuitively obvious to *us*. Whereas the formalists were unperturbed by the presence of non-intuitive concepts like actual infinities in their logical procedures, the intuitionists excluded them by fiat, but both philosophies reject as meaningless the idea of mathematics without mathematicians. One obvious drawback of the intuitionist programme is that the scope of the subject is not well defined. There is no precise definition of what the constructive methods are. We do not know if someone will come along tomorrow and construct results that you believed to be unconstructable by finite intuitive steps. In fact, not long ago some results involving properties of infinite sets proved first by Cantor, and which provoked the original intuitionist revolt, were added to the body of results provable by finite constructive steps.

There exists an indirect connection between the intuitionists' logic and the search for laws of Nature. In Chapter 3 we discussed the dilemma of quantum reality and the issue of whether a neutron 'really' went through one slit or the other, or whether curiosity killed Schrödinger's Cat. The resolution of this logical impasse was the adoption of a radically new picture of reality. An alternative approach to the resolution of the quantum measurement puzzle was the adoption of the three-valued logic of the intuitionists, but not the constructivist methodology that motivated it. In this context it is referred to as 'quantum logic'. Thus, it is no longer necessary to conclude that the neutron went through one slit *or* the other. Schrödinger's Cat is not necessarily dead if it is not alive, and Bell's theorem can no longer be proved. There now exists a further intermediate logical state. The adoption of this quantum logic can provide an 'explanation' of sorts for the world of quantum strangeness, but only at the expense of giving up the logic that applies to everything else. Most physicists regard this as an unacceptable schizophrenia. After all, one has to use ordinary logic to argue for the application of quantum logic.

There is no way of deciding which of these approaches towards mathematics is right or wrong. There is suggestive evidence for some and against others. The best that one can do is to recognize how the adoption of one view or another influences conclusions one might draw about the structure of the Universe. We should also appreciate that the intuitionist position is of a slightly different nature to the others. It is not solely a view of what mathematics is, but an attempt to confine it in a rigid operational fashion. The required definition appears to be quite restrictive, and results in a subject considerably smaller than conventional mathematics. To my knowledge no one has attempted to indicate what would be left of the mathematical sciences if only the intuitionists' corpus of mathematics were regarded as true. The three non-Platonic stances must each come to terms with the effectiveness of mathematics in describing Nature. Why is it that abstract mathematical concepts, devised and explored in the distant past with no apparent application, so often turn out to be key elements in the description of some new area of discovery in physics? The following imaginary dialogue between a Platonist and a mathematician who maintains that mathematics is a human invention covers some of the pros and cons of their respective positions.

*   *   *

Whenever you tell me that mathematics is just a human invention like the game of chess I would like to believe you. It would make things so much less mysterious. But I keep returning to the same problem. Why does the mathematics we have discovered in the past so often turn out to describe the workings of the Universe? Surely this cannot be an accident?

It certainly isn't an accident. When we set out to describe Nature we have to use the only tools that are available to us. For all we know there may be a better language for that purpose than mathematics. We derived most of our mathematics from Nature in the first place, so it would be rather surprising if we couldn't then describe Nature with it.

What about Riemannian geometry? That was developed as a branch of pure mathematics by Riemann and others long before Einstein found that it describes the structure of space–time.

That's an unfortunate example. You see, Riemann's study of the geometrical properties of curved surfaces arose from his interest in a very practical problem: the distortion of sheets of metal when they are heated. The effect is not dissimilar to the distortion of space–time geometry by mass and energy according to Einstein's theory. In fact, some physicists have even used the heated metal sheet as a heuristic to explain Einstein's theory to the general public.

What about group structures and symmetries? The interactions of elementary particles, indeed the *existence* of particular elementary particles, appear to be dictated by mathematical symmetries, and these are described by groups. All these groups were discovered and classified by pure mathematicians more than a hundred years ago, oblivious of modern physics. These properties are exact as well; they predict that there exists a certain number of particles of a certain type and no more. It's not a case of the mathematics being a pretty good approximation. It's either right or wrong, and experience shows us that it's usually right. One might also question your assumption that if some aspect of mathematics is derived from the natural world then this makes it culturally or humanly derived. Quite the opposite, I would have thought. It just seems to reinforce the view that the world is intrinsically mathematical.

I think the world just is. I can't see why I should go any further and say it is mathematical. And I don't think there is anything especially mathematical about symmetry. In fact, mathematicians and physicists were among the last to latch on to the importance of symmetry. Architects and artists had identified and appreciated it long before them. Mathematicians derived it from Nature. Mathematicians' interest in group theory is probably just a sophisticated version of the human attraction to symmetry and pattern, an attraction that was acquired from Nature as a result of natural selection in the first place. I'll give you a good example. The draughtsman Maurits Escher produced designs [Figure 5.1] with no knowledge of mathematics at all, but subsequently mathematicians showed that his pictures contain deep mathematical symmetries and constructions of a complicated sort. The whole Universe could have been fashioned in this way, with no mathematics built into it, only aesthetics. Then along we come with our mathematics and proclaim the Universe to be intrinsically mathematical. You think God is a mathematician, but on the basis of the same sort of evidence you would have claimed Escher was a mathematician: but you would have been wrong. I might also mention that your implication that the usefulness and applicability of mathematics in the real world supports the Platonic view is a bit of a cheat, because your view can't offer any explanation as to why that numinous collection of abstract mathematical objects should have anything at all to do with our everyday physical world. There are all sorts of abstract Platonic entities, like unicorns and centaurs, that are not useful, and which don't seem to exist in the real world: why are the abstract mathematical ones not also of this useless non-existing variety? Lacking an answer to this I think you are begging the question. And anyway, even if I could offer no explanation at all for the usefulness of mathematics it wouldn't strengthen your position one iota.

Figure 5.1   Maurits Escher's 1958 woodcut entitled *Circle Limit I*. This is a conformal representation of the so-called Lobachevskii plane of constant negative curvature. These designs grew out of the interaction between Escher and the Canadian geometer H. S. M. Coxeter but Escher's draughting revealed new symmetries which mathematicians did not discover until some years later.

But how can you defend the idea that mathematics is culturally derived? This seems absurd. Different mathematicians in different cultures living in different circumstances have invariably come up with the same mathematics. Pythagoras' theorem says the same thing whether found by the Greeks, or the Indians, or whoever. It contains a core of truth that transcends individual biases. The whole reason for adopting mathematics as the language of science was, after all, precisely because it was culturally independent and lacking any subjective element. I agree there are fashions in mathematics, national styles even, but they simply dictate the direction in which investigations move or the style in which results are presented; they don't determine what mathematical results will be found true or false. Five is a prime number whether you like it or not!

There must be some contribution by our minds. It's unavoidable I would have thought.

But it's a harmless simplification to ignore this. If you don't then you end up studying an image of a reality which is unknowable by our minds, so why not treat the image as the only reality in the first place? It's the only reality that is relevant.

Then everything is subjective you are saying?

No, I just want to assume our cognition does not distort what is really out there. You don't. That's the crux of the matter.

I think we shall find there is a little more of a distinction between us than that.

If, as you say, we invent mathematics rather than discover it, why do we find it so hard to understand? Why do mathematical objects like sets and groups seem so meaningful to us, if they are just convenient descriptions of some patterns that turn up in our minds—so meaningful, in fact, that we are led to call some of their properties 'true'? Your view must face the same awkward problem that confronts the solipsist. If everything is our subjective creation, why is some of it so hard to understand, and why does it all have so little connection with us in other respects?

Chess is a human invention, but that doesn't stop us failing to solve difficult chess puzzles. These things often defeat us because we are not very smart, that's all.

But I still have this strong conviction that mathematics is altogether different to games like chess. They share common elements of course, but mathematics describes the way the world works: chess doesn't.

Suppose we apply our viewpoints to some other symbolic language—like music, for example. Is a Beethoven symphony invented or discovered? No musician would dream of taking the Platonic view to claim that it was discovered. It is an invention of the composer. It embodies aspects of his personality. Why is mathematics any different?

Although music is a symbolic language it differs from mathematics in a significant way: it does not contain an unbending built-in logic. There are rules of composition, but they are rules to be broken. The reason you regard Beethoven's Fifth Symphony as a creation rather than a discovery is because you cannot believe that somebody else would have written it if Beethoven hadn't. But if

Pythagoras had not discovered his theorem somebody else would have. In fact, somebody else did! There are lots of examples of this multiple discovery: there are no examples of multiple creation in the arts. Newton and Leibniz both discovered the calculus. Gauss, Lobachevski, and Bolyai all appear to have had the same ideas about non-Euclidean geometry. There are not two Hamlets! This feature demonstrates the difference between the sciences and the arts. The products of the latter, being almost totally subjective in form and content, are necessarily unique; but because mathematics is the discovery of something that already exists, it can be—and often is—independently duplicated.

Of course, this fact that very different individuals discover the same mathematical results could be put as evidence for the view that mathematics is determined by a universal human trait rather than particular cultural ones, but I must confess that sounds a bit too much like special pleading.

I agree. It has often been claimed that primitive mathematical notions are innate to the human mind. Poincaré regarded geometries and continuous groups of symmetries as concepts that pre-exist in the human mind. The concept of 'number' is probably the most basic intuition we could attribute in this way. It is often claimed by psychologists that children have the abstract concept of, say, the number 'three' before they ever understand the concrete examples of it—like three little pigs or three kings. However, recently archaeologists have made some remarkable discoveries which shed light upon how abstract mathematical notions evolved amongst ancient Sumerian societies. It appears that around 8000 BC the Sumerians had trading tokens which did not recognize the unity of the concept of a number. They distinguished two sheep from two goats, but they resisted the addition of different types of thing. Lists of their numbers would always include the description of what the numbered things were. However, by 3100 BC they had evolved the concept of number independent of the objects being enumerated. There now existed separate tokens to distinguish numbers of things from the identity of the things being counted.

How can mathematical concepts like 'points', infinitesimally small quantities, or irrational numbers be anything but products of our minds? After all, they do not really exist do they? If these mathematical entities exist 'out there', where are they? It can't be in the space and time of our Universe I'm afraid. You tell me how many dimensions of space there are, and I'll tell you about spatial geometry in twice as many dimensions, even though it doesn't exist.

I'm not claiming that mathematical objects exist in the space and time of our Universe.

You mean these concepts have an existence independent of particular examples of them?

Yes, but I don't know 'where' they exist. All I want to argue is that the Universe is intrinsically mathematical.

Just now you complained at my creating a shadow world of sense impressions of reality one step removed from it. Now you're doing the same. You have created another world stocked with all the mathematical furnishings you require in this one. You can hardly claim that your view that mathematical objects are real is supported by simply inventing another world of real mathematical abstractions.

My other world is unique, and explains why we all see the same mathematical structures, but your other worlds are as many as there are mathematicians, and make it seem distinctly odd that we all detect the same mathematics if it doesn't originate from a common source.

But you don't seem to have an explanation of the effectiveness of mathematics in describing the workings of this world anymore. Your mathematical entities live in some other world, and you have to explain why it just so happens that they describe what goes on in our space and time. And I don't see how you can claim to know anything about your abstract world even if it does exist. You admit that these abstract mathematical entities do not sit in our space and time, but that surely means we cannot know about them. They don't interact with human beings, because our cognition is limited to things in the Universe of space and time.

Well maybe this is just a problem created by our limited picture of what it means to 'know' about something. I'm certainly not claiming that my view is complete yet. It needs a lot of development. And I should add that I am content with the view that the Universe just *is* intrinsically mathematical. I can then avoid all mention of the 'other world' of abstract entities. It might be that mathematics is identical to what you would call logic. Although we think it mysterious that the world is well described by mathematics, I don't hear many people expressing surprise that the world is described and governed by logic.

Maybe intelligent beings cannot exist unless the world is governed by logic and mathematics?

Ah! You've been reading that damned fat square book by Bipler and Tarrow, or whatever their names are. Do you really believe that there are other worlds which are not describable by mathematics? Even the Many Worlds of Everett obey the mathematics of quantum mechanics.

It's not impossible. But I don't think it is going to settle our disagreement either way. It would be enough to know why mathematics describes the structure of any world.

There is one difficulty for us both that I would like to mention. We have both been blithely discussing 'mathematics' as if one word will do to describe the entire thing—whatever it is. Although I believe that Nature is invariably mathematical, I don't think I could claim that *all* mathematics is used in Nature. Perhaps, being British, we must seek a compromise, and maintain that some mathematics is discovered whilst the rest is invented and derived from the former in some way?

That's an interesting idea. It doesn't affect my argument of course, but I notice that it partially undermines yours—and it gives you the awkward problem of deciding where to draw the line between the two different sorts of mathematics.

Well you would say that wouldn't you? Actually, I think it strengthens my position.

You mean because it allows you to shift all the problematic examples which undermine your case into the non-Platonic category? I see a more subtle problem here. Take classical mechanics for example. There are two ways of using mathematics to determine the trajectories of particles moving under the influence of a force, gravity say: either we use differential equations, and determine the future state in terms of the present one in a causal way, or we can use a variational principle which determines the actual trajectory taken by the particle to be that path which minimizes a certain quantity connecting the initial and final state. In this second approach the future partially determines the past. Philosophically, there is a world of difference between these two ideas, but mathematically they turn out to be completely equivalent. So the Platonist has the dilemma of deciding which of the two descriptions is the 'real' one.

Or which is the right one. They may not turn out to be fully equivalent. Incidentally, why do mathematicians regard mathematics as both 'beautiful' and useful?

We are attracted to certain mathematical structures because they are elegant or 'beautiful'. By that mathematicians and scientists mean that they exhibit a deep underlying unity in the face of superficial diversity. They grow into the largest and most intricate logical structures from the simplest beginnings, and exhibit unexpected relationships with other branches of mathematics despite their apparent dissimilarity with them. We might imagine that it is structures like these that are most likely to find application in the real world simply because their applicability requires a minimum of special circumstances to hold.

*   *   *

This little exchange shows the problem we have in determining both the meaning of mathematics and the reason for its special effectiveness in science. There is something emotionally attractive about the Platonic stance. It offers such an enticingly simple explanation for everything. But the arguments cited against it are very persuasive. The final point of the dialogue about different types of mathematics is probably a vital one. We could think of mathematics in the way we think of a language like English. It originates as an expedient way to communicate. It is useful. Sometimes it is even necessary to invent new words. Applied mathematics begins like this. But then along come the grammarians who want to put the whole morass of practice on a firm and logical footing, building their house from the roof downwards. They partially succeed. Next there arise writers and poets who love the language itself. They do not only want to use it for mundane practical purposes. They are attracted by its inherent rhythm and rhyme, and the scope of its grammatical structure for succinct expression. They are conscious of style and form, of the difference between poetry and prose. They are like the pure mathematicians, who pursue mathematics for its own intrinsic structure. They mould the rugged corner-stone into a beautiful sculpture. If this mathematics turns out to be useful so much the better.

Yet this sociological analysis still provides no explanation for the effectiveness of the corner-stones in supporting the scientific description of Nature. The issue is clouded yet further by the fact that there can exist several different mathematical solutions to a physical problem, some of which cannot apply to reality. Thus, some additional correspondence principle must be used to judge whether a piece of mathematics applies to the real world. A good illustration of this ambiguity is provided by the 'Coconut Puzzle' which emerged as a challenge problem in pre-war Cambridge. It can be stated as follows:

> Five men find themselves shipwrecked on an island, with nothing edible in sight but coconuts, plenty of these, and a monkey. They agree to split the coconuts into five equal integer lots, any remainder going to the monkey.
> Man 1 suddenly feels hungry in the middle of the night, and decides to take his share of coconuts at that very moment. He finds the remainder to be one after division by five, so he gives this remaining coconut to the monkey and takes his fifth of the rest, lumping the coconuts that remain back together. A while later, Man 2 wakes up hungry too, and does exactly the same—takes a fifth of the coconuts, gives the monkey the remainder, which is again one, and leaves the rest behind. So do men 3, 4, and 5. In the morning they all get up, and no one mentions anything about his coconut-affair the previous night. So they share the remaining lot in five equal parts finding, once again, a remainder of one left for the monkey. Find the initial number of coconuts.

There are in fact an infinite number of solutions to this problem, but we would like the smallest whole number of coconuts. It is 15621. However, soon after this problem was posed Paul Dirac gave another solution: -4 coconuts! This is clearly a solution. Each time a man arrives at the coconut store he finds -4 coconuts, gives $+1$ to the monkey leaving -5. His one-fifth share of -5 is -1, which he takes, leaving -4 behind either for the next man or the final share-out!

Here we see an example of a perfectly good *mathematical* solution which (since we cannot realize a 'negative coconut') does not have a realist out-working (however, there is a tradition that Dirac's negative solution played some role in his thinking which led to the introduction of the concept of antimatter). In this situation it is easy to exercise a criterion which eliminates Dirac's solution as an unrealistic one, but in more esoteric areas of mathematical physics it may not be so easy to decide upon criteria by which to reject some mathematical predictions as unrealistic.

A shock for the formalists

> God exists since mathematics is consistent, and the Devil exists since we cannot prove it.
> André Weil

By the 1920s the intuitionists' view of mathematics was treated as a heresy popular with a few misguided zealots. Platonism remained the neutral attitude taken unconsciously by scientists, although rarely defended by mathematicians themselves. The dominant contemporary belief was in the formalists' dogma that mathematics is about collections of self-consistent axioms and rules for deriving new statements from those axioms. These statements will be tautologically true but meaningless, in the sense that they are not about anything that really exists. The only necessary property of a collection of axioms is that they be consistent; they do not have to correspond to anything in the physical world. The statements legitimately derived from the axioms using the rules of the system are called 'theorems'. This all seems very satisfactory, and reminiscent of the philosophy behind the presentation of many pieces of elementary mathematics, notably that of Euclidean geometry. When studying geometry one can use the real world or diagrams as a guide to whether your logical steps are likely to be correct. When the axioms of the system do not correspond to familar entities we must be more careful about what we apply our logic to.

First, we need to distinguish between statements *of* mathematics and statements *about* mathematics. Consider a simple example using the English language. The sentence 'my car is dirty' is a sentence in the language. If we were to remove the word 'car' from the page, and arrange the remaining words of the sentence around my actual car, the sentence would still mean the same thing. But the sentence 'car has three letters' is entirely different. It is a sentence about the language. The word 'car' is only a substitute for the phrase 'the word for car'. If we remove the word 'car' from the page, and replace it with a real car, then the second sentence certainly does not retain its meaning. The second sentence differs from the first because it is a statement about the language, as well as a statement in the language. Statements about a language we call *meta-statements*. Thus, meta-mathematical statements are statements about mathematics. For example 'Pythagoras' theorem is true' is a meta-mathematical statement, whereas $2 + 2 = 4$ is a mathematical statement. Although the mathematics of the formalists deals only with symbols and formulae that have no independent meaning, it is meta-mathematics that allows us to make meaningful statements about these meaningless marks.

In 1931 Hilbert and his formalist programme received a dramatic and devastating blow. He and other leading mathematicians had been steadily producing small collections of proofs that appeared to advance them inexorably towards the formalists' goal of capturing all of mathematics in their logical web. A spanner then entered the works. Kurt Gödel, an unknown young mathematician at the University of Vienna, produced a completely unexpected result: that Hilbert's goal was unattainable. No axiomatic system big enough to contain arithmetic can ever be proved consistent. The best that we can achieve is the knowledge that a system is *inconsistent*. This is the curious sort of thing that only mathematicians can do: prove that it is not possible to prove something! Furthermore, no such axiomatic system can be complete: there must exist mathematical statements expressed in the symbols of the system that can neither be proved true nor false using the rules of the system.

Suppose that we have some mathematical or logical system composed of axioms and a set of rules for deducing new statements from those axioms. If we write down some formula, call it F, in terms of the symbols defined in the system, then one of four things can be true of F:

(1) F can be proved true in the system.
(2) F can be proved false in the system.
(3) F can be proved both true and false in the system.
(4) F can be proved neither true nor false in the system.

The options (1) and (2) are obvious possibilities. The result (3) would show our logical system to be inconsistent: if (3) holds, then that system is meaningless because it can be used to show that *any* statement made in the language of the system is true. The possibility (4), that the system is *incomplete*, had not been contemplated by the formalists. This is the situation that Gödel established could occur in mathematical systems. The formalists would be unable to demonstrate the consistency of mathematics. Adding new axioms to an incomplete system never cures the problem. While this may allow previously undecidable statements to be decided (just add them as new axioms for example), it always generates some new undecidable propositions.

The demonstration of the incompleteness of mathematics by Gödel seems to be a totally pessimistic and negative result. Since all our science is built upon mathematical systems, incompleteness must infect these disciplines as well. But, as with the problem of growing old, the alternative is worse still. Suppose we construct a super-computer that can carry out logical operations if we feed it with the axioms and symbolic language of our system together with its rules of deduction. We turn it on and watch it print out all the deductions it can make. Suppose that, after a while, it prints out that 'F is true' and then a little later prints out 'F is not true' (this is the option (3) referred to above). We will discover that our super-computer is just printing out every conceivable statement that can be made using the symbolic language it has been given. Every permutation of symbols and relations will appear, if one waits long enough. What has gone wrong is that the system has been found inconsistent, and the inconsistency allows anything to be proved true within that system. An amusing example of this degeneracy is provided by the story of Bertrand Russell being asked by McTaggart to show 'If twice 2 is 5, how can you show that I am the Pope?' Russell replied at once: 'If twice 2 is 5, then 4 is 5, subtract 3; then $1 = 2$. But McTaggart and the Pope are 2; therefore McTaggart and the Pope are one!' Or, alternatively, find the flaw in the following 'proof' that every statement is consistent: every statement must be either consistent or inconsistent; therefore either they all are or at least one is not. In the latter case the system must be consistent with any statement at all, including the one that it is consistent!

At first sight, Gödel's result might appear to allow us to deduce inconsistent statements in our mathematical system unwittingly. But this is not the case; option (4) is quite distinct from option (3). Gödel's theorem actually protects us from inconsistency, for if we can find just one statement in the language of our system that cannot be proved, then this guarantees that it cannot contain inconsistencies like (3) whose presence would render *all* statements true. Gödel's demonstration of the inevitable incompleteness of mathematics should be seen as a demonstration of its infinite scope and content which cannot be captured simply by the axioms and rules of logic which define it. The mathematical whole is considerably more than the sum of its parts.

One would normally define a 'religion' as a system of ideas that contains statements that cannot be logically or observationally demonstrated. Rather, it rests either wholly or partially upon some articles of faith. Such a definition has the amusing consequence of including all the sciences and systems of thought that we know; Gödel's theorem not only demonstrates that mathematics is a religion, but shows that mathematics is the only religion that can prove itself to be one!

Consequences for physical science

> As far as the laws of mathematics refer to reality they are not certain; and as far as they are certain, they do not refer to reality.
>
> Albert Einstein

In order to prove his remarkable theorem, Gödel exploited the distinction between mathematics and meta-mathematical statements about mathematics which had been introduced by Hilbert. This enabled him to construct explicitly an all-purpose undecidable statement. Each symbol and logical operation of his mathematical system was associated with a prime number (that is, a number like 5 or 17 that is divisible only by itself or by 1). Any statement of the system can then be represented by the product of the prime numbers associated with each of its constituent elements. The motivation behind this step is that any whole number can be decomposed into a product of its prime divisors in one and only one way (for example $2 \times 3 \times 11 = 66$ is the only way to factorize 66 into prime numbers). This means that any given number corresponds uniquely to a particular logical statement composed of the string of symbols corresponding to its prime divisors. Every theorem simply corresponds to a number. This number we call its Gödel number. Hence, each meta-mathematical statement has a unique Gödel number, and so there exists a complete correspondence between arithmetic itself and statements about arithmetic. Gödel then constructed the statement:

The theorem possessing Gödel number X is undecidable.

Finally, he found the Gödel number of this statement, and used it as the value for

X. The resulting statement with this value of X substituted into it is then a theorem which states its own undecidability. This type of argument is good enough to show that undecidability is a disease of all systems that contain arithmetic, since the latter was used to set up the correspondence with meta-mathematical statements via the prime-number decomposition. This is all quite strange for the conceptualist: this construction of our own minds called mathematics has managed to keep a part of itself forever beyond our grasp.

After seeing the way Gödel's theorem is established one cannot help becoming a little sceptical about its real relevance to science. It appears to emanate from the formulation of a rather artificial linguistic paradox. Far more impressive would be the demonstration that some great unsolved problem of mathematics which has tortured mathematicians for centuries *is* actually undecidable, or perhaps, that some very practical mathematical problem, like 'what is the optimal economic strategy?' is logically irresolvable. In 1982 some 'natural' examples of undecidable mathematical statements were discovered in the course of trying to solve a real problem—they are not artificial concoctions.

Suppose we call a set of numbers 'large' if it contains at least as many members as the smallest number it contains. Otherwise we shall call it 'small'. So, for example, the set of numbers (3, 6, 9, 46, 78) is large, but the set (21, 23, 45, 100) is small (because it contains fewer than 21 members). Now it can be shown that if you take a *big enough* collection of numbers, and give each pair of them a label, either black or blue, say, then you can always find a 'large' set within the collection such that the pairs in this set are either all black or all blue. This is not totally surprising, but what is, is the fact that the question 'how big is "big enough"?' cannot be answered using arithmetic: it is undecidable. Several other examples of undecidable questions of a similar type are now known; they are natural in the sense that they arose in the course of trying to solve other mathematical problems.

Another very interesting aspect of Gödel's theorem is its connection with the idea of randomicity. Superficially this connection is rather surprising, yet it turns out to be rather deep. Not only does it transpire that the question of whether a sequence of numbers is random or not is logically undecidable, but posing this question in the right way leads to an illuminating proof of Gödel's theorem which sheds light on the limitations of axiomatic systems.

Suppose we are presented with two lists of numbers, the first entries of which are:

$\{3, 56, 6, 23, 78, \ldots\}$ and $\{2, 4, 6, 8, 10, \ldots\}$

How can we gauge the extent to which these sequences are random? First, for convenience sake, let us convert the numbers into the binary arithmetic read by computers, so the new sequences correspond to lists of zeros and ones. Now ask what is the length of the shortest computer program that can generate each sequence. The length, in computer bits, of this shortest program is called the *complexity* of the sequence. If a sequence is haphazard and contains no special rule for generating one entry from another (as in our first example), then the shortest program can be nothing less than the sequence itself. But, if the sequence is ordered then the required program can be much briefer than the sequence. In the second of our examples the program will just list the even numbers: {PRINT 2N, N = 1, 2, 3, . . .}.

We shall define a sequence to be *random* if its complexity equals the length of the sequence itself. In this case it requires the maximum of information to specify it. So, given any two random sequences of different length, the longer sequence is regarded as the more complex. If you pick a large number of sequences of numbers, say telephone numbers, you will find that most have rather high complexity, and it is quite rare to come across strings of numbers that have low complexity.

Using this notion of complexity, consider giving a computer, whose programs include all the symbols and operations of arithmetic, the following instruction:

> Print out a sequence whose complexity can be proved to exceed that of this program.

The computer cannot respond. Any sequence it generates must by definition have a complexity less than that of itself. A computer can only produce a numerical sequence that is less complex than its own program. We are now able to exploit this quandary to show that there must exist undecidable statements. Simply pick a particular sequence—call it R—whose complexity exceeds that of the computer system: the question

> Is R a random sequence?

is undecidable for the computer system. The complexities of the statements 'R is random' and 'R is not random' are both too great for them to be translated by the computer system. Neither can be proved nor disproved. Gödel's theorem is proved.

The inevitable undecidability of certain statements that this example demonstrates arises because the logical system of the computer, based upon arithmetic, has too small a complexity to cope with the spectrum of statements that can be composed using its alphabet. There is, as a consequence, no way in which one can decide whether the computer program one is using to perform a particular task is the shortest one that will do it.

A computer with a very simple range of logical operations is easily checkmated by being asked to deliver a response of too great a complexity. If its display consists of a single light-bulb that it illuminates to signal 'no' and extinguishes to signal 'yes', then it cannot answer the simple question 'Is your light-bulb illuminated?'

This result poses restrictions upon the scope of any approach to the laws of Nature on the basis of simplicity alone. The scientific analogue of the formalist methodology in mathematics is the idea that, given any sequence of observations in Nature, we try to describe them by some mathematical law. There may be all sorts of possible laws which will actually generate the data sequence, but some will be highly contrived and unnatural. Scientists like to take the law with the lowest complexity in the sense described above. That is the most succinct coding of the information into an algorithm. Sometimes such a prejudice is called 'Occam's Razor' after the medieval philosopher William of Occam, who established the principle that 'entities should not be unnecessarily multiplied'.* We can see that this approach will never allow us to prove that a particular law we have formulated is a complete description of Nature. There will always exist undecidable statements that can be framed in its language. It can never be proved the most economical coding of the facts.

What is truth?

> It is a terrible thing for a man to find out suddenly that all his life
> he has been speaking nothing but the truth.   Oscar Wilde

We have seen that provability is a much straiter gate than truth. In fact, undecidable formal propositions can be generated from linguistic paradoxes. Dozens of these conundrums are known, and some played an important role in motivating careful thinking about the relationship between ordinary language, mathematics, and logic. The most famous example is probably Bertrand Russell's 'Barber Paradox':

> A barber shaves all those individuals who do not shave themselves. Who shaves the barber?

A logical contradiction arises regardless of whether he shaves himself or not.† There can exist no such barber. Another ancient example is the Paradox of Epimenides stated by St Paul in his New Testament Letter to Titus:

> All Cretans are liars. One of their own poets has said so.

Paradoxes like that of Epimenides are resolved by using a definition of truth introduced by the Polish mathematician Alfred Tarski. It makes good use of the idea of a meta-language that we introduced in our discussion of Gödel's theorem. Given some language in which we can write statements, its *meta-language* will be the collection of statements *about* that language (rather than *in* it). For example, if

---

*This is taken to mean that things should be made as simple as possible, but not more so. This dictum has a worrying element of subjectivity. How can we expect to get objective truth from a methodology that relies upon *our* opinion of simplicity?

†We assume that the barber is neither bearded nor a woman.

we talk in French about sentences in English, then French is being used as a meta-language for English. Any meta-language will, in turn, have its own meta-languages; for example, someone might be talking in Spanish about us talking in French about the English language. This hierarchy can continue indefinitely. There exists an infinite hierarchy of logical languages, the subject of each being the one below it.

Logical statements in a particular language cannot be called true or false without stepping outside that language to employ a meta-language. If we want to state that a particular proposition about the real world is true, then we have to make that statement in a meta-language. By this device Tarski proposed an unambiguous way of determining what we mean by saying that a statement is 'true'. For example, take the sentence 'GRASS IS GREEN'; Tarski suggests that 'GRASS IS GREEN' is true if in fact grass is green. That is, the capitalized sentence about the grass is true if and only if grass can actually be demonstrated to be green by replacing the word GRASS in the sentence with a turf of actual grass without altering the meaning of the sentence. So we can talk about the capitalized sentence, decide whether or not it is true, compare it with the evidence; but the statement in capitals alone has no meaning until we do this in the uncapitalized meta-language. This circumambulation pays off in that it removes the linguistic paradoxes. To say 'this statement is false' is to introduce a confusion between the language and its meta-language—we are talking *about* statements in some language, but not making them *in* that language. The Epimenides Paradox makes the same illegal move: it mixes two languages. Without this distinction between different languages logic collapses in inconsistency because one can show that *any* statement would be true, including its negation. Suppose you want to 'prove' that the statement S is true—where S can be *any* statement you like: 'Pigs can fly', 'God exists', 'God doesn't exist', 'Politicians tell the truth', 'Arsenal can win the FA Cup' . . .

Simply consider the statement:

Either this whole sentence is false or S is true.

The whole sentence must either be true or false. If it is false, we see that S is true, and we are done. If, on the other hand, the entire sentence is true, then one of the clauses 'this whole sentence is false' or 'S is true' must be true. Since we are now considering the case where the sentence is assumed true, the first of these clauses cannot be true. Therefore the second clause must be true, and we have shown that the statement

S is true.

is true whatever we choose S to be! The resolution of this crisis is provided by Tarski's remedy: our original sentence is not admissible. It mixes statements in a particular language with statements about that language which belong to its meta-language.

Tarski's definition of a true statement is particularly nice because it applies equally well to logical or mathematical statements written in abstract symbols and to ordinary language. The only thing that differs will be the ways we have available to show that a statement in a lower-order language is in fact true.

Computability

> Computation—whether by man or by machine—is a physical activity. If we want to compute more, faster, better, more efficiently, and more intelligently, we will have to learn more about nature. In a sense nature has been continually computing the 'next state' of the universe for billions of years; all we have to do—and, actually, all we *can* do—is 'hitch a ride' on this huge ongoing computation, and try to discover which parts of it happen to go near to where we want.   Tomaso Toffoli

Way back in 1900, in the age of innocence before the revolution in mathematics and logic brought about by Kurt Gödel, David Hilbert delivered a famous address to the International Congress of Mathematicians when they met in Paris, in which he posed what he considered to be the twenty-three most challenging unsolved problems of mathematics. Several of the conundrums that he set during that address have influenced the development of whole areas of mathematics. Many of Hilbert's problems remain unsolved today. Hilbert's last problem was envisaged as a step towards the completion of the formalist programme for mathematics. It challenged mathematicians to find a systematic method to determine whether or not any mathematical statement is true or false. At that time Hilbert assumed that such a procedure must exist. It was only a question of finding it. One can see that if such a method were found, then it would also serve to define the scope and content of mathematics for the constructivists. However, Hilbert's view that such an automatic method of decision does exist was not universally shared. Some years later, G. H. Hardy expressed the view that

> There is of course no such theorem, and this is very fortunate, since if there were we should have a mechanical set of rules for the solution of all mathematical problems, and our activities as mathematicians would come to an end.

However, Hardy did not seem to base his scepticism upon a belief that such an all-deciding theorem was unattainable in principle, but merely upon the belief that it would be beyond the wit of mortal man to find and use it. Hence, Gödel's discovery that there can exist no method for deciding the truth of all statements was a great shock to mathematicians. Nevertheless, despite this irreducible undecidability at the heart of mathematics, there still remained a remnant of Hilbert's original dream which might be salvaged. Even though some statements made in the language of an axiomatic system must be undecidable, there might still exist a systematic method for finding all the decidable statements and distinguishing the true from the false ones. If so, one might be able to determine the relative degree of incompleteness arising from different sets of axioms.

It was soon demonstrated by Alonzo Church and Emil Post at Princeton, and by Alan Turing at Cambridge, that even these more moderate goals are unattainable in principle. Church found an abstract method for determining whether a particular mathematical operation can be computed for every one of the numbers to which it is applied. He was then able to show that it is possible to construct mathematical formulae which cannot be evaluated or determined true or false in a finite number of steps. *Any* computer would keep running forever in a vain attempt to complete the calculation. For these examples there exists no method for determining whether or not they can be proved. There exist unsolvable problems.

Working alone in Cambridge, Alan Turing established the same conclusion as Church, but by a method that was to have a host of wider implications for the future invention and development of computers. Unlike Church, Turing developed his disproof of Hilbert's conjecture around the conception of a hypothetical *machine* which would decide the truth of statements by a set of well-defined sequential operations. Subsequently, these paradigms have become known as Turing machines. During World War II Turing made important contributions to Allied cryptography by pioneering the construction of real mechanical devices which would try vast numbers of combinatorial alternatives in the search for a correct decoding of intercepted German communications.

A Turing machine is the essence of any computer. It consists of a memory tape of indefinite length, and a processor which manifests the current state of the machine. That current state is determined by a combination of the previous state and the last instruction which told it how to change. A real computer will possess all manner of fancy accoutrements—monitors, graphics, software, keyboards, and so forth—but these play no role in the logical capability of the machine. They are aids to its use. No real computer possesses greater problem-solving capability than that of Turing's idealized machine.

All a Turing machine can do is take a list of natural numbers and transform it into another list of natural numbers. The operation of the machine just matches a number in the first list with one in the second list. If the operation of interest is multiplication by two, then all the numbers in the second list will be double those in the first. Unfortunately, there exist operations which are infinitely more complicated because there exist infinite sets which are infinitely bigger than the infinite set of natural numbers (for example all those numbers, called *irrational* numbers, like 'pi' and the square root of two, which cannot be expressed exactly as vulgar fractions—one natural number divided by another). They cannot be systematically paired with the natural numbers. Such infinite sets are called *uncountably* infinite.

Turing concurred with Church by showing that there exist mathematical problems which cannot be decided in a finite number of logical computations by one of his idealized machines. Such unperformable operations are called non-computable functions, and their existence is a consequence of the existence of uncountably infinite sets. As an example, suppose the function $G(N)$ is applied to the natural numbers $N = 1, 2, 3, \ldots$ and chosen either to be equal to the value of the Nth computable function plus one, or to zero if the Nth computer program fails to stop with an answer in a finite number of steps after N is inputted. We see that G cannot be a computable function, because if the Nth input program did compute it then we would have the impossible result that

$$G(N) = G(N) + 1.$$

Another set of non-computable examples are the so-called 'Busy Beaver' functions $B(N)$, defined as the largest number produced in the output of any program less than N bits in length. The Busy Beaver function grows in size faster than any function that could possibly be computed. Hence, it is not computable. In practice, it is the hardest mathematical problems that are most easily shown to be unsolvable, because they can most easily defeat the computer.

Although it cannot compute everything that is fed to it, most computer scientists believe that a Turing machine is capable of computing anything that can be solved in a finite time by any physical sequence of realizable operations: that is, a problem is solvable if it is solvable by a Turing machine. This is called the Church–Turing hypothesis. Until recently this hypothesis was regarded as telling us something about mathematics, of the sort that all the possible ways that one could dream up of computing things are at root equivalent because they can all be reduced to the capability of a Turing machine. Yet, it is clearly telling us something deeper about the structure of the physical world, and the fact that we find mathematics so beautifully adapted to its workings.

David Deutsch, an Oxford physicist, has recently argued that the defining feature of the computable functions which the Church–Turing hypothesis maintains can be computed by a Turing machine is that they can be realized in Nature. If we had two 'black boxes', one containing some real physical process and the other an idealized Turing machine, then identical outputs are possible from the two boxes for the same inputs. We could not tell from the result alone in which box it had been computed. This idea draws the link between mathematics and the laws of Nature very tight. The laws of Nature happen to allow us to build exact physical models which execute the arithmetical operations of addition, subtraction, division, and multiplication. There is no known reason why Nature should simulate the rules of arithmetic; no reason why the laws of Nature should be constrained by, or linked to, the computational ability of mathematical algorithms. It is fortunate that Nature does run in this simple fashion. Were it not so, then no physically realizable electronic device of silicon and metal could compute the operations of addition, subtraction, division, and multiplication.

These simple arithmetical operations would be non-computable functions which we would be unable to perform or employ in constructive proofs. All we could achieve by our mathematics would be non-constructive existence proofs that told us that they existed, in much the same manner that we can deduce the existence of undecidable propositions. For example, if our computers could do no more than carry out the geometrical constructions which we can achieve on paper by the use of a straight edge and a pair of compasses (that is, drawing straight lines and arcs of circles), then we would be unable to trisect an angle computationally, even though the concept of a trisected angle would be familiar to us and the existence of some line which did trisect an angle might be provable non-constructively. The other side of this apparent match between Nature and arithmetical computation is that it is unexpected and unproven, except by experience, and so we should not be surprised, although we certainly would be, if there were found a physical process in Nature which simulated the computation of a function that could not be computed by a Turing machine.

This correspondence between physical realizability and computability seems to require something like the quantum picture of reality to be true. Certainly, the classical world of Newtonian physics does *not* possess the Church–Turing property. Energy does not come in discrete countable quanta in the non-quantum view of the world. The continuum of states that exist for every classical physical system prevents the existence of a one-to-one correspondence between computability by a countably infinite sequence of operations and the laws of Nature. However, Deutsch has argued that it may be that all finite systems in Nature can be simulated by a 'quantum computer'. The prescription for such a hypothetical device can be drawn up in general terms, and it appears to have the potential to carry out computations faster than any Turing machine, although it cannot compute any function that cannot be computed by a Turing machine. Most striking of all, quantum computation appears to require an objective quantum reality to exist. This is only possible in the 'Many Worlds' interpretation of the quantum theory, and offers the hope that there may one day exist an experimental resolution of the problem of quantum ontology.

Inherently difficult problems

> Many physical systems are computationally irreducible, so that
> their own evolution is effectively the most efficient procedure for
> determining their future.
>   Stephen Wolfram

It may be disappointing to learn that there are problems which computers cannot solve, so let us return to those which they can. We would expect there to exist some hierarchy of difficulty associated with the solvable problems, or, what is the same thing, with the computable functions. A simple dichotomy—solvable or unsolvable—computable or uncomputable—is too coarse a classification to be helpful in practice. Furthermore, there may exist functions which are uncomputable when all possible inputs are considered, but perfectly computable for some particular sub-collection of inputs which are of interest in practice.

In real problems we are primarily interested in the length of time it would take to solve a problem. Any problem which takes twenty billion years to compute its solution on a Turing machine would be solvable in principle, but unsolvable for all practical purposes. This pragmatic approach to solvability is the philosophical basis of modern codes. No practical code is unbreakable in principle, but it suffices if it requires billions of years to crack by systematic computation using the other side's fastest computers. Cryptographers have developed a collection of coding operations called 'trapdoor functions' which operate very simply when applied in one direction to transform a message into code, but which would require billions of years to decode them in the other direction. A simple example of such operations which are slow in one direction but fast in the other (hence the name 'trapdoor') is to take a collection of four large prime numbers having, say, a thousand digits in each, and multiply them together. This is a simple operation for a computer. But show a computer the answer, and ask it to find those four prime numbers which divide into it exactly, and the computer is faced with a problem that could take a hundred lifetimes to solve.

A practical measure of the difficulty of a problem is the rate of increase in running-time of the shortest computer program required to solve the problem as the size of the input is increased. Suppose we have a program which must be run on an input consisting of N different numbers. If the fastest program which computes the output runs in a time that increases only as a power of N as we increase the size of N, for example as $N^2 (= N \times N)$, then the problem is said to be solvable in polynomial time. Such problems are labelled *polynomial*, or P, problems. Harder problems, requiring a computational time that increases more rapidly than a power of the input size, say as 2^N or N^N, are designated *non-polynomial*, or NP, problems. In practice, only P problems are regarded as being tractable. In Figure 5.2 we show the increase in execution time for a variety of P and NP problems as the size of their input increases.

It has never been shown that there exists some deeper and more fundamental difference between P and NP problems other than computational running time. The inherently difficult NP problems divide into classes of comparable difficulty. What binds the members of each class together is the feature that, if any problem in one of these classes is solved, then it supplies a universal algorithm to solve every other member of its class. The solution of any NP problem would, therefore, be a gold-mine.

Real mathematical problems whose computational execution time grows faster than a power of N are well known. The fastest known chess programs grow as 2^N, but there may exist unfound algorithms that are polynomial, or even linear, in N. In addition, there exist mathematical problems for which it can be proven

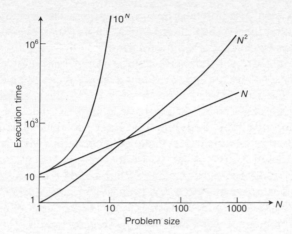

Figure 5.2 The growth of computation time with problem size for different types of problem.

that no polynomial program exists in principle. Such problems are unsolvable in practice.

An application of this practical measure of computational difficulty to the familiar system of arithmetic which we have inherited from the Arabs is instructive. This system of Arabic numerals and arithmetical operations is beautifully compact and efficient compared with other laborious notations like the Roman numeric system (just try doing multiplication using Roman numerals!) If we want to 'compute', either in our brains or on a commercial calculator, the function which converts X into X^2, then if X is written in our system of decimal arithmetic it will possess $N = 1 + \log_{10} X$ digits (for example if $X = 10$ then it has $1 + \log_{10} 10 = 2$ digits). Hence, when we compute X^2 by using the notation of multiplication, the number of computations performed increases with N as N^2, but if, instead, we had to compute it by performing repeated additions, then the number of computational operations needed becomes proportional to X, and so to 10^N. The Arabic system of notation that we employ has the effect of making some NP problems into P problems. This is the intrinsic simplification that we find when we carry out multiplication in the Arabic system. The formalism seems to have built into it some logic about the operations that are conducted with it.

In the study of natural phenomena there arise many problems which are rendered intractable by the length of computer-time that would be required to resolve them. A characteristic of these problems is that there does not exist any simplifying procedure or approximation scheme which might serve to reduce the complexity of direct computational attack. The only way of studying such problems is by explicit simulation or, in effect, by observing them as they occur. In the absence of any short-cuts we can do nothing but watch the evolution of a model of the system programmed into a computer. The running-time of the program will be no shorter than the enactment of the real process—and no more instructive in the absence of other information.

Unfortunately, there still exist other fundamental limitations on the use of mathematical algorithms and equations to describe the workings of Nature. Our discussion so far has been concerned with the logical limitations to the computational process and with the time required for the computation of a mathematical formula by an ideal computer. But we might legitimately ask 'what is an ideal computer?' Computers are made out of materials which are themselves subject to the laws of Nature. What constraints, if any, do laws of Nature place upon the scope and speed of computers? What is the best possible computer?

This is a problem to which Nature has been slowly finding progressively better answers over millions of years of 'experiment'. Human and animal brains are complicated computers, but because they have been honed by the process of natural selection they contain many ingredients which we would not be concerned with if we were setting out to build a super-computer in the laboratory. The human brain contains all manner of special programs which are associated with the 'hardware' of flesh and blood in which it is embodied and on which that program called 'intelligent life' runs; these control the movement of limbs and muscles and other functions which are only indirectly associated with the computational power of the brain. Yet, there are other physical restrictions upon the evolution of our brains and bodies which are so fundamental that they would also play a role in determining the limits of any man-made computer.

In most birds and vertebrate animals, brain size increases systematically with their total body size, L, roughly as

$$\text{Brain size} \propto L^{9/5}$$

These warm-blooded creatures survive by ingesting food. This provides their bodies with a source of energy. But animals are continually losing heat to their surroundings if those surroundings are cooler than their body temperature. The rate at which they can generate heat from food is proportional to their volume, that is to L^3, but the heat-losses to their cooler surroundings will be proportional to the surface area of their bodies, that is to L^2. Thus, we see that small animals are in trouble in the cold: they cannot ingest food at a high enough rate to maintain a constant body temperature. This is why we find polar bears rather than mice in the Arctic, and why the average size of birds increases as one moves from the equator towards the polar regions. Large brains require large animals, and in order for large animals (and humans are the largest two-legged animals) to maintain a balance between the internal heat generation by their beating hearts and their heat losses, their pulse rate, p, must be inversely proportional to their size L because,

$$\text{Heat production rate} \propto \text{pulse rate} \times \text{volume} \propto p \times L^3$$

and,

$$\text{Heat loss rate} \propto \text{surface area} \propto L^2.$$

So, larger animals have stronger, slower hearts, and as we grow from childhood to maturity our pulse rates fall accordingly. What has all this got to do with computers?

A computer processes information at a some rate that might be called its 'pulse rate'. The heat generated in its electronic circuits must be lost by the computer or its temperature will steadily increase, and eventually melt its circuitry. The bigger the computer grows, and the faster it processes information, the harder it is to remove the waste heat. One could try to build a computer that increased its surface area by much more than one would expect when its volume increased. There are many animals that do this: sponges and flat-fish are common examples. The convoluted surface of a sponge has much more surface area exposed than a smooth solid ball of the same diameter. But if one attempted to beat the cooling problem in a computer by this means one could only get so far before the length of circuitry required to run around all the superficial crenellations would become prohibitively long. This would mean that any signal, even one moving at the speed of light, would take too long to travel from one part of the computer to another, and there would be a consequent drop in the speed at which it could carry out computations. These simple considerations indicate that there may exist a limit upon the size and power of 'brains', whether they are based upon carbon chemistry like our own, or upon the silicon chemistry of the micro-electronics industry.

The ultimate limits Nature places upon the capability of computers have yet to be fully worked out. We can foresee some known laws of Nature which must play a role in determining these constraints: the restriction that the transmission of information cannot occur at a speed faster than that of light, for example. But what of the quantum and thermodynamic constraints? For a long time it was believed that the version of Heisenberg's Uncertainty Principle which forbids the simultaneous measurement of a time interval and an energy with an accuracy less than Planck's constant divided by 4π must enforce another fundamental limit upon the ultimate computer. If the time interval concerned was the time for a logic gate or switch to change in the computer, and the energy uncertainty was associated with the minimum energy expended as waste heat in the switching process, then it would appear that the maximum information processing rate of any computer must be limited absolutely by the Uncertainty Principle. However, this is not necessarily so. There is no principle of the quantum theory of Nature that requires the energy-spread in the energy–time Uncertainty Principle to be associated with a quantity of dissipated heat generated in a physical process. The computational process can be carried out with less than one quantum of energy dissipated at each step. An event of duration T can be measured with an expenditure of energy E which is less than T times Planck's constant divided by

4π, so long as the time is measured by some external standard that is not intrinsically defined by the physical system itself. Thus, the Heisenberg Uncertainty Principle does not place any fundamental restrictions upon the computational process in principle. The problem of waste heat we mentioned above boils down, at the most fundamental level, to the question of whether the logical operations of a computer are necessarily irreversible. If so, then in accord with the Second Law of thermodynamics they must generate heat and entropy. Ordinary arithmetic certainly seems to be irreversible. If we consider the sum $2+2=4$ then there is no unique inverse. The final total 4 is generated equally by the sums $1+3$ and $0+4$, and so given the number 4 there is a loss of information or a generation of entropy in arriving at it by the addition of two numbers. This irreversibility is similar to the general principle we explained in Chapter 3 when discussing the relationship between the Second Law of thermodynamics and time-symmetric laws of Nature. There are always more ways of decomposing the total 4 into two numbers than the one total that results from the sum. An asymmetry exists, even though the basic operations of arithmetic are reversible. But perhaps there are other ways of carrying out computations which *are* reversible?

Some years ago Edward Fredkin showed that a logic switch exists whose operation *is* reversible, and hence, in principle, it can process information with no generation of entropy and waste heat. The Fredkin gate is illustrated schematically in Figure 5.3. The practical limitation upon these devices is that one must have a way of getting information in and out of the computer. The processes which couple the computer to the outside world *are* limited by the laws of thermodynamics, and are irreversible.

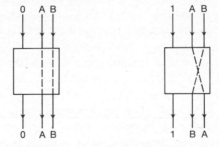

Figure 5.3 The Fredkin gate permits reversible computation. The gate has three inputs and three outputs. The left-hand channel is transmitted unchanged in each case but its binary value (0 or 1) causes the other two inputs either to remain unchanged or to exchange. This computational operation creates no information loss and is its own inverse; hence the operation is reversible.

We have described some of these ideas of modern computer theory to show that there is growing interest in the ultimate limits imposed upon computer capability by the laws of Nature. As yet the fundamental limits upon the capabilities of *quantum* computers have not been examined in any detail. They will undoubtedly be able to carry out tasks that no current computer can approach. These considerations as to the ultimate practical limits of computation would seem to be heading towards the calculation of some new fundamental constants of Nature. There would appear to exist a hardest problem that could be computed in the time since the beginning of the Universe until the present. If the Universe is finite in size, and will exist only for a finite period of cosmic time, there may exist problems which are computable in principle (in some Platonic world) but not in our Universe.

The dilemma of ignorance

> The greatest contribution of medievalism to the formation of the scientific movement [was] the inexpungable belief that every detailed occurrence could be correlated with its antecedents in a perfectly definite manner.
>
> A. N. Whitehead

The subtleties of mathematics and the ultimate limits of computational processes reveal mysterious connections between the inner world of mathematical logic and the physical processes that comprise the Universe. We have found there to exist limits both of principle and of practice to the scope of mathematics as a description of the laws of Nature. In all of these deliberations we have been considering the non-quantum world to be an idealization in one vital respect: we have assumed that whatever is knowable in principle can be known with perfect accuracy. This is clearly a false assumption, but is it a harmless simplification, or the thin end of a dangerous wedge that will upset the mechanical world-view that still conditions our approach to the non-quantum world? What are the consequences of imperfect knowledge of the laws of Nature? This is the problem to which we shall now turn our attention.

Maxwell on determinism

> In the fairy tales the cosmos goes mad, but the hero does not go mad. In the modern novels the hero is mad before the book begins, and suffers from the harsh steadiness and cruel sanity of the cosmos.
>
> G. K. Chesterton

When Maxwell returned in 1871 to take up a position at Trinity College, Cambridge, he resumed his membership of an exclusive group of Cambridge dons who would hold regular conversaziones to discuss the philosophical and religious relevance of their subjects. This perpetual group of twelve members called itself the Apostles (but should not be confused with a much later, infamous Cambridge group of the same name!). In years to come it would recruit such luminaries as Bertrand Russell and G. E. Moore. Many of Maxwell's closest friends were numbered amongst its membership. Although Maxwell grew up in the Church of Scotland, he underwent a dramatic religious conversion in 1853, and became what we would now call an Evangelical. Maxwell believed that the domain of science was the arena of universal and shared experience, while religion was a part of that quite different personal world: the 'I am' of that personal world, he argued, 'cannot be used in the same sense by any two of us, and, therefore, it can never become science at all.' Maxwell used the Apostles' informal discussions about science and religion as a sounding-board to develop his thinking about the relationship between these two sets of different ideas. Like other Apostles he was sceptical of the over-confident opposition to Darwinian evolution offered by the conservative religious thinkers of the day, and he was also unsympathetic to the logic of William Paley and his Deistic successors, whose opposition to evolution was grounded in the belief that every contrivance of Nature witnessed to the benevolence of a Divine Creator.

When the Apostles met on the evening of 11 February 1873, Maxwell had been asked to present some ideas on the subject of free will. His brief was to introduce the motion 'Does the progress of science lead one to favour a belief in the existence of free will rather than determinism?' Before seeing what he had to say, it is worth digressing to consider the Apostles' motivation for such an esoteric discussion.

Darwin's revolutionary suggestion that our human physiology and mental attributes were the result of chance events, preferentially selected over huge periods of time by their ability to survive, had created an apparent conflict with determinism. Several notable nineteenth-century physicists had argued that Darwin's theory offered no more explanation for the existence of living creatures adapted to their habitats than did the older notion that they were just miraculously created that way and have remained so ever since. They believed that the rigid determinism of the Newtonian mechanical world-view ensured that the present state of Nature was a unique and direct consequence of a particular state at the beginning of the Universe. The existence of intelligent human beings and the harmonious world around them then becomes no more (and no less) mysterious than the existence of that special starting state, regardless of whether any evolution ever took place. However, faced with the physicists' faith in the determinism of the mechanical Newtonian world-model, many theologians, while finding Darwin's ideas distasteful, also found themselves impaled on the horns of a dilemma. They liked to use the fabulous success of Newtonian physics as evidence that the world was fashioned like a watch, and hence deduce the existence of a Divine Watchmaker. But at the same time they wanted to appeal to the traditional evidence of and for miracles—that is, *departures* from the clockwork predictability of the Newtonian world—in support of their religious faith.

With regard to Darwinian evolution, Maxwell was to argue that the existence of 'molecules' (actually what we now term atoms) as identical microscopic building-blocks in Nature showed there to be a limit to the influence of natural selection. Molecules had definite properties that were not subject to change. The fact that they were all *identical* was for him the hallmark of their Divine design. In his famous address to the British Association in 1873 he would argue that:

> No theory of evolution can be formed to account for the similarity of molecules, for evolution necessarily implies continuous change, and the molecule is incapable of growth or decay, of generation or destruction.
>
> None of the processes of Nature, since the time when Nature began, have produced the slightest difference in the properties of any molecule . . . They continue this day as they were created—perfect in number and measure and weight; and from the ineffaceable characters impressed on them we may learn that those aspirations after accuracy in measurement, and justice in action, which we reckon among our noblest attributes as men, are ours because they are essential constituents of the image of Him who in the beginning created, not only the heaven and the earth, but the materials of which heaven and earth consist.

Maxwell admitted the possibility that there might have once existed other varieties of atom that proved less well fitted for the constitution of solid bodies than those which we now see, but brushed it aside in the absence of any fossil remnants of a maverick population of defunct atoms with non-uniform properties.

In his private presentation to the Apostles on the subject of free will and determinism, Maxwell had little to say on the vexed subject of free will. Instead, he focused upon some popular misconceptions about the deterministic character of Newtonian physics, and made a profound observation concerning the apparent conflict between free will and deterministic science that so readily springs up in our minds. Those persuaded towards determinism, he argued, have been swayed in their judgement by the fact that physicists, and especially their public spokesmen, always focus attention upon problems that reinforce the image of the clockwork universe. Their examples invariably possess one very special property: *a small change in the starting conditions creating some motion produces only a small change in the final state that results.* The false view that all motions exhibit this neat property is what has led to 'a prejudice in favour of determinism'. He suggests instead that:

> Much light may be thrown on some of these questions by the consideration of stability and instability. When the state of things is such that an infinitely small variation of the present state will alter only by an infinitely small quantity the state at some future time, the condition of the system, whether at rest or in motion, is said to be stable; but when an infinitely small variation in the present state may bring about a finite difference in the state of the system in a finite time, the condition of the system is said to be unstable.
>
> It is manifest that the existence of unstable conditions renders impossible the prediction of future events, if our knowledge of the present state is only approximate, and not accurate . . . It is a metaphysical doctrine that from the same

antecedents follow the same consequents. No one can gainsay this. But it is not of much use in a world like this, in which the same antecedents never again occur, and nothing ever happens twice . . . The physical axiom which has a somewhat similar aspect is 'That from like antecedents follow like consequences'. But here we have passed from sameness to likeness, from absolute accuracy to a more or less rough approximation. There are certain classes of phenomena, as I have said, in which a small error in the data only introduces a small error in the result . . . The course of events in these cases is stable.

There are other classes of phenomena which are more complicated, and in which cases of instability may occur . . .

Maxwell recognized that there exist many physical situations in which any uncertainty, however small, in our knowledge at one time leads to increasing uncertainty in determining the future state. Even if we had the perfect laws of Nature in our possession, they might be of no use in predicting the future in such circumstances.

Suppose we take a bowl and some marbles. Let a marble drop from a point on the inside surface of the bowl, and it will eventually come to rest at the lowest point of the bowl. If we release the marble from a slightly different starting position it makes no difference to the result: the marble still finishes its motion resting at the bottom. Clearly, a tiny uncertainty in our knowledge of the starting position of the marble does not prevent our determining its future behaviour. The uncertainty is effaced by the overwhelming tendency to fall towards the stable end-state.   Now consider the precarious situation pictured in Figure 5.4. In state (a) the initial positions of two cones differ by only an infinitesimal amount, but the result (b) of the tiny difference turns out to be enormous.

If you are unable to determine the initial position of a physical system with sufficient accuracy then Newton's laws of motion (even if they *perfectly* described the working of the world) will not be of any use whatsoever in determining its future behaviour. Thus the Newtonian watch-world is not deterministic in practice, even if it is deterministic in principle, because we may be unable to pin down the state of the world 'now' with sufficient accuracy to exploit the predictive power of Newton's equations. The equations are not to blame. They find themselves in the position of the computer blamed for sending some poor little old lady 20,000 gas bills. It is the accuracy of the information fed into equations that determines the reliability of what comes out—'garbage in, garbage out'!

This breakdown in predictability also occurs in important practical situations. It is the basic reason why it is impossible to predict the future behaviour of a complicated system like the weather. Meteorologists' inability to determine the *present* state of the weather accurately enough is why their forecasts so often go awry.

This undermining of determinism in practice has nothing to do with the undermining of determinism in principle effected by quantum theory, although

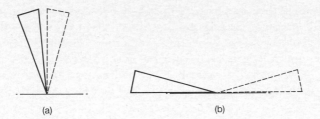

Figure 5.4 An infinitesimal difference in starting state of an unstable system of two identical cones infinitesimally displaced from their states of unstable equilibrium balanced on their points, (a), results in vastly different final states (b).

quantum uncertainty tells us that there must always be some finite uncertainty in our knowledge of the starting state of any physical system. The lack of absolute precision in our measurement of the present state of things denies us the type of determinism that Pierre Laplace, writing in an age whose recognition of the necessity of reason had given way to a belief in the sufficiency of reason, had in mind when he claimed that:

> We may regard the present state of the Universe as the effect of its past and the cause of its future. An intellect which at any given moment knew all the forces that animate nature and the mutual positions of the beings that compose it, if this intellect were vast enough to submit its data to analysis, could condense into a single formula the vast movement of the greatest bodies of the universe and that of the lightest atom: for such an intellect nothing could be uncertain; and the future just like the past would be present before its eyes.

We might mention that, although Laplace is always credited with introducing an explicit statement of determinism, he was preceded by Leibniz who had stated that*:

> Now each cause has its specific effect which would be produced by it . . . If, for example, one sphere meets another sphere in free space and if their sizes and their paths and directions before collision were known, we can then foretell and calculate how they will rebound and what course they will take after the impact . . . From this one sees then that everything proceeds mathematically—that is, infallibly—in the whole wide world, so that if someone could have sufficient insight into the inner parts of things, and in addition had remembrance and intelligence enough to consider all the circumstances and to take them into account, he would be a prophet and would see the future in the present as in a mirror.

*This description highlights another defect of the conventional deterministic view of Newtonian mechanics. If collisions take place, then it is necessary to know the elasticity with which the rebounds occur in order to predict the change in speeds of particles involved, and hence uniquely determine the future course of the dynamics. This involves physical information which lies outside the remit of mechanics. In modern computer studies of the dynamics of many moving bodies the interactions are 'softened' in an artificial way in order to prevent actual collisions being realized.

One might think that the answer to this challenge to determinism is increased accuracy of observation. The more accurate our present knowledge of the world, so the more reliable will be our future predictions. Unfortunately, increased accuracy does not really help us, because the uncertainty about the future grows so fast that it very easily overcomes our paltry attempts to improve our specification of the present. Here is a simple, but realistic, example of how one is chasing the wind.

We will locate a marker on the circumference of a circle by the angle its position makes with the vertical (see Figure 5.5). Now, we shall prescribe a law of motion which will move the marker to a new position angle by an amount equal to twice the value of its old position angle. Therefore the new position (in degrees of angle) is always the angle that remains after dividing twice the old angle by 360 degrees. This rule is mathematically precise, and the ensuing motion of the marker will be precisely determined after any number of steps. If we know the starting position exactly then we will know *exactly* where the marker lies after any number of moves. But what happens if, as will always be the case in practice, we can only ascertain the starting position to within some finite uncertainty: say we know that it is at an angle of 30 degrees to within one hundredth of a degree. After one move

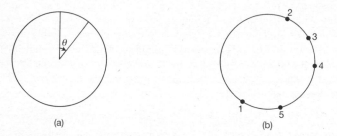

Figure 5.5   The circle algorithm described in the text. The algorithm moves the point on the circumference on to a new point situated at twice the value of the angle marking the previous position. All angles are determined to be the remainder after division by 360 degrees. Thus if the first position is 190 degrees, the second is at 380 minus $360 = 20$ degrees, the third is at 40 degrees and so on. The first five positions are shown.

the uncertainty will be doubled, and after N moves it will have snowballed to $2 \times 2 \times 2 \times \ldots$ (N times)$= 2^N$ times the starting uncertainty. This means that after only sixteen moves the uncertainty becomes larger than 360 degrees, and we can conclude nothing about the subsequent position of the marker on the circle. All one can then predict is that there will be an equal probability of it being found anywhere on the circle, no matter where it began. We can postpone the time when our deterministic law becomes useless in this way by specifying the starting position more accurately, but so rapid is the growth in uncertainty that, even if we specified the position with an accuracy equal to the size of a single atom, it would take only 38 moves to amplify it up to 360 degrees. In practice, greater accuracy can only put off the evil day by an insignificant amount. Unpredictability always catches up with you. There has emerged a fundamental problem of principle in our problem: sensitive dependence upon initial conditions. If you were unaware of this subtlety then you would go on blithely predicting the position of the marker with complete faith in the exact law for computing it. But after just a handful of moves your precise predictions as to its position would be entirely meaningless.

Chaos

> It is impossible to study the properties of a single mathematical trajectory. The physicist knows only bundles of trajectories, corresponding to slightly different initial conditions.
>
> Leon Brillouin

Physical phenomena which exhibit this sensitive response to tiny changes in their starting state are called *chaotic*. They are by no means of purely academic interest. They are all around us, and an understanding of their behaviour is crucial if we are to make progress in understanding the most complicated phenomena in Nature. However, as our previous example shows, things do not need to be complicated for chaos to occur.

What could be more deterministic than the motion of billiard balls on a billiard table? So straightforward and predictable did such a situation once appear that the term 'billiard-ball universe' was used as a byword for the deterministic mechanical world-view of Newton. We have just seen that Leibniz used the example of such collisions as an exemplar of determinism. However, cue games like billiards and pool exhibit that extreme sensitivity and instability highlighted by Maxwell. We know from experience of these games that a catastrophic error results from the slightest miscue. Suppose we ignore the effects of air resistance and friction between the balls and the table (in practice, it is these that cause the ball eventually to stop rolling after it has been struck). If we could know the starting state as accurately as the quantum Uncertainty Principle of Heisenberg allows, then this would enable us to reduce our uncertainty as to the starting position of the cue-ball to a distance less than one billion times the size of a single atomic nucleus (this is totally unrealistic in practice of course, but suspend all practicality for one moment). Yet, after the ball is struck, this uncertainty is so amplified by every collision with other balls and with the edges of the table that after only fifteen such encounters our irreducible infinitesimal uncertainty concerning its initial position will have grown as large as the size of the entire table. We can then predict nothing at all about the ensuing motion of the ball on the table using Newton's laws of motion.

It is curious how long it took for the significance of these simple ideas to be appreciated. In 1929, long after the insights of physicists like Maxwell and Poincaré, the influential French mathematician Jacques Hadamard was widely followed when he laid down three criteria which he believed were traditionally required for equations to be introduced as mathematical descriptions of natural phenomena:

(i) Solutions to the equations must exist to the future of where the initial conditions are specified; that is: *solutions must exist.*
(ii) A set of initial conditions must not lead to more than one solution of the equations; that is: *solutions must be unique.*
(iii) A small change in the initial conditions must produce only a small change in the solution at later times; that is: *solutions must be stable.*

Why others had wished the laws of Nature to be so restricted is not clear. All the important laws of Nature are described by equations which exhibit chaotic phenomena, and so violate Hadamard's third criterion. We do not have any theories that violate his second criterion if the initial conditions are taken to be precisely known. However, quantum theory obeys this criterion only in the academic sense that the Schrödinger equation which governs the evolution of the wave function in space and time satisfies the criteria (i) and (ii), but of course the wave function is not observable. The quantities that we can observe are statistically, but not deterministically, specified by the Schrödinger equation, and it is possible for two identical experiments on a quantum system to yield different *observed* results. Einstein's theory of general relativity disobeys Hadamard's first criterion. Solutions describing the behaviour of sufficiently large aggregates of mass do not exist for all future time, but evolve into a 'singularity' at which the physical variables become infinite and space and time are no longer mathematically defined. This pathology may be a physical consequence of the attractive nature of gravity rather than some peculiar unrealistic property of Einstein's equations which renders them physically inadmissable.

The study of chaotic phenomena has become a growth industry over the last ten years. In order to deal successfully with chaos a number of fundamental issues have to be faced. The identification of chaotic phenomena in Nature brings with it deeper problems than a dilemma about uncertainties over starting conditions. Imagine that we want to describe mathematically a very complicated process like the flow of water issuing from a roaring tap or a cascading waterfall. This fluid mess we usually call 'turbulence'. The problem of the laws that govern it has been a stumbling-block to physicists for a very long time because of its enormous complexity. If one watches the complicated flow from a bath-tap then one can see how parts of the flow which leave the tap close together end up being mixed far apart as the falling water becomes mixed-up with water already in the bath. A good way to observe this phenomenon in slow motion is to introduce a few drops of ink or dye into a glass of stirred water. The tendrils of ink are rapidly twisted into a complicated pattern. Parts that begin close together are muddled irretrievably into different places. In the face of complicated sensitive phenomena like this, our confidence in traditional mathematical physics is dented. If a slight mistake in the specification of the starting state of the system under study can result in totally incorrect predictions concerning its subsequent state, even when the equations governing the behaviour are perfectly known, what if they are imperfectly known? The consequences of slightly altering the rule for moving our marker around the circle in Figure 5.5 to multiply the existing angle by 2 and one-tenth rather than by two would be enormous after a few moves. The slightest error in any formula we use to describe chaotic natural phenomena renders that formula quite useless in practice. Given that any theory we formulate will always contain some imperfections, approximations, or external perturbing factors that we have ignored or not suspected to exist, we are clearly in trouble when faced with using equations to describe chaotically complicated phenomena.

But what if we were to abandon the hopeless search for *the* one true equation which precisely describes the chaotic phenomenon in question and concentrate instead upon discovering the common properties of *all possible equations*?

This ambitious suggestion is the one that has been taken up by physicists in the recent past. The search for the perfect mathematical model of a physical situation has given way to a search for the properties of all possible mathematical models satisfying some very broad restrictions sufficient only to delineate the type of problem under study. In order to appreciate how this is done we must first see what the equations used in the study of Nature are like.

Equations

> Physics is mathematical not because we know so much about
> the physical world, but because we know so little: it is only its
> mathematical properties that we can discover.
>
> Bertrand Russell

In practice, laws of physics mean differential equations. These creations are algorithms for determining the future state of a physical system from its present configuration. To 'solve' the differential equation means to find an expression which tells us what all the future states ensuing from any starting state could be, and when and where they will arise. Differential equations are the particular dialect of the language of mathematics that most effectively describes how Nature works. This fluency can be traced to their three disjoint ingredients:

(i) The *algorithmic structure*; that is, how one determines the future state from the present.

(ii) The starting state, or *initial conditions* as they are usually called.

(iii) Various constant quantities which are unchanged by the application of the algorithm. These are the quantities that we call *constants of Nature*.

As we have mentioned in earlier chapters, various principles of symmetry allow us to predict and deduce the general form of the algorithms (i), which are admissible as laws of Nature. They must obey certain restrictions in order to be consistent with existing observations of the world and apply universally. The situation with regard to the qualities (ii) and (iii) is quite different. Initial conditions and constants of Nature are the pregnant properties of the world which are not determined by its laws as we now understand them. We have no way of determining initial conditions. They are presented to us. Einstein's theory of general relativity may provide a good description of how the Universe changes with time, but we have no law which tells us what the initial state of the Universe was like. There have been suggestions of 'laws' of initial conditions which we shall discuss in Chapter 6, but they are little more than speculations.

The freedom to specify initial conditions to which laws of Nature can then be applied is a peculiar and two-edged feature. In applications to mundane and particular problems like the flight of a rocket, the independence of the law of motion for the rocket from the allowed starting states is what gives the law its universality. It is applicable to the description of rockets launched with any velocity. If the law worked only for a particular launch velocity it would be little more than a description of a particular rocket's motion. The independence of initial conditions from the laws of Nature is a measure of their utility. However, in the cosmological problem this universality becomes an embarrassment. There is only one Universe, but the equations of Einstein which govern its evolution allow the initial conditions to be freely specified. There should, it seems, exist some principle which restricts these starting conditions in some way.

The third ubiquitous ingredient in our equations is the constants of Nature (iii), which arise as proportionality factors. Unfortunately, at present we have no 'law' that determines what these must be either. In Chapter 2 we discussed Newton's law of gravitation. It states that the force of attraction between two masses is directly proportional to the product of their masses, and inversely proportional to the square of their distance apart. The constant of proportionality we call the gravitational constant. Newton's law implies that it should be found to have the same value for all pairs of masses whatever their separations; but the only way its numerical value can be found is by measuring the gravitational forces between various pairs of masses with different separations. One of the great goals of fundamental physics is to find the reason for the precise numerical values of the constants that appear in the equations that prescribe the laws of Nature. Many physicists believe that, eventually, we will be able to calculate the values of the fundamental constants, and specify the initial conditions of the Universe by some principle of internal consistency that reveals there to be one, and only one, logical choice for all of them. We are a long way from being able to do such a thing, and even farther from seeing how we would ever know that it had been done correctly and completely.

The three ingredients (i), (ii), and (iii) have one very nice property which is one reason why differential equations are so useful to scientists. They neatly separate our knowledge about the world from our ignorance. We can deduce the form of the law (ingredient (i)) accurately, despite the fact that we may know nothing about the initial conditions (ingredient (ii)), and have only rough measurements of the universal constants involved in a problem (ingredient (iii)).

A sceptic might well worry that the concepts of 'initial conditions' and 'constants of Nature' have arisen out of our preference for formulae that possess the three disjoint ingredients (i)–(iii). For instance, there may be something odd about the notion of initial conditions. When we use Newton's laws of motion to predict how a body will move, we must specify its starting position and its starting velocity as initial conditions. But there is something peculiar about the idea of an 'initial velocity', because the concept of velocity in the differential equation involves the notion of the position of the particle at some infinitesimally small future time as well as at the initial time.

Initial conditions and constants of Nature may be just names for particular aspects of the world that are not specified by the form we have chosen to express its laws. They may not have quite the fundamental significance we have chosen to endow them with.

Law without law

> Ignorance of the law is no excuse.
>
> A judge

What hope is there that we might be able to find properties of all possible equations that might be employed to describe a complicated chaotic process? At first sight the answer seems to be 'none at all'. Anything that could ever happen will be the solution of some equation, so there cannot exist useful information about *all* possible equations. Anything we could say about the solutions of all equations would be so weak that it would be of no real interest. However, the fact that we are not interested in situations which only arise with tiny probability enables us to restrict our attention to those equations which are in some sense *likely* to be chosen from the collection of all possible equations. This rules out situations akin to a needle being dropped and landing perfectly balanced upon its point, and remaining in this precarious equilibrium ever after. There is nothing to forbid such an event in principle, but it is extremely unlikely to be observed, because the slightest perturbation of the situation destroys the perfect balance. It is unstable. The slightest change completely changes the outcome. If we eliminate equations which are very special in an analogous sense, because an infinitesimal change turns them into another type of stable equation, then we are left with representatives of what we will call *almost every* equation. This collection possesses the property that, if one of its members is slightly changed, then its solutions retain qualitatively similar properties to other members of the collection. If a non-member is slightly changed, then it behaves like a member of the first collection as well. Remarkably, there are general properties of the collection of 'almost every' equation which are not trivial. In particular, it has been found that when they get beyond a certain complexity they exhibit chaotically unpredictable behaviour.

This approach has led to dramatic advances in our understanding of turbulent and chaotic phenomena in Nature. These phenomena are typical in the sense that they are realized in Nature. They do not require a precarious balance of special circumstances to maintain (just turn on your tap). This means they will be described by an equation that is a member of our family of 'almost every' variety of equation. Hence, without knowing the exact physical law governing turbulent motion, we might hope to describe some of its properties. This subject is of considerable current interest because it appears that one can divide the 'almost every' set of equations into groupings which contain many apparently different members. Yet, all the members of a grouping possess definite properties in common. These groupings are called *universality classes*. Their existence may prove a godsend in the solution of very difficult problems, because if a superficially intractable equation lies in the same universality class as an easy one, then we can confidently confine our attention to the easy one in the knowledge that the solution of the harder one will have similar general properties.

The most striking universal property possessed by a large class of equations is the sequence of events by which they degenerate from ordered behaviour into chaos. Imagine that a flow of liquid is proceeding in a smooth, regular fashion, but we can adjust a jet to make the flow run faster or slower. As the jet is gradually turned up, a definite series of changes occurs. At first the flow oscillates into waves which repeat in a time period T; then, as the jet is steadily increased, the flow becomes a little less regular, and takes a time $2T$ to repeat itself; then it repeats only after a time $4T$, then after $8T$, and so on until a state of unpredictable chaos is reached for a particular setting of the jet. In this particular state the time required for the flow pattern to repeat itself is infinite. It contains no cyclic predictability at all. This sequence of events has been found to lie at the root of a large class of chaotic natural phenomena. Regardless of the details of the physical situation in which the final chaos appears, this 'period doubling' transition can often be found. But most remarkably, the rate at which the period doublings occur on approach to the critical value where the period becomes infinite is always the same. If the nth doubling of the period occurs when the external control of the jet has the value k_n, then as n is allowed to increase k_n approaches its critical value k^* in accord with

$$k_n - k^* \propto D^{-n}$$

where $D = 4.669\,201\,609\,102\,9 \ldots$ is a pure number that is always the same. The simplest example of this 'universality' phenomenon is the algorithm

$$x_{n+1} = Ax_n(1 - x_n) \tag{5.1}$$

where A is a number between 1 and 4 which we can treat as a variable control. This algorithm allows the $(n+1)$th value of a quantity x to be determined completely from its nth value. It has the characteristic form of any process in which there is competition between births (the first, Ax_n, term on the right) and deaths (the second, $-Ax_n^2$, term on the right). If you pick a starting value for $n = 1$, and calculate the sequence of x values that the algorithm produces for different choices of A, one finds a definite pattern.

When A is near 1 the successive values $x_1, x_2, x_3, \ldots$ are periodic. If A is chosen with a larger value, then the period increases from 1 to 2 to 4 to 8, successively doubling until we pick the critical value $A = 3.5699$. At this value the output becomes aperiodic, and as A is increased to 4 the chaos continues. The probability distribution of the output from the algorithm settles down to a definite form after a large number of iterations. This is shown in Figure 5.6 for $A = 4$.

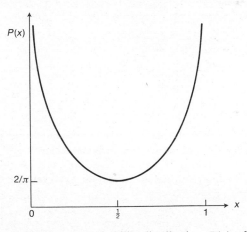

Figure 5.6 The steady long-run probability distribution, $P(x)$, of the repeated outputs of the algorithm in equation (5.1).

The steady growth of complexity in chaotic systems displaying period doubling is associated with the fact that, for each value of the control parameter at which a period doubling occurs, two distinct types of future evolution of the flow are allowed from the same current state. As the critical value is approached, the number of distinct behaviours grows dramatically. This proliferation is curious because a small steady change creates a discontinuous change in something like the number of solutions that there exist to an equation. Yet, this type of thing can occur in a very simple manner. Imagine that the British economy possesses stable states X which are determined by various partially controllable factors like inflation, foreign exports, unemployment, and so forth. Let us call them A, B, and C. Suppose, for the sake of illustration, that these variable factors determine the equilibrium state X through the quadratic equation

$$AX^2 + BX + C = 0. \tag{5.2}$$

The number of different solutions to this equation for X will give the number of different equilibria that the economy can have. However, that number will vary according to the relative size of B^2 and $4AC$ as shown in Figure 5.7. If at one time

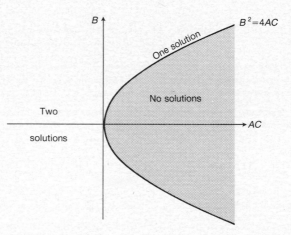

Figure 5.7 The different number of real solutions of the quadratic equation given in equation (5.2) as the three control parameters in the equation A, B, and C vary. The number is determined by the relative values of B^2 and AC and changes discontinuously between zero, one, and two even when the quantities A, B, and C change smoothly.

we have a state in which B^2 exceeds $4AC$ very slightly then there will exist *two* separate solutions. But if environmental changes occur so that B^2 becomes equal to $4AC$ there will exist only *one* equilibrium state, and if a further changes occurs so that B^2 slips just below $4AC$ in value, there will exist *no* possible equilibrium economy at all. A very small continuous change in the control parameters produces a discrete change in the number of solutions to the equation. Likewise, in the route from order to chaos there can arise discrete and sudden changes in the qualitative behaviour of a system when an environmental influence changes just a little. Notice that this non-uniqueness of the future state of a system following the prescription of a definite starting state is the very antithesis of what was envisioned as 'natural' by Hadamard's first and second conditions.

The discontinuous phenomenon exhibited in Figure 5.7 is a fairly common event in physical systems. The study of how these discontinuous changes arise has become something of a cult subject following its espousal as a panacea by the French mathematician René Thom in his classic book *Structural Stability and Morphogenesis*. At Christopher Zeeman's suggestion it has become known as *Catastrophe theory*. To some extent it is an extension of the philosophy we have been describing here, whereby properties of all possible laws of Nature are elucidated independent of their detailed form. Catastrophe theory is built upon a number of precise mathematical results, which prove that if a smooth surface is described in an unknown way, but controlled by a fixed number of parameters which can vary arbitrarily, then the behaviour of the system near any discontinuity must have a particular form. This has led to the invocation of Catastrophe theory to model all manner of complicated processes, from prison riots to mental illness, where no underlying mathematical model is known. By assuming that the phenomenon in question is controlled by a small number of control parameters it is possible to predict the presence of catastrophic discontinuities and the nature of changes in their vicinity. The results, however, are clearly only as well founded as the assumptions they are built upon—the number of control parameters involved in the unknown model for example: Nevertheless, this elucidation of the generic discontinuous change has shed light upon many optical phenomena where caustics and diffraction occur.

This type of emphasis illustrates the trend of modern mathematical thinking about complexity in Nature. In the past it was traditional to pick on a particular natural process and seek an exact equation to describe it. Now there is emphasis upon those properties which will be shared by the most general equations and the most complicated phenomena; phenomena whose intrinsic complexity defies the scientist to find some exact equation which describes their workings precisely. In the past, when faced with natural phenomena of great complexity physicists rushed to adopt a statistical approach. Turbulence would be attacked as a statistical process. This stochastic aspect was admitted just for convenience in arriving at a description. There is nothing intrinsically non-deterministic about turbulence. In recent years there has been a move towards reinstating a *deterministic* approach to complex turbulent phenomena in which their chaotic appearance is traced back to the existence of intrinsic sensitivity to slight changes. Such phenomena are entirely deterministic in principle, but because our information regarding their exact state at any instant of time must be incomplete, we adopt a statistical description that gives information about what will be seen over a long period of observation. Observation must be repeatable to be useful, and repeatable as often as one wishes. The outcome of a whole series of observations will always give us a probability distribution. In some cases, where the system is not chaotic, the 'distribution' will be trivial, with every observation yielding virtually the same value, but in others this will not be the case. Many chaotic systems eventually settle down to a form of equilibrium that is described by a definite probability distribution of events, like that shown in Figure 5.6 for the $A = 4$ algorithm. The equilibrium distribution of outputs is adhered to with greater and greater accuracy as the evolution continues, even though the individual events show no sign of approaching any particular limiting value.

However, there is more to chaos than a probability distribution. An instructive example is supplied by the following algorithm, first studied by the French astrophysicist Michel Henon:

$$x_{n+1} = y_n + 1 - 1.4x_n^2$$

$$y_{n+1} = 0.3x_n.$$

(5.3)

The formulae show how the $(n+1)$th values of the variables x and y are calculated deterministically from their nth values for $n = 0, 1, 2, 3, \ldots$ Hence, if $x_0 = 0$, $y_0 = 0$ then $x_1 = 1$, $y_1 = 0$, $x_2 = -0.4$, $y_2 = 0.3$ etc. This algorithm is chaotic in the sense we have been discussing. A small uncertainty in the starting state is magnified very rapidly. If we set the starting values x_0 and y_0 equal to 0.631 and 0.189 respectively, and then plot all the subsequent values of x and y that are generated by repeated use of the algorithm, then a strangely structured diagram results (Figure 5.8). After a large number of iterations the distribution of the output takes on a characteristic form. Areas where the population of output points is densest indicate regions of the x–y plane where the probability distribution is peaked. However, the picture hides a remarkable property. If we look at a blow-up, (b), of one of the sections of the picture, (a), where the points are strung out in narrow bands, then we find that the magnified image of a single band reveals further substructure, with bands within bands within bands [compare (c)]. If we continue to examine the pattern of the output with a finer and finer resolution, we will always find a new level of structure and complexity in the output pattern. The content of the chaotic output never fades away into uniformity. Although there may exist simple universal laws which govern the rate at which a new structure appears as the number of output points increases, there is irreducible complexity on every microscopic scale that no finite description can exhaust.

The essential component of the never-ending levels of structure in the Henon mapping is something that mathematicians call a *Cantor set*. It is a curious creation hovering on the brink between nothing and something. In order to construct one we can start with a straight line of length 1 lying between 0 and 1 and including both those two endpoints.

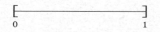

Now, remove the middle third of this line (between 1/3 and 2/3) leaving its endpoints (1/3 and 2/3) behind. We are left with two sub-intervals

We now repeat this subdivision *ad infinitum*, in each case throwing away the middle one-third of each piece while retaining its two endpoints. At the end of this

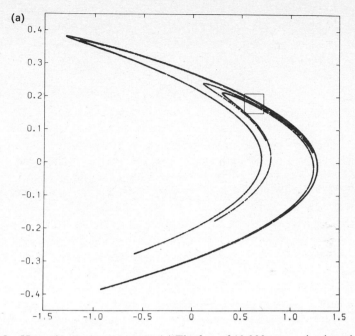

Figure 5.8 Henon's strange attractor. (a) The fate of 10,000 successive iterations of the algorithm given in equation (5.3), starting with values $x=0.63135448$ and $y=0.18940634$; x values are plotted horizontally and y values vertically. (b) An enlargement of the boxed region in (a) showing an increase in the amount of structure present. The number of points computed has been increased to 100,000 to help delineate the structure. (c) An enlargement of the boxed region in (b) showing further substructure. The number of plotted points has been increased to one million. Examination reveals a hierarchical structure of levels within levels. At each level the global structure is similar when the overall size has been allowed for. This 'self-similarity' is characteristic of the 'Cantor set' discussed in the text. [Reprinted by kind permission of the author from Henon, M. (1976). A two-dimensional mapping with a strange attractor. *Communications in Mathematical Physics* **50**, 69–77.]

operation the 'dust' of scattered points that remain forms the Cantor set. One might have supposed that this remnant would consist of nothing at all, but this is not the case. The remnant Cantor set has zero length because the total length of line thrown away is the infinite sum of the geometric progression of lengths

$$\frac{1}{3}+\left(2 \times \frac{1}{9}\right)+\left(4 \times \frac{1}{27}\right)+\left(8 \times \frac{1}{81}\right)+\cdots=1$$

Yet, it contains an uncountable infinity of points, and as a result its dimension is not zero like that of a single point which also has zero length. More information is necessary to generate it than is needed to fix a single point, but not as much as is required to specify a line. This informational measure of dimension (called the

(b)

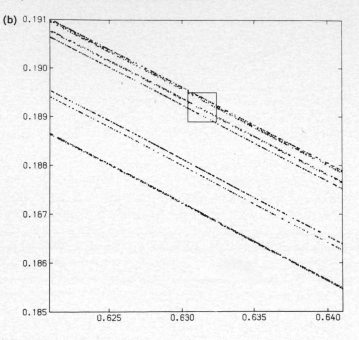

(c)

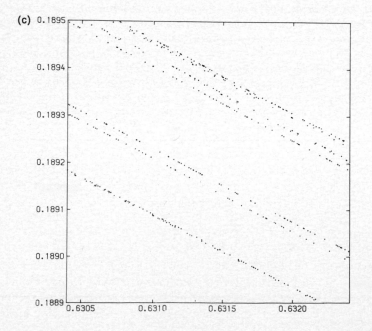

Hausdorff dimension) need not be a whole number, and in this case is equal to $\log(2)/\log(3) = 0.6309 \ldots$ The final states of Henon's mapping form a structure which is described by a line multiplied by a Cantor set (recall that a plane results when a line is multiplied by a line, and a cylinder results when a line is multiplied by a circle). The result is something intermediate between a line and a plane from the point of view of the amount of information needed to generate it, although it can be represented on the plane pictorially because its geometrical dimension (the number of co-ordinates necessary to fix the position of each point uniquely in space) is still two.

The 'worlds within worlds' structure of the output from Henon's algorithm is characteristic of many complex algorithms which exhibit chaotic behaviour. Another example was provided by the fractal shown on p. 220. In that case the mapping was illustrated by drawing the boundary which separates those points which remain mapped within a finite distance of the origin from those which are mapped off to ever-increasing distances after many iterations. The boundary of the black region in Figure 4.10 contains a bottomless pit of complex, connected structures. There are an infinite number of small disc-like structures attached tangentially to the heart-shaped component; these in turn spawn a further infinity of even smaller discs connected by thin tendrils, and so on *ad infinitum*. Remarkably, the whole of this complex boundary can be generated from any finite piece of it by iterating the algorithm sufficiently often. We see that a very simple arithmetical rule can store within itself an infinite amount of complex information that reproduces itself upon the simple act of iteration. This gives us some clue as to how Nature manages to evolve such complexity from simple beginnings.

Are the laws of Nature computable?

> In thinking and trying out ideas about what is a 'field theory' I found it very helpful to demand that a correctly formulated field theory be soluble.
>
> Kenneth Wilson

We have stressed the role of *algorithms* in the application of mathematics to the physical world. In practice, we require computable algorithms in order to compare the predictions of a physical theory with what can be observed or measured. We may have laws of Nature written in terms of partial differential equations, but we have to select some computational algorithm in order to solve those equations. This algorithm may be implementable with pencil and paper in simple cases, or it may require a super-computer. Usually, we can separate the concept of a physical theory from its implementation. However, one can envisage a situation in which a law of Nature cannot be distinguished from its implementing algorithm. There may exist no way of abbreviating certain sequences of events; the simplest representation of them will be the sequence of events themselves. A random sequence has this property of course, but it is also possible for non-random sequences to possess such intractability. In this case there will exist no algorithm for the practical application of the mathematical theory in question.

We have already seen how Alan Turing and others have shown that non-computable numbers exist. The question of interest to the physicist is whether there exist laws of Nature, framed in mathematical form, which predict the values of measurable quantities that are not computable numbers. If an observable number is not computable, then there can exist no mechanical program which will approximate the number with arbitrary accuracy in any finite number of computational steps. With a law of Nature of this type one would be faced with the situation of having numerical predictions which could not be checked to high accuracy against the observed facts, because the checking process would require the construction of an algorithm able to carry out comparisons between the predicted number and the measured value to arbitrary accuracy. Since the predicted number is not computable, no such comparison algorithm can exist. Each new level of accuracy would require qualitatively new principles of approximation for its attainment. Non-computable numbers are defined by a bottomless pit of novelty. This is why no mere mechanical algorithm can exhaust their content.

The concepts and constants of classical physics involve numbers (like 'pi') and dimensionless physical constants (like the ratio of electron and proton masses) which are both measurable and computable in the context of theories of atomic physics. Quantum physics need not give rise to decidable or computable predictions. Given two arbitrary quantum states of a physical system possessing an infinite number of degrees of freedom, then the question of whether they are macroscopically distinguishable cannot be decided by any computational algorithm. The infinite number of degrees of freedom plays a key role here, because if that number were finite all the possible macroscopic states associated with each of the two quantum states could just be enumerated in two finite lists and compared. Another, different sort of non-computable physics problem has recently emerged in the investigation of quantum gravity. The mathematical theory under investigation predicts that a particular quantity (call it X), which in principle determines measurable numbers, is defined by summing a series of terms each of which is evaluated for a member of the set of all four-dimensional compact manifolds. However, there can exist no algorithm which can systematically list and distinguish the manifolds required to evaluate the sum. Such a listing is a non-computable function. A distinct and qualitatively new (not derivable from the last step) idea is required to evaluate each term in the sum. Whereas this example does not prove that the observable X is not computable—for there may exist other mathematical expressions for it which avoid the conduct of a non-computable operation—it does offer a warning that the deep mysteries of any 'ultimate' theory of the Universe may well have at their heart a mathematical structure that is non-computable. This might even be a natural expectation to have of a theory which applies only to a unique object like the Universe. It need have no implementation except that Universe.

These two examples of non-computability in physics are rather far from home. The last example we shall give is more fundamental. It has been shown that if we take the archetypal ordinary differential equation

$$dy/dx = F(x, y)$$

where F is a continuous function which cannot be differentiated twice (this means that a graph of F, although it can be drawn without taking the point of the pencil off the page, can contain creases and cusps like those found at the point of a cone), then even though the function F may be computable there need be no computable solution to the equation. This problem exists for partial differential equations as well. If the wave equation is examined for waves moving in more than one dimension of space then it suffers exactly the same problem. If the initial profile of the wave is prescribed in space by a continuous but not twice-differentiable function G, then even though G is a computable function there may exist no computable solution of the wave equation. In one-dimensional wave propagation this cannot happen: computable starting conditions are sufficient to guarantee computable solutions. At first, this result appears profoundly worrying. Wave equations and ordinary differential equations of the sort exhibited above include just about all the varieties of differential equation employed in describing laws of Nature—brain waves, water waves, quantum wave-functions, gravitational waves, all are included it seems. However, it appears that the lack of smoothness required of the functions F and G is absolutely essential for the non-computability. If F and G are twice-differentiable functions then the solutions of the differential equations are always computable. It has been traditional to regard such a level of smoothness as a necessary condition for F and G to be physically realistic, but how do we know that this is correct? At the molecular level (and below), the flow of a liquid is the concatenation of innumerable collisions which exhibit discontinuities at the moments of collision when viewed classically. What happens in the quantum view we do not fully understand. Our differential equations are averaged descriptions of a myriad of individual algorithms describing individual encounters. Even if they are computable it is still possible that the discrete descriptions which underlie them are not.

The cosmic code—a final speculation

> This ... theory [of Shannon's] is so general that one does not need to say what kinds of symbols are being considered— whether written letters or words, or musical notes, or spoken words, or symphonic music, or pictures. The theory is deep enough so that the relationships it reveals indiscriminately apply to these and all other forms of communication.
>
> Warren Weaver

If we think of the process of scientific investigation as the decoding of a message that is conveyed to us, then the classic work of the American information theorist Claude Shannon has taught us that, somewhat surprisingly, it is possible to find a transmission code such that the decoding of the message can be done with accuracy as good as one cares to choose, even in the presence of noise and interference which distorts the signal. Different recipients can all get the same *message* as accurately as they wish even though they may not receive the same sequence of signals, or even use the same method of communication. Of course, in practice, information signals will only have this resilience against noise if there is a lot of redundancy built into the communication with far more symbols being used than would be necessary to send the message in the absence of noise. Furthermore, in order to realize this promise of arbitrary high signalling fidelity the message must be encoded in a special way. In some strange metaphorical way Nature seems to be 'encrypted' in one of these expedient forms. There are many observers and interpreters of Nature who at different times and in diverse places have gathered similar and various facts and observations—'information'. All these investigators are beset by forms of 'noise'—whether it be personal prejudice, imperfect or incomplete data, error, the Kantian categories of thought innate to the human mind, experimental selection effects, or cultural bias. Yet, according to this analogy with Shannon's discovery, it is still possible for certain types of message to be discerned with arbitrarily high accuracy in the face of all these distortions if the source has certain properties. If the 'message' is associated with the laws of Nature, we might interpret our faith in the harmlessness of filtering the observations of the world through human cognition as a belief that we have determined the message regardless of the presence of noise which partially corrupts the signal. The redundancy required to ensure high fidelity is available through the repeatability of experiments and observations, and the way in which similar structures appear to be deployed in different aspects of Nature. Perhaps the evidence that the laws of Nature give a consistent picture regardless of so many superficial distortions and in the light of such a perplexing diversity of approaches is analogous to them possessing an 'optimal coding' in some sense. Could this be a reflection of the unexpected usefulness of mathematics as a description of the physical world?

6

Are there any laws of Nature?

There is no law except that there is no law.
John A. Wheeler

Heretical notions

Heresy is but the bridge between two orthodoxies.
Francis Hackett

Often, chaos leads spontaneously to order. The atoms around us move haphazardly, each in its own way, yet there is comparative order over a large scale. The market 'forces' that move an unconstrained economy are made up of the random and independent decisions of millions of individuals, but they can produce a coherent overall direction. There is no 'law' of evolution which directs the course of biological development over aeons of time: only the statistics of the independent results of adaptation and survival. Yet the flora and fauna of our planet exhibit what we would have to admit as 'order'. This impressive state of affairs makes it reasonable to ask if it is not possible that there is also no 'unseen hand' at work in the Universe at large. Could it be that there are no laws of Nature at all? Perhaps all the order we see is a manifestation of that peculiar type of total lawlessness and independence that leads to predictability?

In the past there has often been recognition of the fact that chance lies at the root of many regularities of Nature. Normally, this idea is used only to explain the average uniformity arising from very large numbers of events, each acting individually in conformity with deterministic laws. Of this situation, Schrödinger writes,

> The individual process may, or may not, have its own strict regularity. In the observed regularity of the mass phenomenon the individual regularity (if any) need not be considered as a factor. On the contrary, it is completely effaced by averaging millions of single processes, the average values being the only things that are observable to us. The average values manifest their own purely *statistical regularity*.

The fact that we do not observe *individual* microscopic events means that the laws of molecular motions are not of immediate observational relevance. What matters for our impression of Nature is the average behaviour of large collections of individual motions. And the larger a collection of objects, the smaller is the minimum level of random fluctuation that characterizes it. None the less, despite the appearance of large-scale order out of microscopic disorder in the processes of the sort Schrödinger cites, he is not trying to argue that these individual motions are not each subject to strict laws of motion. They only have the appearance of disorder because of their complexity, and as we discussed more fully in Chapter 3, it is our inability to handle this complexity mathematically that leads us to seek a statistical description, not the presence of some intrinsic randomness in molecular motion. Furthermore, like deterministic laws, statistical regularities have the potential to be tested.

Here we would like to entertain the more radical idea that the underlying laws governing those individual phenomena are themselves of statistical origin. This has the implication that they cannot be exact. They might have arisen as an asymptotic state after a long period of evolution. The appearance of regularity, then, need be evidence of nothing more than the extremely long time that has existed since the origin of the Universe.

Although the Christian West has inherited a faith in the immutability of the laws of Nature, it has always been tempered by the belief in the possibility of 'miracle'. Rather than being seen as a contradiction, the co-existence of these two views would have been explained as evidence of belief in the laws of Nature as properties divinely imposed on the Universe from outside, rather than being holistic or immanent within Nature. In other traditional religions the scope for deviance from the canonical laws of Nature is usually much wider, because there is no a priori or analogical reason to associate the immanent character of Nature with lawful consistency and regularity.

Most ancient pagan beliefs place more emphasis upon the non-uniformities of Nature than the regularities. Rival deities battle to impose their wills upon the world. Marduk slays the mother of the gods and the spawner of monsters. Her body is torn asunder, and from it the heavens and the Earth are fashioned. In such stories the world changes its character in unison with that of its transient overlords.

We have already discussed the extent to which the dominance of one or other of these belief-systems played a role in incubating particular attitudes to the predictability and lawfulness of Nature. Whereas the ancients seem to have had a propensity for the melodramatic birth of the order of Nature out of the clash of titans, the medieval Europeans had evolved a curious fear of what they could not understand, and of lawlessness in any shape or form. A macabre example that has recently been documented is the fact that throughout the Middle Ages it was common for animals (and even insects!) to be tried and punished for 'crimes'. They were remanded in custody, represented by legal counsel, sometimes acquitted—all at substantial cost to the state. What could have induced such absurd behaviour? Presumably, a deep-seated feeling that law should govern everything in the natural world: that nothing and nobody should be exempt from its jurisdiction and penalties.

In the early days of the Church Fathers the apologist Arnobius argued that, because the laws of Nature 'initially established' had not been disrupted by the advent of Christianity, this proved that the faith that he preached did not possess some terrible and unnatural ingredient which would subvert Nature if embraced by mankind. The scientific culture that arose in Western Europe, and of which we are the inheritors, was dominated by adherence to the absolute invariance of laws of Nature, which thereby underwrote the meaningfulness of the scientific enterprise and assured its success. This attitude is an extreme one. It may be the right one. But what if it is not? The opposite extreme of a chaotic and lawless world that characterized early mythopoeic thinking appears to be in total conflict with the world that we experience. But is it? Could there be no laws of Nature? Or could the laws that we assume to hold yesterday, today, and forever be slowly changing with time, or be different in distant parts of the Universe? Do the laws of Nature hold everywhere in the Universe? We have already encountered the cosmologists' arguments that the Universe had a beginning in time where the laws of Nature must break down. Could there also be sites in the Universe today where no laws of Nature hold?

Between a rock and a hard place

> A really naked spirit cannot assume that the world is thoroughly intelligible. There may be surds, there may be hard facts, there may be dark abysses before which intelligence must be silent, for fear of going mad.
> George Santayana

If there are real laws governing the bedrock of Nature, could they not be changing with the passage of time, varying from place to place in the Universe like the fields of matter and energy they govern? This is a common speculation, the basis of many a science-fiction story. But what could it mean? Again, we must be careful to distinguish between the two different ways in which we talk of the laws of Nature. On the one hand we might be referring to our current description of the Universe and its internal workings. This description is something that is constantly changing and evolving. Some would claim this provisionality to be a necessary property of any scientific (rather than religious) description of the world. At any moment parts of our description may, unbeknown to us, be totally incorrect. It is a map being drawn by explorers. Some parts of the territory it traces will be well trodden by many investigators, while others will have been glimpsed only by scouts unsure of their bearings or confused in their account. But this map is quite unlike any we might use when trekking across the countryside, because it tells us not only what there is to be seen but how the different sights are linked together. More than that, it gives predictions as to what might be found around the next corner or on the next map sheet. More impressive still, its microstructure is strangely bound up with the way in which it is consulted. It is more like an interactive computer adventure-game than an Ordnance Survey sheet.

This descriptive map of the laws of Nature can, and clearly does, change with time. Indeed, different scientists even disagree as to what the best edition of the map is at any one time. But what of the underlying reality of which the map is but an imperfect representation: can that be changing with time and place? After all, we have learnt that the entire Universe is in a state of dynamic change, with galaxies rushing away from one another at ever-increasing speeds, and we have learnt that the very structure of space and time are not fixed and immutable. Their properties are determined locally by the changing patterns of matter and energy residing within them. Perhaps the laws of Nature change at the rate at which the Universe expands, or are determined locally in the same way as the geometry of space and time. We can give a definite 'No' in answer to this seductive suggestion. If there are laws of Nature which change with time and place, then either there is a rule governing how they change or there is not. If there is such a rule then we have found a more fundamental unchanging law, but if there is not then there are no laws of Nature. There are either constant laws or there are no laws. It does not matter whether we happen to know what they are or not. If the laws we believe to be invariant are found to change, it demonstrates our ineptitude not Nature's inexactitude.

It is interesting to note that even Maxwell was mistaken in his evaluation of this question. He claimed that all laws of Nature were time-invariant in the sense that they must be described by differential equations which do not contain the time variable explicitly. However, one can always perform mathematical operations on an equation that contains the time variable explicitly to transform it into another equivalent equation that does not. The new equation gives the invariant law governing the time changes in the original one. Maxwell's requirement does not eliminate any differential equation, because any law of change can be derived from the constancy of some other quantity.

If there really were to exist no laws of Nature, how could we know? We can only question the truth of a particular law if we assume the truth of others to provide the framework for rational comparison. Undeniably, human beings have a habit of perceiving in Nature more regularities and patterns than really exist there, and of extrapolating unjustifiably without noticing the fact. This is an understandable by-product of an activity that sets out to codify and organize our knowledge of the world. During the last twenty years, however, physicists have made progress as much by overthrowing bogus laws of Nature as by discovering new ones. Many quantities which were once believed to be unchanging have been found to possess tiny variations. Traditional conservation laws have been questioned; many apparent symmetries of Nature have turned out to be merely 'almost' symmetries upon closer scrutiny; elementary particles of Nature, like the neutrino, once routinely believed to possess no mass, are now equally routinely believed to possess a small but non-zero mass. It would probably be true to say that if one had attempted to publish a research paper ten years ago that assumed neutrinos possessed a mass one would have had almost as much difficulty getting it accepted by the editorial board of a scientific journal as one would today a paper which assumed neutrinos to have zero mass! There is undeniably a trend from more laws to less, and a desire to believe that everything that is not explicitly forbidden by a physical theory is compulsory. Maybe this is a healthy sign that we are working towards some ultimate unified 'theory of everything' which contains no superfluous elements: the ultimate in logical and conceptual economy. Laws now regarded as distinct and unrelated would be synthesized within a single unique description. This is what most physicists working on such fundamental questions believe. But it may be that our undoing of the catalogue of Nature's laws will take us down a different road, which will lead us to the recognition that there is no such ultimate theory of everything: no law at all.

At first sight such an idea seems outrageous stuff and nonsense. The observation of order in the physical world was what led humans to begin scientific inquiry for its own sake in the first place. How could the result of that inquiry be the revelation that it was based upon an illusion? But Nature is more subtle. We have seen that the ordered behaviour of large numbers of objects, be they molecules or stars, is the consequence of many 'chance' encounters. It is because each is oblivious of all the others that there is an underlying statistical trend towards some types of uniformity, just as the continual tossing of a coin produces a closer and closer convergence between the number of 'heads' and 'tails'. Paradoxically, chance lies at the root of most of the uniformities of the world we are familiar with. Complete chaos possesses statistical uniformities. Completely deterministic laws possess definite uniformities also. It is the in-between world of partial disorder involving neither millions nor just one or two objects that is so difficult to describe. In Chapter 3 we encountered the ubiquity of the Gaussian distribution of outcomes. As the number of outcomes of independent random events gets larger, so the better is the pattern of frequency with which different outcomes arise described by the bell-shaped curve (p. 118).

Consider a different type of example: if you stir a barrel of thick oil it will rapidly settle down to the same placid state, no matter how it was first stirred. Some physicists have explored whether such phenomena may have analogues in the way the laws of Nature arise. We could imagine an initial chaos in which the behaviour of electricity and magnetism was dictated by a combination of all the possible symmetries—a state of affairs that would in practice amount to having no symmetries at all. As the Universe expanded and cooled the effect of more and more of these symmetries would become negligible, and by the present one would dominate over all the others and give the impression of being unique. In this way the evolution of the Universe from a hot, dense Big Bang to the present rarefied and cool expanse could dictate the environment within which chance acts. In such a world there would appear to be no laws at all in the inferno of the Big Bang when all the symmetries would be on an equal footing, rather like living in a new country which had adopted the legal systems of every other state simultaneously. But, with the passage of time, dominant trends of behaviour would win out, because in a random environment it would be too much of a coincidence to expect all the different symmetries to die away at exactly the same rates as the Universe cooled.

Such a speculative scenario is only attractive if one can show that the symmetries which we observe in the present-day Universe, and thereby the particular laws of Nature associated with them, arise inevitably no matter what the initial chaos was like. It is interesting to note that such an idea is actually quite the opposite to that envisaged by many elementary-particle physicists today. 'Grand Unified gauge Theories' (GUTs) present a picture in which symmetries are increasingly restored as the temperature of the Big Bang gets higher and higher, and we approach the beginning of the Universe's expansion history. The change in the effective strengths of the different forces of Nature produces a convergence of their effects at high energies and a divergence at low energies. The alternative we propose is to imagine that symmetries disappear at high energies and are restored as the temperature falls. In fact, we know of certain circumstances which, if they prevail in Nature, prevent the philosophy of 'Grand Unification' being realized. In Chapter 4 we discussed how the ceaseless activity within the quantum vacuum leads to the phenomenon of *asymptotic freedom* on which the possibility of Grand Unification is built. In the case of the strong force between coloured quarks, asymptotic freedom means a weakening of the strength of the effective interaction between them in high-energy environments. This occurs only if the number of different types of quark is no greater than eight. At present we know of only three families. If, at very high energies five more exist then, above those energies, the strong interaction would start to become stronger rather than grow asymptotically weaker. There could then be no point of meeting between the force strengths of electromagnetism and colour charge interactions. In such a situation we would look for an explanation of the low-energy world in the shedding of symmetries as the Universe cooled out of lawless chaos.

This alternative speculation, that symmetry may arise as a sort of 'survival of the fittest', is more reminiscent of attempts by the 'chaotic cosmologists' to show that, no matter how the Universe began, it would inevitably end up looking pretty much like the one we see to today, because of physical processes that occur within it as it expands and cools. Again we recognize this persistent dichotomy in scientific study of fundamental questions: on the one hand there is an attempt to show that what exists now does not depend on how things began, because all initial situations inevitably evolve towards something similar to what is now seen. Hence, the unknowable initial conditions play an inessential role in determining the structure of the world we see. On the other hand there are those who wish to place the onus for explaining the present primarily upon the initial conditions, and deny the role of any temporal evolutionary effects: 'the first morning of creation wrote what the last dawn of reckoning shall read.'

There is one further scenario. Suppose that the Universe did indeed begin in a state of chaotic anarchy. If the evolutionary emergence of particular symmetries depends upon the local temperature and density, then they will emerge at different rates in different parts of a chaotic universe. Only in places where dominant symmetries, and hence close approximations to invariant laws of Nature, exist will the fundamental building blocks of complex systems of matter and energy be able to arise, and only on these oases of order can those complex systems we call living beings eventually evolve. If there are places in the Universe where Nature displays no close adherence to laws, then we would not expect to be living in these sites. If the phenomenon of inflation occurred in such a heterogeneous Universe then one could find that our entire Visible Universe evolved from the accelerated inflation of a single microscopic domain possessing similar laws and symmetries. Outside this domain things might, literally, be unimaginably different.

Although we can conceive of gauge symmetries and their associated conserved quantities arising as a result of the evolution of the Universe from very high energies to the quiescent conditions that exist today, this is not really sufficient to do away with laws of Nature. We recall from Chapter 4 that, while gauge invariance was a powerful condition which was able to impose the form of the forces of Nature and the way they act, it fails to determine how many species of particle can exist. Similarly, the chaotic scenario for the origin of gauge symmetry is faced with explaining the extraordinary fact that matter is composed of a relatively small number of apparently *identical* elementary particles. What could be less random a state of affairs than that?

Too many laws?

> 'We actually made a map of the country, on the scale of a *mile to a mile!*'
> 'Have you used it much?' I enquired.
> 'It has never been spread out, yet,' said Mein Herr: 'the farmers objected: they said it would cover the whole country, and shut out the sunlight! So we now use the country itself, as its own map, and I assure you it does nearly as well.'   Lewis Carroll

It is worth reflecting a little further on the image of our quest to understand the structure of the Universe as that of creating a map. It is a common allegory, but a bad one. If we are involved in nothing more than a *description* of the Universe lacking any contact with an underlying bedrock of reality, but merely a useful guide to finding our way in it—a sort of upmarket 'Hitch-hikers Guide to the

Metagalaxy'—then science is faced with the depressing goal of discovering only a more and more detailed description of what is, with no hope of foreseeing what is to be; the creation of a model whose scale is getting closer and closer to full size; an enterprise doomed to the fate of Borges' imperial cartographers:

> In that empire, the art of cartography achieved such perfection that the map of one single province occupied the whole of a city, and the map of the empire, the whole of a province. In time, those disproportionate maps failed to satisfy and the schools of cartography sketched a map of the empire which was the size of the empire and coincided at every point with it. Less addicted to the study of cartography, the following generations comprehended that this dilated map was useless and, not without impiety, delivered it to the inclemencies of the sun and of the winters. In the western deserts there remain piecemeal ruins of the map, inhabited by animals and beggars. In the entire rest of the country there is no vestige left of the geographical disciplines.

There is an analogous fate awaiting if we are doomed merely to keep track of the map of events in time. We would be keeping a catalogue of sequences of events. In the view of Poincaré,

> Suppose we have been able to embrace the series of all phenomena of the universe in the whole sequence of time. We could envisage what might be called the *sequences*; I mean relations between antecedent and consequent. I do not wish to speak of constant relations or laws, I envisage separately (individually, so to speak) the different sequences realized.

The scientific method is not just interested in collecting sequences. It is committed to classifying those sequences which are alike, and understanding the interrelation of similarities and differences.

There is a form of chaos that is caused by too much order; too many laws; too much complexity. Our confidence in the charming 'simplicity' of Nature may be misplaced. Nature may look simple only because we have unlocked so few of its secrets. As we dig deeper into the microscopic structure of matter and space–time, we may strike a seam of great complexity created by the simultaneous interplay of an enormous number of factors. Such a situation might appear as lawless as pure chaos.

Spontaneous order

> The universe is full of magical things patiently waiting for our wits to grow sharper.
>
> Eden Phillpotts

John Wheeler paints a striking picture of how order can emerge from disorder by a subtle process of collective self-interaction. He asks us to consider a parlour game of twenty questions, in which the guesser is banished from the room while the others decide upon a word that must be guessed by eliciting the answers 'yes' or 'no' to no more than twenty questions when the guesser returns to the room. But, on this occasion, the party enters into a conspiracy. When we return to begin our interrogation . . .

> we find a smile on everyone's face. We innocently start our questions. At first the answers come quickly. Then each question begins to take longer in the answering—strange, when the answer is only a simple 'yes' or 'no'. At length, feeling hot on the trail, we ask, 'Is the word "cloud"?' 'Yes', comes the reply, and everyone bursts out laughing. When we were out of the room, they explain, they had agreed not to agree in advance on any word at all. Each one around the circle could respond 'yes' or 'no' as he pleased to whatever question we put to him. But however he replied he had to have a word in mind compatible with his own reply—and with all the replies that went before.

The point of this little game is that it conveys something of the concept of observer-created reality which we discussed in Chapter 3 in connection with the interpretation of quantum theory. If Bohr's extreme idealism is true, and the act of observership actually 'creates' the phenomenon that is measured, then the orderly world of law and regular processes that we have come to associate with the rule of law in the Universe must be a form of observer-created reality in which the acts of observation and participation combine to manufacture the picture of reality by the 'game' of quantum observations. The order is generated spontaneously. If Everett's Many Worlds exist then there are forking paths through a succession of 'multiverses' which any observer could trace to compose a history that told of an erratic lawless universe. Our path does not seem to be like that. At first sight we might think this a strong argument against the Everett picture, because there are so many more disordered paths than ordered ones to follow through the labyrinth of world-splitting. But we must temper this by the worry that it may be impossible for biochemically complex intelligences like ourselves to evolve after fifteen billion years in any but the ordered realities. We may necessarily inhabit an ordered branch of quantum reality that is atypical of the whole ensemble of other worlds.

Chaotic non-linear systems of the sort we discussed in the last chapter display remarkable propensities to generate ordered structures spontaneously, despite their superficially chaotic behaviour. Typically, such systems describe physical situations where there exists some connection with the outside environment, either in the form of a continuous throughput of energy, or a consistent small perturbation or disruption of the main system. As the outside influence is slowly altered a series of sudden changes occurs, during which the behaviour of the local system changes in dramatic fashion. For example, if the speed of a flow of water from a tap is increased steadily there is a sudden transition to turbulent behaviour which exhibits new types of order. This common phenomenon of 'bifurcation' into qualitatively new types of behaviour when laws of Nature have non-linear aspects teaches us important lessons. First, it shows how sudden changes are to be expected rather than slow and gradual evolution to new equilibrium states. Second, it shows how transitions can occur which move a complex non-linear system into a regime where qualitatively new types of law dictate what occurs. These laws of organization are not inconsistent with the basic laws of physics which govern the gross aspects of the physical phenomenon under study, but in no sense are these organizational laws reducible to fundamental laws governing the basic forces and elementary particles of Nature. The new structures which appear at the bifurcation points are complementary to the spontaneous appearance of new types of complex-ordering principle which emerge when a particular threshold of complexity is crossed. The third lesson we learn from these spontaneously ordered phenomena is the importance of understanding 'chance' correctly. The intuitive view of random behaviour would persuade us that it is improbable that ordered behaviour should arise spontaneously in Nature. But that assumption is based upon our intuitive feeling for the Gaussian 'law of large numbers', which is predicated upon the assumption that many *independent* events constitute the realm of possibilities. However, when physical systems approach bifurcation points such an assumption is no longer valid. The long-range correlations within the system and the coupling to the outside world drive the system far from its equilibrium state in the absence of external influences, and make the appearance of ordered structures probable, in complete contrast to the situation when the system is close to thermal equilibrium where the 'law of large numbers' applies and ordered structures are improbable.

This departure from the conventional reductionist story has many interesting parallels and extensions. In general, we see that there can exist layers of laws which govern physical situations, not all of which are hierarchically reducible to one law. For example, the word-processor on which I am writing this page combines several complementary levels of law. There are the laws of quantum electrodynamics, which control the basic atomic and subatomic structure of all the components of my personal computer's electronics. Then there is a set of software programming 'laws' which have been imposed upon that circuitry to produce the operating system of the computer. Then there are the rules of grammar, or in the case of a computer-game, the rules of play which have been coded into the program disc. In no sense are the latter two sets of laws, which dictate the rules by which information can be organized, reducible to the quantum laws of Nature which govern the electromagnetic interactions of Nature. Such reductionism is logically impossible. Thus we see how highly ordered systems give rise to new propensities for assuming organizational laws that are 'novel' in the sense that they are not predictable in terms of the laws of physics.

Does life transcend the laws of Nature?

> The concepts 'soul' or 'life' do not occur in atomic physics, and they could not, even indirectly, be derived as complicated consequences of some natural law. Their existence certainly does not indicate the presence of any fundamental substance other than energy, but it shows only the action of other kinds of forms which we cannot match with the mathematical forms of modern atomic physics . . . If we want to describe living or mental processes, we shall have to broaden these structures. It may be that we shall have to introduce yet other concepts.
>
> Werner Heisenberg

> Can these bones live?
>
> Ezekiel

Life and consciousness should be recognized as phenomena that emerge when a particular level of complexity is attained. This complexity level is beyond our present ability to visualize or simulate. The most powerful computer that exists has information-storage and processing capacity less than between ten to one thousand times that of the human brain. A high level of complexity is necessary but by no means sufficient to produce such intricate effects as 'thought', for we know that even at the chemical level very special environments and biochemical candidates must be present to perform the complex functions. Life is like a form of software that runs on certain complex biomolecules. As such it cannot be 'explained' or reduced to the laws of physics that govern the forces of Nature, any more than can the operation of *Pac-Man*. A structure like the human brain is more complex than the underlying laws governing the chemical and atomic forces of Nature. It operates because of the way in which components are organized, just as a computer operates because of the way it is hard-wired together, and hence a knowledge of the individual operation of each of the brain's myriad of nerves and cells will not tell us how it works collectively, any more than the subject of human anatomy could be used as a basis for predictive sociology. A knowledge of the alphabet is necessary but far from sufficient to produce the collective effects that Shakespeare achieved. There can be no laws of thought and action which are described by equations containing the constants of Nature.

We should stress that we are not suggesting any form of vitalism—the discredited notion that living matter differs from all other matter by possessing some peculiar ingredient or *élan vital*. The manifest difference between living and non-living systems is not in basic atomic components but in the attainment of particular thresholds of complexity where new self-organizing principles can spontaneously come into play.

The *aficionados* of artificial intelligence in its strongest form take an operationalist view of the mind, and define it to be nothing but the algorithmic aspects of its information-processing—that is, a piece of 'software' that produces particular outputs for each input. As such it can be mimicked or simulated by a man-made computer, which can in turn be defined as intelligent if no operational procedure can distinguish a human's responses from those of the computer.

We should first remark that even this goal could not be attained if the human mind carries out procedures which are *non-computable* mathematical functions, and hence not within the scope of the action of a Turing machine. This could conceivably be true if quantum processes play some role on the scale of the human brain, but there is as yet no positive evidence for such a view. If we lay to one side this tantalizing possibility we might question the operationalist view that intelligence is nothing but an algorithm. Such a simple view seems to fail to distinguish between the processing of information which our brain carries out subconsciously, or when we dream, and *understanding* which we associate with conscious information processing. Whether this is a distinction without a difference is the question that the pursuit of artificial intelligence may one day answer. But perhaps the prospects of unravelling the problem are not as promising as many naïvely think, because even if the artificial intelligentsia are successful in their quest, they will face curious new problems of understanding. If a complex mind were to be fabricated, then the amount of information necessary to specify and understand it would be prohibitively high. Moreover, it would exhibit disconcerting properties like those that we call irrationality, free-will, subjectivity, and probably the occasional statement of disbelief in human intelligence. A full understanding of such a machine could be impossible—and it would be a job for the psychologist and psychiatrist as much as for the computer scientist.

The attribute of self-reference plays some role in complex systems both at a 'hardware' and a 'software' level. This need not be very subtle. We could devise an 'expert' system capable of upgrading computer hardware and R(ead) O(nly)M(emory) circuits, which is attached to a robotic motor system able to perform mechanical and electrical manipulations on the expert system itself.

Biological systems possess teleological aspects, not in the form of some grand plan or final cause to which the whole course of the evolutionary process is heading, and which determines their ultimate form, but by virtue of the fact that organisms surpassing a critical level of complexity exhibit purposeful behaviour which can alter the way in which they subsequently evolve. Human beings are no longer entirely at the mercy of environmental forces and natural selection because we can imagine and simulate the effects of these pressures, we do not have to learn only by experiencing them. Our minds allow us to imagine many plausible futures and we can alter the environment to maximize our survival probability.

It is important to distinguish three types of *reductionism*: *Ontological reductionism* maintains that there is no *élan vital*. All the material content of the world can ultimately be reduced to elementary particles and forces of the sort studied by physicists. Most scientists assume this to be true.

Methodological reductionism holds that all explanations must be deterministic, and cast in the language of mathematical physics. We should search for explanations of the complex at lower levels of complexity, and ultimately at the level of elementary particles where the most general and powerful laws of Nature act transparently.

Epistemological reductionism claims that 'laws' formulated in one area of science can always be reduced to special cases of laws in other areas of science—that all psychology can be reduced to biology, all biology to chemistry, and all chemistry to physics, for example.

From our discussion it should be clear that, while we are happy to believe in the reasonableness of ontological reductionism, there is no convincing reason to believe in methodological reductionism, and every reason to maintain that epistemological reductionism is false. Large and complicated systems which exhibit chaotic behaviour and stable statistical properties argue against the methodological reductionist. The existence of laws of organization which are independent of the underlying physical laws, the existence of teleological behaviour in living systems, the requirement that anthropomorphic considerations be introduced in order to understand our observation of the Second Law of thermodynamics, and the observed outworkings of cosmological models of the inflationary variety: all witness to the error of the epistemological reductionist and the shifting sands upon which the methodological reductionist is erecting his house. There are no fundamental laws of human history, no theorems about human behaviour, no laws of thought, rather:

> There are more things in heaven and earth, Horatio
> Than are dreamt of in your philosophy.

Accidental symmetries

> Accident, *n*. An inevitable occurrence due to the action of
> immutable natural laws.   Ambrose Bierce

Throughout the post-war period there has been a steady erosion of many supposed laws of Nature and the symmetries imagined to underpin them. This trend should not be interpreted merely as a reflection of the ineptitude of scientists in prematurely adding candidates to the roster of laws of Nature before they have been adequately scrutinized. Rather, it reflects the extraordinary extent to which the Universe possesses 'almost' symmetries which fail to be exact by tiny margins. Only after the development of very sensitive technological probes was it possible to uncover the inexactitude of some of these almost symmetries.

There are a number of interesting examples which fall under this heading. For example, it was long believed that there should exist complete symmetry between matter and antimatter in physical processes. That is, for any interaction involving elementary particles there exists an interaction possessing the same rate, but in which all particles are exchanged for antiparticles. For many years it was believed that this symmetry existed in Nature, but eventually experiments revealed that there are situations in which it is violated by a tiny amount. We believe that it is very fortunate that this is the case for, were the symmetry exact, the Big Bang would have given rise to an equal abundance of particles and antiparticles. In the enormous densities of the early moments of the Big Bang the result would have been a catastrophic annihilation of matter and antimatter into radiation. The abundance of matter surviving this holocaust would have been miniscule. The average density of material in the Universe today would be more than ten billion times less than what it is, and too sparse for galaxies and stars ever to emerge. By contrast, the tiny asymmetry that Nature possesses between matter and antimatter meant that the material which we (by convention) call 'matter' decayed and annihilated just a little more slowly than the material we call 'antimatter', so that one proton of matter survived for every billion or so proton–antiproton pairs that annihilated to produce photons of light.

'Almost' symmetries like the slight imbalance between matter and antimatter in Nature are accidents which can appear either large or small depending upon the type of consequences that one focuses upon. The actual difference in rates between physical interactions involving particles rather than antiparticles is tiny, but on the cosmic scale the consequences of that imbalance are overwhelming.

In Chapter 4 we reflected upon the problem of determining what it was about the Universe that required an explanation, separating those things which are the consequences of physical laws from those which are random outworkings of those laws. The 'almost symmetries' of Nature probably owe their existence to symmetry breakings that fell out in one way rather than another because of chance events rather than invariant aspects of Nature. The tiny margin by which they fail to be exact invariances is slightly worrying. For a long time we were oblivious to their lack of perfect invariance. In the future many of our other assumed invariances might well follow the same lead. For some deep reason it appears that Nature has an inordinate fondness for lawlessness that is slight, for symmetry that is almost, but not quite, perfect.

Places where the laws of Nature break down

> General relativity contains within itself the seeds of its own
> destruction.
> D. W. Sciama

We have seen that we must choose between having unchanging laws at some level or no laws at all. Can there exist some hybrid of these extremes? For example, could there exist laws of Nature everywhere except for special places in the Universe where anything can happen: where the laws of physics governing the world outside do not apply? Remarkably, this may well be the case. Einstein's law of gravitation is a law that has the unusual property of predicting that there can arise states in which its jurisdiction does not apply. It predicts that it cannot predict. These singular points where breakdowns occur demarcate the edge of the Universe. We have encountered them already in the context of the Big Bang singularity which we discussed in Chapter 4. Here we wish to examine their meaning in more detail, and discuss their existence in situations other than the Big Bang. Let us first amplify our earlier discussion.

The distant galaxies are flying away from one other in a state of universal expansion. If one extrapolates this state of affairs backwards in time then we appear to encounter an event of infinite density at a finite time in our past: a *singularity* that has become known as the Big Bang. Superficially, it marks both the beginning of our description of the Universe and its laws. To begin with, the prediction of this singular state was regarded as a pathology of the extremely idealized mathematical pictures of the expanding universe being used. At first, Einstein thought that it was only a consequence of a model which expanded at exactly the same rate in every direction. Make it a little less symmetrical, and one would expect that as the expanding material was traced backwards in time it would not all arrive at one place at one time, and the singularity would be defocused. Moreover, Einstein proposed that the inclusion of realistic pressures could resist the state of compression, and avert the prediction of a singular state in the past, just as the gas pressure within a balloon reacts against our attempts to squash it to infinitesimal size.

Unfortunately, both of Einstein's objections failed to work as he had expected. In fact, far from preventing the creation of a singular state in our past they appeared to make it worse. The asymmetrical models would have experienced it more recently in our past and the universes containing conventional types of pressure ended up creating a worse singularity. The reason can be traced to another of Einstein's discoveries: the formula '$E = mc^2$'. This reveals that all forms of energy, E, including pressures, are equivalent to a mass m. As a result, they exert gravitational attractions of the sort recognized by Newton. Hence, as we go backwards in time and approach the singularity, the rising pressure also contributes an extra gravitational compression that more than compensates for the resisting pressure. Trying to avoid the singularity is like trying to pull yourself up by your own boot-laces.

Later, a more subtle objection was raised. What if the indication of a singular state in the finite past was solely an artefact of your way of mapping the Universe? For example, on geographers' globes of the Earth we use a grid of latitude and longitude lines to label positions on the Earth's surface uniquely. We call this grid a system of co-ordinates because it enables us to co-ordinate different positions. As we move away from the equator towards one of the poles the lines of longitude begin to converge, and they all intersect at the Poles. At these points the system of mapping co-ordinates we have employed to describe the earth's surface have developed a 'singularity'. But if we care to pay a visit to the Polar regions we may readily confirm that there is no rupture in the surface of the Earth at the places where our map co-ordinates become singular. If we were travelling near the North Pole we would use maps that employed a different and more convenient choice of co-ordinate grid that was perfectly well-behaved at the Pole. How do we know that the Big Bang singularity is not also of this benign variety, the consequence solely of an inappropriate way of describing the Universe, but of no physical significance?

At first this seems very difficult to ascertain. Suppose you look at one system of mapping co-ordinates after another, and the Big Bang singularity arises in each one of them. Does that really prove anything? No, because there are an infinite number of candidates to check, and you do not know whether there exist as-yet untried co-ordinates in which things are non-singular. Likewise, the discovery of a co-ordinate system in which the singularity disappears leaves you having to decide why you should regard that as the decisive one rather than the others in which the singularity appears.

It was these awkward dilemmas that forced cosmologists to define very carefully what they mean by a 'singularity' in the Universe, in an effort to avoid all the complications of pressures, co-ordinates, and asymmetries in the structure of the Universe. The first step was to abandon as fundamental the traditional picture of the Big Bang singularity as a place where the density or the temperature, or indeed anything else, becomes *infinite*.

It is easy to see why the concept of an infinity occurring at a point in the Universe is not a good concept to take as primary. A description of the Universe achieved by solving Einstein's equations, or those of some other theory, gives us a space–time map of the Universe. Let us now examine this map for the 'singularities' where the density becomes infinite, and cut them out of the map. What remains is a non-singular map of a universe. We would probably object to this sleight of hand, because the perforated non-singular universe seems to be, in some vague sense, *almost* singular near the perforations. Worse still, if we were able to find a non-singular mathematical model of the Universe how would we know whether our method of finding it or describing it had not implicitly performed a surgical removal of the singularities?

A *singularity* is defined to occur when the path of a light ray, or that of a particle, comes to an end. If this happens, then on reaching the end of its path the particle disappears from the universe because it runs out of space and time. What could be more singular? Conversely, particles could come into being out of nothing at the end of one of these finite space–time paths (although, of course, no mechanism or reason for this is supplied by our definition). This definition is nice because if our paths through space and time encounter holes that have been pruned out of our model universe because they contain points where the density is infinite, then they come to a stop at the perforation edge just as surely as when they hit a point of infinite density where space and time are destroyed. In both cases the inextendable path would signal that the Universe was singular.

We are now left with a number of questions to answer: What is the meaning of the end of a path through space and time—what is a singularity? Can such points occur in the real world, and if so, where might they be found, and can they be observed?

The collection of end-points of paths through space and time is the edge of space–time. Strictly speaking this boundary is not part of the Universe. At this edge the Universe would come to an end, since no particle or light ray could continue to exist in space–time. If we encountered the boundary we would cease to exist in our Universe. This sounds peculiar. Why should the Universe come to an end there? Cosmologists have devoted much effort to the task of proving that Einstein's theory of gravitation ensures that the density of matter, or some other physically measurable quantity, always becomes infinite on the singular boundary of possible universes. Very often this is the case, and the arena of space and time comes to an abrupt end because it is destroyed—torn apart by infinitely strong gravitational forces. But it has not been proved that this is always the case. If it were, then all singularities would be of the intuitive 'Big Bang' variety. Einstein's theory of gravity is known to contain examples of possible universes in which the edge of the universe is not accompanied by any physical quantity becoming infinite. What is not completely decided is whether these situations are likely to occur. All the existing evidence indicates that they are not: the smallest perturbation of the known examples of special benign singularities, or 'whimpers', transforms them into ones of the Big Bang variety which contain physical infinities.

Mathematicians have proved a number of precise theorems which tell us that if certain things are observed to be true then the Universe must contain singularities—places where the laws of Nature break down, and space and time cease to exist. The most powerful of these theorems was that constructed by Roger Penrose and Stephen Hawking, in that its preconditions for the existence of an edge to the Universe can all, in principle, be tested by observation. What are these conditions that the Universe must uphold? We discussed them first in Chapter 4; let us recapitulate briefly. First, general relativity holds and gravity must act upon everything as an attractive force. Second, time-travel must be impossible. Third, there must be a certain amount of material in the Universe. These are not unreasonable assumptions. Certainly, every type of matter we have ever encountered feels the attractive force of gravity. Most people would regard the possibility of time-travel, and the accompanying overthrow of the laws of cause and effect, as something a good deal worse than a singularity.

Having found the assumptions of the singularity theorem to hold true in that part of the Universe which we observe today, we are forced by mathematical logic to conclude that there exists a singularity in our past. That is, if we follow the possible past histories of all the particles and light rays we can see in the Universe today backwards in time, at least one of them must come to an end after a finite amount of time measured by someone moving along them. It is usually believed that this singularity would be universal, and in fact all the paths would come to an end together at a Big Bang. We are only at liberty to disbelieve the inevitability of a singularity if we disbelieve one of the assumptions which led to its deduction.

Although the assumptions are found to be true in our experience, such experience is really rather limited by cosmic standards. We live in quiescent conditions on a pleasant planet some fifteen billion years after any Big Bang could have occurred. Our terrestrial investigations into the behaviour of matter and energy under extreme conditions have not approached the extremities that Nature routinely creates for herself. If the Universe emanated from a state of unlimited density and temperature then all sorts of undreamt-of phenomena might arise. We do not know whether gravity will always remain attractive under such conditions, and even if it did, there could arise new repulsive force-fields in the Universe to oppose the attraction of gravity. All we can say is that if such strange new forces of Nature do exist to prevent the need for the expansion of the Universe to have issued from a Big Bang singularity, then they must arise at densities and temperatures that are so high that a merger of the great theories of general relativity and quantum fields will be necessary to describe them adequately. But we should not forget that the singularities which general relativity renders inevitable do not, as far as we know, necessarily require the presence of extreme physical conditions of density and temperature. Maybe one day we will be able to show that they all do, but if they do not there is little ground for expecting that new force fields or anti-gravitating forms of matter will save us from their reality.

Black hole ontogenesis

> The black holes of nature are the most perfect macroscopic objects there are in the universe: the only elements in their construction are our concepts of space and time.
>
> S. Chandrasekhar

The Big Bang is not the only place where astronomers can encounter the edge of space–time and a concomitant breakdown of the laws of Nature. It is possible for the conditions required to produce a singularity to be met within a finite amount of material in space. If sufficient mass is attracted into a small enough region by the pull of gravity, then the gravitational field that it creates can become so strong that nothing can escape—not even light. To visualize this it is useful to recall Einstein's picture of space and time as a rubber sheet whose shape is moulded by the presence and motion of matter upon it. In places where there is a large concentration of matter in a small region, so deep a ditch can be made in the space–time geometry that a part becomes 'pinched off', and separated from the rest (see Figure 6.1). Anyone caught within the trapped region would be incommunicado with events in the space and time outside it. This state of affairs gives rise to what astronomers call a 'black hole'. The possibility of such dense cosmic light-traps was appreciated long ago, first in 1783 by John Michell, one-time professor of earth sciences at Cambridge, and subsequently, in 1798, by Laplace. They both noted that one could conceive of a spherical astronomical body in which so large a mass was confined within so small a radius that the escape speed—that speed that must be attained on launch in order to escape completely from the gravitational field of the body—was equal to the velocity of light. This state of affairs was in accord with Newton's theory of gravity if one regarded light as composed of little particles. Because the conditions required were so extreme, and the picture of light as a wave was more readily subscribed to then, this idea did not produce any further scientific investigations at the time—although Michell's paper seems to have attracted some attention at the Royal Society where it was first read. The President, Sir Joseph Banks, actually mentions it as the most interesting of recent scientific ideas in his letters to American scientists of the period.

Stars are vast nuclear reactors stabilized by gravity. A star is defined to be a body in which the central pressure is high enough to raise the temperature to the level required to initiate spontaneous nuclear reactions. For most of their subsequent lives the inward pull of gravity is balanced by the outward pressure sustained by the nuclear reactions burning at the core of the star. The lifetime of the star is determined by the time required to burn all its hydrogen fuel into helium and heavier elements. The most massive stars live the briefest lives because they attain the highest central pressures and temperatures, and so their nuclear reactions consume their fuel most rapidly. When stars that are more than

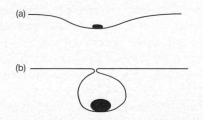

Figure 6.1 A schematic picture of the distortion of spatial geometry by mass. A heavy ball is dropped into a block of foam rubber. In (a) a small mass generates a small curvature of space which we interpret as a weak gravitational field. In (b) the mass of the ball is so great, and the depression in the rubber so deep, that it closes up around the ball, cutting it off from the outside world. Analogously, the presence of a sufficiently great mass within a sufficiently small region of space results in that region being closed-off from the outside world. We say that a *horizon* appears. The distortion of space and time is so great that light cannot pass to the outside world. A black hole has been formed.

three times the mass of our Sun finally exhaust their reserves of nuclear fuel they have no means of opposing the inward pull of gravity. Their fate is to continue collapsing to a denser and denser ball of matter. During this process a sufficiently large mass should be compressed inside so small a region that a *horizon* will always form. Nothing dramatic happens locally to signal that material has passed the point of no return when it falls within that horizon. Only if a rocket attempted to retrace its path back into the distant Universe would it discover that it was within a trapped region. A trapped region of space–time concentrates such a large quantity of mass within a small volume that the speed of light would have to be exceeded in order to escape from it. The surface of no escape defines the boundary, or 'horizon', of a black hole.

A black hole is not a solid object, only a surface in space and time bounded by the horizon. The horizon is like a one-way membrane. Nothing falling in through it can pass back out. Observers outside the horizon cannot receive signals from within the horizon. The only things they can know about the material trapped within the black hole horizon is its total mass, its total angular momentum, and its total electric charge. These are quantities preserved in global conservation laws, and this tells us that the laws of physics retain the invariances reflected by these conservation laws even when black holes are involved in natural processes. There is nothing else to be known about a black hole but its mass, electric charge, and angular momentum. It has no other properties. Two black holes possessing the same values of mass, angular momentum and charge could not be distinguished in any way *from the outside*. This does not mean that the material inside the horizon ceases to have any other properties, only that none of the others are accessible to outside observers. In this sense black holes are the simplest objects in Nature: they do not possess the millions of idiosyncratic defining properties characteristic of even the simplest everyday objects.

It is worth stressing that, although at first glance this entity resembles the one envisaged by Michell and Laplace, it is actually far more extreme. In general relativity and other related theories of gravitation the black hole's horizon is an absolute barrier to outward travel or communication. In the Laplace–Michell object, which is based upon Newtonian gravitation, one can escape as far as you wish from the object. The escape velocity may be equal to the highest attainable velocity (i.e. that of light), but the Newtonian escape velocity is the speed that must be attained to escape from the gravitational field completely, and that means travelling *infinitely* far away. For some launch speed less than that of light one can get as far away from Laplace and Michell's 'black hole' as one wishes. This cannot be done in the case of the relativistic black hole.

Let us first dispel some myths about black hole horizons. If we crossed the horizon of a very large black hole, say with a mass one hundred million times that of our Sun (of the sort suspected to reside at the centre of many large galaxies), we would experience nothing unusual—at first. Conditions would be very mild. The average density of matter within the horizon would be about that of air. We could all be living inside a very large black hole at this moment without noticing anything amiss. But eventually, as the pull of gravity drove us closer and closer to the centre of the black hole, if we were a body with a finite size (that is, not an idealized mathematical point of zero extent in which case we would feel nothing at all, and would be in the same 'weightless' state we experience when falling freely under gravity, say, from a swimming pool diving-board) the tidal forces upon us would become unbearable. We would be torn apart. These forces increase inexorably as the centre of the black hole is approached. What awaits us at the centre? According to the singularity theorems the centre should reveal a singular point: a part of the edge to the Universe where matter is squeezed to an infinite density, and the associated pull of gravity is sufficient to rend the malleable sheet of space–time asunder. As in the case of the Big Bang singularity we do not know whether some unknown force of Nature will allow that fate to be avoided. But it is interesting to examine what happens if a singularity really does occur inside a black hole, since there are probably billions of them in the visible universe.

Cosmic censorship

> It is one of the little ironies of our times that while the layman was being indoctrinated with the stereotype image of black holes as the ultimate cookie monsters, the professionals have been swinging round to the almost directly opposing view that black holes, like growing old, are really not so bad when you consider the alternative.
>
> Werner Israel

All the laws of Nature will come to an end at a singularity. Nothing appears to govern what can emerge from them, any more than there is a known law governing the coming into being of the entire Universe out of the Big Bang singularity, if such there was. What is most curious about such a situation is that it emerges as a prediction of Einstein's theory of gravitation that it must break down somewhere. Such a law of Nature as Einstein's contains within itself the seeds of its own destruction. It predicts that there must exist places in the Universe where *anything* can happen. These points are part of the singular boundary of the Universe. But there is something very unusual about their situation inside a black hole. There, they are invariably surrounded by the horizon surface of the black hole. They are not visible, nor able to exert any influence whatsoever upon the world outside the black hole. If the laws of Nature do cease to exist at the singular points inside black holes, and unpredictable things emerge from those singularities at their centres, just as they might from a Big Bang, then their consequences are trammelled up inside their horizon surfaces. Now, although we *suspect* that a black hole horizon will form whenever stars die and collapse to high density, this is not proven. It is conceivable that a concentration of matter and energy at high density could give rise to a singularity in space and time that was not surrounded by a horizon. Such a 'naked singularity' would be able to influence us in a completely unpredictable fashion. It would be like watching the beginning of the Universe. The breakdown of predictability and scientific inquiry that this would allow is so unsavoury to contemplate that it has been suggested that there exists a law of Nature to the effect that all space–time singularities must be surrounded by horizons. This proposed natural legislation against 'naked singularities' is graphically dubbed 'cosmic censorship' by Roger Penrose, the Oxford mathematician responsible for first proving a singularity theorem and for initiating the study of singularities. It is still not known whether the hypothesis of cosmic censorship is true or false.

Very simple physical reasoning suggests that cosmic censorship, in some form, is true in Nature. We know that energy appears to be conserved in Nature, and moreover that it comes in two forms—kinetic energy of motion and potential energy—and it can be shuffled between these two forms during physical changes. Thus, when water falls from the top of a dam it begins with little energy of motion but enormous potential energy (defined to be the work that you would have to do to raise it up to that starting position); when it hits the hydroelectric turbines at the bottom of the waterfall then all that potential energy has been converted into energy of motion that can be transformed into electric power. If naked singularities did exist in the Universe unguarded by horizons, then any material that fell into the singularity from outside would have to lose an *infinite* amount of potential energy *en route* to a point singularity, and that energy would be radiated back into the Universe in various forms. Our own existence seems to indicate that this does not happen. Something either stops a singularity being reached—and so keeps the amount of radiated potential energy finite—or a horizon forms around the singularity and traps the radiation, or both. This nasty 'heat death' by the consequences of unstoppable gravitational collapse is what would result from the Newtonian 'black holes' of Laplace and Michell.

If naked singularities do occur in Nature then the laws of Nature do not allow us to predict the future. Anything can emerge from a naked singularity as far as our understanding of gravitation is concerned. If naked singularities do not occur in Nature then we could still observe the totally unpredictable emanations from a naked singularity, but we would have to be inside a black hole horizon in order to do so. We would be unable to send any information about what we had discovered to our colleagues back on planet Earth outside the horizon, and we would be doomed to take our knowledge with us to the oblivion of the central singularity.

There is only one situation where the inside story of a black hole is the only story that matters. The Big Bang, if it really occurred, was a naked singularity as far as we are concerned. We are directly influenced by it. If the Universe is 'closed' and destined to recollapse in the future, then one can think of the entire Universe as the interior of a black hole to which there is no exterior. Everything we see around us is the product of that naked singularity: matter, time, space, the laws of Nature. These things tempt one to believe that there is some degree of order behind what comes out of a naked singularity—a meta-law governing the behaviour of singularities, and so *inter alia* of cosmic initial conditions.

In modern cosmology the role of the Creator is essentially assumed by the naked Big Bang singularity. However, a super-civilization could, in principle, create a local naked singularity by moving enough material into a small enough region of space. Their 'theological' status is somewhat undermined by the realization that they could be 'man-made'. This local construction would allow its perpetrators both to create and destroy space and time. This is undoubtedly the hard way to make a Universe. If we could discover how to effect a transition between different energy states of elementary particle matter of the sort that stimulates the period of accelerated inflationary expansion near the Big Bang, then we might be able to 'create' another expanding sub-universe within our own. There have been some attempts to investigate whether it is possible to create a universe 'in the laboratory' in this way. At present it appears that this is impractical, because it is necessary to have a singularity present in order to effect the inflation of a microscopic region locally. But this view may change.

It is not known whether there are laws governing the behaviour of singularities, but there is one interesting speculation as to the type of law that could exist. It appears that the Big Bang singularity in our past was comparatively orderly rather than completely chaotic. But if the Universe one day recollapses to a Big Crunch singularity in the future, it is inevitable that this singularity will be far more chaotic. Irregularities are magnified in the process of gravitational squeezing. There are many ways in which things can become chaotic, but only one way of staying ordered. This type of thinking brings back memories of the discussion of the Second Law of thermodynamics, which codifies our recognition that disorder ('entropy') increases in physical processes. It has been suggested that there may exist a measure of the entropy of singularities and of the Universe as a whole, and that this entropy also obeys the Second Law of thermodynamics. If true, this has a number of consequences. We would expect the initial Big Bang singularity to be an ordered state, and its emanations would not be random. As the Universe expanded, so matter could cluster into irregular structures like galaxies as a manifestation of the gravitational entropy increase. A chaotic final 'Big Crunch' singularity could be the end-result of this increase in cosmic disorder. Attractive as this story sounds, it tells us very little in addition to the facts that were used to motivate it in the first place. The reason it is regarded as a possibility is that an analogous 'gravitational' entropy to that which it proposes does exist for unchanging gravity fields like those of black holes with horizons.

In 1974 Stephen Hawking showed that black holes are, thermodynamically, black bodies. They obey the laws of equilibrium thermodynamics. They have temperatures and entropies given by their gravitational fields and surface areas. It had been known prior to 1974 that the total area of event-horizons of black holes involved in any process cannot decrease, and there was a formal resemblance between the rules obeyed by the surface area of a black hole and by entropy in thermodynamics. Hawking showed that when the influence of quantum mechanics upon black holes is considered this analogy becomes a reality. Quantum black holes radiate particles with a temperature and entropy actually determined by the gravity and area of the horizon, and after a finite time this radiation could result in the complete evaporation of the black hole. The end-result of this evaporation process is not yet clear. It may well be a naked singularity. If so, this would show that 'cosmic censorship' is untrue when quantum theory is included in the behaviour of gravity. On the other hand, the end-result of the evaporation may well be some new type of object—a superstring perhaps, or an inert elementary particle. If the evaporation leaves some massive remnant, however small, the formation of a singularity will be avoided.

These exotic properties of black holes could be true. We say 'could' because we do not know whether some new quantum gravitational force stops the singularities ever forming in the first place. They highlight a trend we have seen already in other areas of modern physics: the importance of distinguishing phenomena from *observable* phenomena.

Gravity imposes limits upon what knowledge of Nature is attainable by observers of it, and these limitations depend upon where the observers are located. The existence of black hole horizons partitions the Universe into causally disjoint regions. No prediction regarding the events on the inside of black hole horizons can ever be verified by an outside observer. Does this mean that the inside is not an area of 'scientific' investigation for those on the outside? If so, then what must be the judgement of observers inside the horizon regarding the nature of the outside world from which they have passed? What would the inside observer make of the idealists' claim that the notion of a horizon is just a mental creation, or of the instrumentalists' argument that it is but a device for understanding the world? These questions make one doubly suspicious of any attempt by philosophers of science to 'define' science or scientific method. Any such definition needs to know of the extent and depth of the subtlety of Nature and the objects within the Universe. Notions like falsification, verification, and operationalism seem like gloves that ill-fit the hand of Nature, however fashionable they may appear on the hands of idealized scientists.

Can we probe a singularity?

> What doth gravity out of his bed at midnight?
>
> Shakespeare

Let us suppose, just to make life more interesting, that singularities of the sort predicted by Einstein's theory of gravitation really do occur in Nature. They may arise at the centres of black holes shrouded by horizons that prevent them affecting the outside world, or they may exist in a naked form. In the first case we could only be influenced by the singularity's anarchic tendencies if we were already inside the horizon ourselves. In the other we could, in principle, observe and record the output from the singularity, and still depart to report it back to Earth. But what if we wanted to reach the singularity itself, or at least send a probe all the way into the singularity—would this be possible? If, as seems likely, a singularity is accompanied by ever-increasing densities of matter and energy, then we are in something of a quandary. Any probe must be made of some material and have a finite size. As it approaches closer and closer to a singularity of the high density sort it will be subject to ever-increasing tidal stresses, and eventually will be torn apart; first into pieces, then into atoms, then into quarks, and so on. In some sense the singularity prohibits close scrutiny. The enormous forces in its vicinity act as a security system to guard its innermost secrets. The more detailed and precise the information one seeks to extract from the singularity, so the more sensitive and sophisticated is the type of probe that must be sent into it, and the harder it becomes to deposit it close to the singularity in working order. If the singularity is a very high-density environment it is guarded by another general feature of Nature which makes it increasingly difficult to extract information from it. Any light signals that our probe transmits back to us must escape from the gravitational pull of material in the neighbourhood of the singularity. This will be possible if the singularity is naked, but only at some cost. The light signals must use energy to overcome the gravitational field just as you or I must do if we wish to walk uphill. This loss of energy is called the 'gravitational redshift', and can be observed using very sensitive measuring devices when light rays move in the Earth's gravity field. It means that the strength of the signal is degraded and its information content decreased when it reaches us far away from the singularity. Something like this is also associated with the singularity at the beginning of the Universe, as William McCrea has pointed out. The closer the source of our information to the beginning of the Universe, so the more is that information degraded by the time that it reaches us here and now. McCrea speculates that there may exist a fundamental restriction upon how much we can know about what happened at some time in the distant past because of the degrading of information in the form of light signals (or indeed signals of any sort). This restriction would also limit our scrutiny of a naked singularity.

Staccato time

> There is a time for everything.
>
> Ecclesiastes

There is one further question to pose concerning the approach to a singularity of high density: how long does it take to get there? In the case of the Big Bang singularity this amounts to asking how old the Universe is. The traditional answer would give the time measured by a clock that fell freely under gravity into the singularity, if followed over its history. This measures the *proper time* interval we discussed in the context of special relativity. But why should this apply to the Universe as a whole? The Universe is governed by general rather than special relativity. How do we know that there is not some other more fundamental measure of time associated with the Universe as a whole?

Consider the fate of a clock as it approaches the high density environment of a singularity. Gradually it starts to be crushed in one direction and then stretched in another until it is broken to pieces. Immediately, we counter this by switching to another, more robust, chronometer, but very soon this too is destroyed. We switch to clocks that use atomic oscillations, but they too are torn apart. And so we go on. What, we must ask ourselves, do we mean by proper time when there are no clocks left to measure it? In this extreme environment we must look around for a way of measuring time that is tied to the only thing that does survive: the curvature of the sheet of space and time. It has been found that Einstein's equations predict that some singularities make the space and time around them behave in a very characteristic fashion that provides a sort of natural clock. The squeezing of matter to higher and higher density as the singularity is approached has an undulatory character which vibrates faster and faster as the singularity is approached. But whereas a finite amount of proper time elapses before an hypothetical clock hits the singularity, an *infinite* number of these oscillations of space would occur during this finite interval of proper time. In 'oscillatory time' the Universe would be judged to be infinitely old. This is not a trick of the Zeno paradox variety, for an infinite number of distinct physical events occur before the singularity is reached. We reject Zeno's argument that it takes forever to walk across the room because there are an infinite number of sub-intervals to be traversed, because those intervals do not physically exist. By the same token, as Charles Misner has stressed, we must hold this oscillatory Universe to be *infinitely* old because an infinite number of physically distinct events happen according to a clock defined by the changing geometry of space itself. If the idea of an infinite number of things happening in a finite amount of time seems odd then consider a simple example, shown in the Figure 6.2 where we have displayed the graphs of the functions $y = \sin(x)$ and $y = \sin(1/x)$ over a range of values for x. As x approaches zero the graph of $\sin(1/x)$ undergoes more and more oscillations. An infinite number of them must occur before $x = 0$ is reached, however close to zero one starts out from on the x axis.

Thus, if one were a being (or a computer) whose subjective time 'ticked' at the same rate as these oscillations, one would experience 'living' forever. One would never reach the singularity where the laws of Nature do not exist: they would remain a 'well at the world's end'. The question of whether there exists some unique absolute standard of time defined globally in the Universe by its intrinsic geometry is a major unsolved problem of cosmology. Until we know the answer

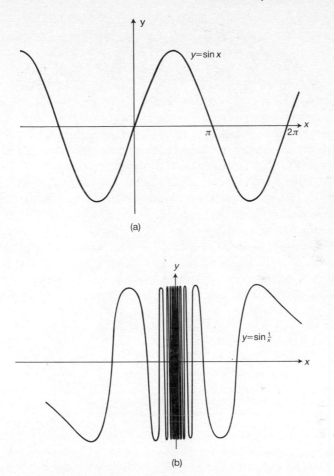

Figure 6.2   (a) The graph of $y=\sin(x)$. (b) The graph of $y=\sin(1/x)$ which undergoes an infinite number of oscillations on approach to the origin at $x=0$.

we will not really know what physical meaning to associate with space–time singularities, or how to answer those who ask whether the Universe is of finite or infinite age (see also Figure 6.3 for another example).

This approach to the definition of time will be recognized as an operationalist one. It regards time as defined by the process whereby it is measured, and is a close cousin of the constructivist philosophy of mathematics. The singularity theorems of Hawking and Penrose which we have described, both above and in Chapter 4, are non-constructive mathematical proofs which establish the existence of some singularity by contradiction, and hence would not be admissible as valid truths to the constructivist mathematician nor, presumably, to an operationalist physicist like Bridgman. For them the issue of the

Figure 6.3 Maurits Escher's 1960 woodcut *Circle Limit IV–Heaven and Hell*. This is a conformal representation of the Lobachevskii plane for which 'infinity' is at the circular boundary. There are an infinite number of interlocking devils and angels diminishing in size as they radiate outwards from the centre. An infinite number of drawings appear in principle on approach to the boundary. A clock that 'ticked' every time it encountered one would take an infinite amount of its own time to reach the boundary which none the less resides at a finite Euclidean distance from the centre. This type of design is sometimes called the Poincaré plane after Henri Poincaré who also suggested that it might give a picture of a finite universe with a practically unattainable boundary.

singularity's existence under the assumptions of the theorem would remain undecidable (it is an interesting question to find out what additional assumptions would be necessary to prove the singularity theorems constructively). Such a view of the singularity 'theorems' would be consistent with the staccato time we have just discussed. It is only in the particular type of universe model which exhibits the infinite sequence of oscillations that no singularity is encountered after a finite number of operationally outworkable steps. In other universe models this type of operational definition of time may not exist, and only a finite number of physically realizable things may occur before a singularity is reached.

Constants of Nature

> Suppose that instead of 60 chemical elements there were 60 milliards of them . . . uniformly distributed. Then every time we picked up a new pebble there would be a great probability of its being formed of some unknown substance; all that we know of pebbles would be worthless for it.
> H. Poincaré

In our mathematical equations that purport to describe the workings of Nature there arise certain quantities to which we ascribe a special status, and which we have come to call 'constants of Nature'. They arise naturally in our equations as constants of proportionality between different variable physical quantities. As such, they can be determined only by measurement. Although the form of the relationship between the physical variables will be dictated by some symmetry principle or invariance, the principle will tell one only that, for example, energy is proportional to mass. It cannot give the value of the constant of proportionality. The first such quantity to be discovered was the gravitational constant of Newton. It arose from the deduction that the force of gravity between any two masses was proportional to the product of those masses and inversely proportional to the square of their separations. This constant of proportionality is called the Newtonian gravitational constant. Although the existence of such a parameter is inevitable in Newton's scheme, and also in Einstein's theory of gravitation that superseded it, its numerical value cannot be determined by the theory. It has been ascertained only by observation of the actual strength of the gravitational force.

The fact that these constants arise as proportionality constants in equations describing natural phenomena might ultimately turn out to be misleading. They may be artefacts of our particular mode of description. Nevertheless, this possible bias is a small price to pay for the convenience of a way of describing the world which is able to separate automatically those parts which are unknown from those parts which may be dictated by symmetry considerations.

A constant of proportionality in a physical equation will only be given the epithet 'constant of Nature' if it appears in a law of Nature which is believed to be *universally* true. Thus, Newton believed that his law of gravitation applied to everything, whether it be on the Earth or the heavens, and so the constant of gravitation which it implies lays claim to the status of a universal constant of Nature. The elasticity of your bicycle tyre is also a proportionality constant in an equation describing the stretching of the tyre under pressure, but it is hardly a universal constant because it is associated with a particular piece of rubber which differs from every other piece in some way. We would, however, be interested in attributes of individual objects all of which were identical throughout the known universe; for example, the amount of electric charge on an individual electron, since we believe all electrons to be identical in every respect.

Real advances in our understanding of the physical world always seem to involve either:

(i) The discovery of a new fundamental constant of Nature,
(ii) A formula showing how the value of one constant of Nature is determined only by the numerical values of others, or,
(iii) The discovery that a quantity believed to be a constant of Nature is not constant.

For example, the introduction of the quantum theory by Planck, Einstein, Bohr, and others brought with it the new fundamental constant known as Planck's constant which dictates the quantum limit on direct observation which we discussed in Chapter 3. Einstein's theory of special relativity gave universal status to the velocity of light in vacuum, and Einstein showed it to provide the link between the concepts of mass and energy. Towards the end of the nineteenth century Maxwell's combination of the theories of electricity and magnetism into the unified picture of electromagnetism was also mediated by the fundamental status of the velocity of light.

To most physicists the ultimate goal of their subject is nothing less than the *determination* of the numerical values of all these universal constants: the demonstration that they can possess only one possible set of self-consistent values, and that this requirement of self-consistency, together with a minimum of symmetry principles, will be enough to determine the structure of the world uniquely. This requires the development of theories in which the role of constants is deeper than that of mere proportionality constants. We require the existence of the proportionalities themselves to be contingent upon the value of the proportionality constant. It is interesting to see the extent to which this view has been continually echoed by different scientists in slightly more specific ways.

Einstein thought that with regard to the constants of Nature:

> I would like to state a theorem which at present cannot be based upon anything more than upon a faith in simplicity, that is, intelligibility of nature: there are no *arbitrary* constants of this kind; that is to say, nature is so constituted that it is possible logically to lay down such strongly determined laws that within these laws only rationally completely determined constants appear (not constants therefore, whose value could be changed without destroying the theory).

Or, the recent views of the Nobel prize-winning particle physicist Steven Weinberg:

> Quantum mechanics and relativity, *taken together*, are extraordinarily restrictive, and they therefore provide us with a great logical machine. We can explore with our

minds any number of possible universes consisting of all kinds of mythical particles and interactions, but all except a very few can be rejected on a priori grounds because they are not simultaneously consistent with special relativity and quantum mechanics. Hopefully in the end we will find that only one theory is consistent with both and that theory will determine the nature of our particular universe.

Both these writers would like to believe that the world is uniquely determined by the requirement of self-consistency. This is a modern notion, and it is interesting to recall the opposition of Newton to the concept that Nature possessed a necessary structure. Through the pen of Samuel Clarke, he wrote to Leibniz that

> From this fountain [the free will of God] it is that those laws, which we call the laws of Nature, have flowed, in which there appear many traces of the most wise contrivance, but not the least shadow of necessity. These therefore we must not seek from uncertain conjectures, but learn them from observations and experiments. He who is presumptuous enough to think that he can find the true principles of physics and the laws of natural things by the force alone of his own mind, and the internal light of his reason, must either suppose the world exists by necessity, and by the same necessity follows the laws proposed; or if the order of Nature was established by the will of God, that himself, a miserable reptile, can tell what was fittest to be done.

Newton's views were not coloured entirely by expediency of scientific practice. They reflect the 'voluntarist' theology of his day (which had been codified into 'nominalism' by Occam during the thirteenth century), with its belief that there were no limits imposed upon Divine power by natural laws. These laws only held 'nominally', but they could be temporarily abrogated in 'miraculous' events.

The development of many new fundamental theories possessing débutante constants of Nature is the mark of science passing through a vibrant adolescent phase. The result will be many theories each successful within its own sphere of application, but partially, or even completely, disjoint from the others in the range of phenomena to which it applies. General relativity and quantum theory are good examples. Each has been brilliantly successful within its own domain, giving both successful predictions and satisfactory explanations. But these two theories are incompatible with each other in their present forms; at least one of them, and almost certainly both, must undergo substantial revision if they are to be wedded together. This marriage is necessary if a complete picture of the earliest moments of the Universe's evolution are to be reconstructed. It is only during those first instants of the Big Bang that gravitational forces are strong enough to warp the curvature of space and time so greatly that the wave-like character of the masses involved become important. It is only in this exotic environment that we come face to face with the conjunction of influences of gravity upon the microscopic world of elementary particles. Elsewhere, and in particular within the parts of the Universe which we observe today, the forces of gravity are so weak compared with those of electricity and magnetism that the quantum world is, for all practical purposes, decoupled from that of general relativity. Nevertheless, we would expect any legitimate union of quantum theory and general relativity to provide links between the different constants of Nature which characterize the theories. Such a link would, in effect, reveal one of the fundamental forces or constructs of Nature to be derivative from others. Alternatively, it might be that both these theories must suffer a change into something richer and stranger than either before a synthesis of their essential content is possible. We recall how the theories of static electricity and magnetism once looked so different, and were unified into a single theory of electromagnetism only when their dynamic aspects were incorporated. The unusual features of superstrings have singled them out as prime candidates to unify and explain the different strands of Nature. We have already discussed (in Chapter 4) some of the hopes and fears of this latest quest for an all-encompassing, all-explaining theory of Nature. It is quite possible—even likely—that a successful theory of this type will ultimately predict the numerical values of constants of Nature, but this would still leave undetermined the cosmological initial conditions and the values of the 'almost' symmetries of Nature that arise by quasi-random symmetry breakings during the history of the Universe.

Weights and measures

> 'Tis as if they should make the standard for the measure we call a
> foot, a Chancellor's foot; what an uncertain measure would that
> be? One Chancellor has a long foot, another a short foot, a third
> an indifferent foot.   John Selden (1689)

If we are to appraise the problem of determining and interpreting the fundamental constants of Nature, we must point to one important aspect of the quantities which we regard as the most fundamental. We are used to measuring things in terms of various systems of weights and measures. There are many such systems. None is sacrosanct. Each is designed for convenience in particular circumstances. Many were derived anthropomorphically from the dimensions of parts of the human anatomy. The 'foot' is the most obvious unit of this sort. Others are no longer familiar. The 'yard' was the length of a tape drawn from the tip of a man's nose to the farthest fingertip of his arm when stretched out horizontally to one side. The 'cubit' was the distance from a man's elbow joint to the farthermost fingertip of his outstretched hand, and varies between about 17 and 25 of our inches in the different ancient cultures that employed it. The nautical unit of length, the fathom, the largest distance-unit derived from the human anatomy, was defined as the distance between the fingertips of a man with both hands outstretched horizontally to the side.

The most obvious problem with such units is the fact that different men are different sizes. Which man do you measure as your standard? The king is the obvious candidate (and one unlikely to be faced with a rival in ancient times) but, even so, this results in a recalibration of units every time the throne changes hands. One interesting response to these problems was that devised by David I of Scotland to define the 'Scottish inch' early in the twelth century: he ordained that it was to be the *average* drawn from measurements of the width of the base of the thumbnail of a 'mekill' [big] man, a man of 'messurabel' [moderate] stature, and a 'lytell' [little] man.

The modern metric system of centimetres, kilogram, and litres, and the traditional 'Imperial' system of inches, pounds, and pints are equally good measures of lengths, weights, and volumes. They were defined originally in terms of some standard object kept at one place in the world, with copies of it elsewhere. For example, the kilogram was the mass of a special cylinder made of an alloy of platinum and iridium stored inside a vault at the French Bureau of Standards in Sèvres. This procedure for defining standards in an objective manner divorced from human standards was first introduced by the French in 1799.

Likewise, we can measure temperatures on the Celsius or Fahrenheit scales. These are conventions. Clearly it would be nice if we could express the basic constants of Nature in a way that is independent of the system of units being used. This we can do by expressing the fundamental quantities as ratios between quantities of the same type. Thus, the ratio of the electromagnetic force between two protons and the gravitational force between them is a pure number; it does not depend upon how one chooses to measure the units of force, or mass, or anything else for that matter. The answer is close to

$$10^{39}$$

Take another example: the ratio of the mass of the proton to the mass of the electron is also a pure number. Its value is roughly

$$1836$$

The strength of the electromagnetic force between two electrons separated by the wavelength of their quantum waves can also be represented by a pure number; the result is roughly the fraction

$$1/137$$

And so we could go on. Stripped of all its superficial and secondary complexities, our most basic description of the essential character of the physical world reduces to a catalogue of pure numbers of this sort. They are the ingredients necessary to turn our mathematical equations and invariance principles into a working description of *our* world rather than some other hypothetical one. If all our existing physical theories were perfectly correct in every deductive principle we would still need to measure these numbers in order to make use of them. The general structure of the theories that we develop, and the fact that they give rise to constants of Nature, could have anthropomorphic aspects, but the values of these constants cannot.

Varying constants?

> God [could] vary the laws of Nature, and make worlds of several sorts in several parts of the universe.   Isaac Newton

We began this chapter by warning against too careless an appeal to the idea of changing laws of Nature. When working scientists speak of such a concept they invariably have something more specific in mind. They are questioning the standard dogma that the quantities, like those of the last section, which we expressed as pure numbers are really the same everywhere and everywhen. If a constant of Nature is not a pure number, but has some units of measurement associated with it—metres, feet, or inches, say—then one cannot give any meaning to the idea that it may be changing in space or time because one can make it appear so to do by changing the system of units used to measure it. But a quantity that is a pure number is oblivious to the systems of units used to evaluate its components, and a change in its value would have a real and observable meaning. It is the fact that these numerical constants of Nature appear to possess the same values to very high precision everywhere we have looked in the Universe that forms the corner-stone of our belief in the constancy of the laws of Nature. Furthermore, because of the finite speed of light-signals travelling through space, our observations of very distant objects are also observations of events which occurred in the distant past. The light which we are *now* receiving from distant quasars was emitted in the distant past. In the spectrum of the light from distant astronomical objects we are therefore observing the consequences of the numerical values of the constants of Nature as they were more than twelve billion years ago; long before human life existed on this planet.

There is a striking terrestrial example of the constancy of the constants of Nature which emerged in 1976. In a uranium mine situated in the African state of Gabon there exist abundances of two isotopes of the rare element samarium (isotopes are versions of the same element where the atomic nucleus contains the same number of protons but a different number of neutrons). In ordinary samarium the ratio of these two types of samarium is about 9:10, but in the sample from the Oklo mine the ratio is dramatically reduced to about 1:50. The reason for this is that the conditions in the mine conspired, over billions of years, to produce a natural nuclear reactor which has transformed one isotope into the other. However, the transmutation of one samarium isotope into the other needs a very special set of coincidences to obtain, which involve the relative strengths of the electromagnetic, radioactive, and nuclear forces of Nature. The fact that the reactor did go critical two billion years ago means that the values of the fundamental constants of Nature which combine to create the coincidence necessary for the special nuclear reactions to occur must have existed then just as they do now. This places amazingly strong limits on the amount of variation that could have occurred in the values of the pure numbers which characterize the strengths of the fundamental electromagnetic, weak, and nuclear forces of Nature. Over the fifteen-billion-year age of the observable Universe these constants can have changed by no more than one part in a million, one part in fifty, and one part in fifty million respectively. The weak force, as its name suggests, has such minor effects compared with the other two forces of Nature that it is allowed to change quite a lot before there arise the sort of adverse consequences that obtain from relatively small changes in the other two forces.

Astronomical observations are well disposed to give powerful constraints upon any change in the accepted constants of Nature with space or time. We recall that the enormous distance from us of stars, galaxies, and quasars means that the light we receive from them today left long ago, and so we are seeing the distant parts of the Universe as they were many millions, or, in the case of galaxies and quasars, even billions of years ago. Thus, it is possible to compare the physics of the radiation from distant astronomical sources with that on Earth here and now to check whether there exists any difference. Considerations of this sort enable us to conclude that any changes in the values of the constants of Nature, if they exist, are by less than one part in a hundred over the fifteen-billion-year expansion history of the Universe.

The question of variations in space is more problematic from a conceptual point of view. Variations of, say, Newton's constant of gravitation—or some dimensionless combination of it in conjunction with other dimensionful constants of Nature—appear to require the existence of 'preferred' places in the Universe where a particular constant would be a maximum or a minimum. However, philosophical objections notwithstanding, a universe in which constants of Nature varied in space would be an interesting place. Suppose that only the 'fine-structure constant', $\alpha = 2\pi e^2/hc$ (e is the electric charge of the electron, h Planck's constant, and c the velocity of light in vacuum), which controls all the electromagnetic interactions of Nature, were to fall in value as one moved outwards in space from some particular place.

This variation in the strength of electromagnetism would influence both chemistry and physics. In this hypothetical world there would exist an 'annulus of life' at some special distance from the centre of the universe, where α was close to $1/137$, in which atoms could exist. In places where α was larger there would exist no atoms, because the electrons would get pulled into the atomic nucleus. In sites where α was smaller atoms would be too weakly held together to survive. Observers like ourselves could only arise in the annulus of life. If, instead of supposing that the fine-structure constant falls off steadily in size with distance, we were to postulate an oscillatory variation of a sinusoidal sort, then there would be many (infinity if the Universe stretched all the way to spatial infinity) annuli of life. The reader who knows a little physics can pass an amusing evening piecing together what this sort of Universe would look like to one of the inhabitants of one of the 'annuli of life'.

There is no observational evidence for the variation in space or time of any of our traditional constants of Nature, but there exists a curious unexplained coincidence which might be telling us that the constants of Nature have values which are determined statistically, and are changing in space and time in a manner that is bound up with the evolution of the Universe as a whole. It has long been known that the dimensionless number which characterizes the strength of the force of gravity between two protons is equal to about 10^{-39}. This is a curiosity in itself, but one notices the 'coincidence' that this is roughly equal to the inverse square root of the number of protons (about 10^{78}) in the visible Universe today, and this inverse square root is the expected statistical fluctuation in a random collection of objects. This implies that the gravitational constant arises in some way as a statistical manifestation of the total number of atoms in the visible Universe. This is, of course, very speculative (the statistical relation is only suggested if one compares the number and gravitational force between *protons* in the Universe. When other, possibly more fundamental, elementary particles are used in the argument the inverse-square-root coincidence no longer exists). We shall have more to say about the numerical coincidence of Nature on which the statistical coincidence is based in the next chapter.

A window onto extra dimensions?

> 'Gracious!' exclaimed Mrs Snip, 'and is there a place where people venture to live above ground?' 'I never heard of people living *under* ground', replied Tim, 'before I came to Giant-Land.' 'Came to Giant-Land?' cried Mrs Snip, 'why, isn't everywhere Giant-Land?'
>
> R. Quizz

For many years the idea that the dimensionless constants of Nature might change slowly with the passage or time, or from one place to another, was one that was not tied into the main body of physics. It was a baseless speculation that relied upon completely *ad hoc* ideas as to how the variations occur. But in recent years it has emerged how observed variations might be expected.

We have stressed that if there exist any laws of Nature there must exist unchanging constants—a manifestation of an *invariance*. These constants will codify unchanging aspects of the three dimensions of space in which we live. But what if space possesses more dimensions than the three with which we are familiar? Suppose it were four-dimensional. The extra dimension could be imperceptibly small, with no overt effects save in the world of elementary particles. We would be in the position of ants crawling upon the surface of a ball with no conscious experience of the extra dimension of space that exists away from the surface. The real laws and their unchanging constants would exist in the four-dimensional space, and our three-dimensional section of it would not necessarily have unchanging constants associated with it. If the strength of the electromagnetic force of Nature is constant in the four-dimensional world, then it will be found to change in strength in the three-dimensional world in proportion to the extent of the fourth dimension. If the extra dimension is getting bigger then we should observe our forces of Nature to be getting weaker. If the extra dimensions are shrinking in size then our fundamental forces should be getting stronger. We can suppose the Universe to possess additional spatial dimensions, and calculate how any change in the extent of these extra dimensions is manifested through changes in our three-dimensional 'constants' of Nature. For example, the charge of the electron will decrease inversely as the mean radius of any extra dimensions of space. The gravitational constant of Newton should fall inversely as the volume of the extra dimensions of space. These predictions enable the precise observations which witness to the constancy of the constants of Nature to be interpreted as direct evidence that any extra dimensions of space (if they exist) are not changing with time. If the 'universe' of extra dimensions is expanding like our observed Universe, or even contracting, then it is doing so at a rate that is about a billion times slower than that of the observed three-dimensional universe. Such evidence for the static character of any additional dimensions of space is quite consistent with the expectation of particle physicists. We recall that they believe any extra dimensions should be confined to a microscopic scale, and not be in the macroscopic state of expansion exemplified by the three dimensions that we inhabit.

The lesson of this development is that we may be living in a small slice through a larger Universe. The true laws of Nature and the most compelling symmetries will, perhaps, then be found only when the whole Universe is appreciated, and not just our slice of it. But, somewhat surprisingly, we have found that the presence of extra spatial dimensions can have observable consequences within our observable three-dimensional Universe. With such a broadening of our horizons as to what determines the forms of the laws and constants of Nature around us, we would be wise to regard the fundamental laws of physics that we use so successfully as provisional. When we understand so little about something as we do the origin of symmetry and uniformity, it behoves us to be tentative as to the true degree of its permanence, and to recall Sydney Smith's aphorism, 'when I hear any man talk of an unalterable law, I am convinced that he is an unalterable fool'.

Selection effects

> Procrustes, you will remember, stretched or chopped down his
> guests to fit the bed he had constructed. But perhaps you have
> not heard the rest of the story. He measured them up before they
> left next morning, and wrote a learned paper 'On the Uniformity
> of Stature of Travellers' for the Anthropological Society of
> Attica.
> A. S. Eddington

Patterns in the trees

> Let every student of nature take this as his rule that whatever the
> mind seizes upon with particular satisfaction is to be held in
> suspicion.
> Francis Bacon

On London underground trains there exists a particular type of advertisement
which I can even remember seeing as a child. It informs you that if you supply
the next entry in several sequences of numbers, or pick the odd one out of a
collection of superficially similar shapes, then you could land a well-paid job in
the wonderful company whose telephone number you can find listed below *et
cetera*. Advertisements like this illustrate how readily our society has come to
equate the ability to spot patterns or relationships, usually those of a
mathematical or geometrical nature, with 'intelligence'. Indeed, most IQ tests
place a very heavy emphasis upon these mental abilities. Whether or not they
measure anything as definable as 'intelligence' is usually irrelevant to those
who set these tests. They are interested primarily in that specialized ability of
pattern recognition and isolation. In this sense, intelligence is defined as being
what intelligence tests measure. It is likely that our very existence as a species
owes much to this ability to delineate patterns. Evolution may have made us a
little too adept at spotting them. Our propensity to see patterns where none exist
at all is witnessed by ancient Man's enthusiastic identification of ploughs and
hunters, crabs and scales tracing out the patterns of stars we call the
constellations, or by modern Man's enthusiasm for Martian canals. But this
disposition to identify patterns where none exist is a better one to be saddled with
than a failure to perceive patterns that really do exist. If you keep telling your
family that you see tigers in the trees when there aren't any they will merely think

you are a little paranoid, but fail to see tigers in the trees when they really are there and you're dead! Over-sensitive pattern recognition tends to survive.

Today, tigers are not such a problem. But we have inherited an ability to perceive patterns that in some sense are not really there. The fact that this propensity varies from individual to individual is something that psychologists have attempted to exploit by using the Rorschach ink-blot personality test to evaluate mentally disturbed patients. It is interesting to undertake some simple experiments to see how the eye and brain operate. The picture drawn below in Figure 7.1 consists of many concentric circles of dots. Each circle contains the same number of dots, and the dots on one circle lie half-way between those of the circles on either side of it. So the dots all lie along straight lines passing through the centre of the system of circles. Thus the patterns which 'really' exist in the sense of having been built into the picture deliberately, are circles and straight lines. But what does the human eye perceive? Close to the centre of the picture we do see circular rings, but towards the outside the eye picks out crescent-shaped petals. It does this because the brain is adept at drawing imaginary lines from a point to its nearest neighbour. Near the centre of the picture the dots are closely ringed, so that the nearest neighbour to any particular dot is one next-door to it on the same circular ring. As we look farther from the centre the dots on any circle become more spaced-out, and the nearest neighbour is to be found on an adjacent circle. The crescents are the lines the eye most readily draws in your mind between these closest neighbours. As a further experiment you might like to tilt the book, and look at the picture along the page. You will see a new set of apparent patterns because the different perspective creates a new set of nearest neighbours for the points. The patterns change according to your angle of view. Figure 7.2 reveals the confusion that your brain experiences when it cannot decide upon a pattern. It continually changes its fix, and this creates a peculiar dynamic effect. There are many influences at work in producing a percept of Figure 7.2, not least of which is the positioning of prominent visual cues at the centres of circular arrays of points traced by nearest neighbours. The eye identifies the symmetrical patterns most readily.

This little game of illusions has a serious side to it. At the moment astronomers are trying to determine whether significant intrinsic patterns of lines and cells really exist in the observed distribution of galaxies in the Universe. Patterns certainly do 'appear' to exist, but it is not clear whether they are just chance effects highlighted by the eye's peculiar fondness for patterns, or attributable to features intrinsic to the galaxy formation process.

Psychologically we find pattern, symmetry, and order appealing. Throughout the arts of ancient cultures we find symmetrical patterns of great sophistication developed for purely decorative purposes. Subsequently, some of these patterns and symmetries have been found to possess sophisticated mathematical properties. In modern times many examples of the draughtsmanship of Maurits Escher (see for example Figure 5.1) have turned out to exploit very subtle

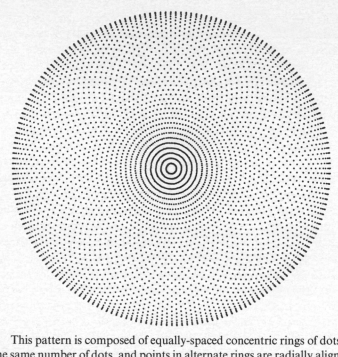

Figure 7.1 This pattern is composed of equally-spaced concentric rings of dots. Each ring contains the same number of dots, and points in alternate rings are radially aligned. Despite these built-in patterns of circles and straight lines, the eye perceives three dominant patterns: concentric rings near the centre, crescent-shaped 'petals' farther out, and radial straight lines near the perimeter. These impressions are dictated by the tendency of the eye to trace out imaginary lines between nearest neighbours. Near the centre of the figure the nearest neighbour of any point lies on the same circle. Farther out the nearest neighbour is to be found on an adjacent circle, until there is crowding near the periphery. If the reader tilts the page and views the Figure at an angle it will be found that the pattern has completely changed. This change reflects the fact that the nearest neighbours are altered by the projection effect of the inclined viewpoint.

mathematical symmetries which he perceived visually with no knowledge of mathematics at all. Consequently Escher seems to have been adopted by mathematicians as their cultural attaché for the arts. All this must trouble the scientific realist very deeply. It appears that the human mind has evolved an ability to recognize geometrical patterns where none exist. What else might it be recognizing that does not really exist?

One of the realist's axioms of faith which we listed in Chapter 1 held our separation of natural phenomena from the perception of them to be a harmless simplification. Maybe this is not true. Our perception of Nature as governed by particular geometrical factors and predictable regularities may be an illusion. Rather, such orderly trends may be the only aspects of Nature we are any good at

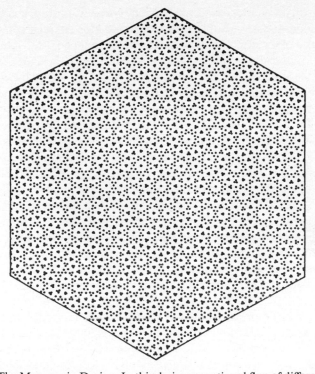

Figure 7.2 The Marroquin Design. In this design a continual flux of different patterns is evident. The eye has a tendency to seek out larger and larger circular patterns which eventually dissolve to be replaced by the distinctive local centres of the pattern. This design also contains a more subtle pattern which many viewers seem unable to discern. Within each of the largest of the circles that the eye easily picks out there is inscribed a twelve-sided 'Swiss' cross and this pattern is repeated periodically throughout the design. [From Marroquin, J. L. (1976). Human visual perception of structure. Master's thesis, MIT by kind permission of MIT.]

discovering (look again at Figure 7.2, and try to see the network of twelve-sided Swiss crosses that cover it). This is not to deny, as some anti-realists might, the reality of the laws of Nature we perceive and employ, but simply leaves open the possibility that this mathematical watch-world we have mapped out may fall short of the whole. Our cross-examination of Nature may have elicited only a particular type of evidence because we have hit upon a successful line of enquiry, and by the direction of our questioning we are able to determine that part of the whole truth that comes out.

This type of subjectivity has worried modern philosophers and historians of science. There is a traditional way of writing the history of science that has died out amongst the historians, but it can still be found in the writings of many scientists interested in the history of their subject. It is also the perception of the ordinary person in the street as to what history is about. Indeed, it is how we were taught it at school: dates, people, events; how we arrived at the present. In this story of the rise of science we might trace the forerunners of the 'right' answer we know today. Who can we find in the history books who thought that the Earth went round the sun before Copernicus wrote about it? These individuals from cultures widely separated in space and time we link together as forerunners of the heliocentric view. We draw an imaginary thread through the ages to chart the course that we judge to be the 'correct' one. All wrong views are ignored. This approach was dubbed the 'Whig' theory of history by Herbert Butterfield. The name derived from those past historians who treated history as a record of events that culminated in the political system dear to their own hearts: the liberal democracy. Scientists' pictures of their subject's history usually suffer from this delusion. It is rather like taking their realist approach to Nature out of its useful context and applying it to the course of science, believing that there is one true history which is either right or wrong. There is a true record of events, of course, but only one that includes all of them. Erroneous ideas and incorrect measurements also played a role in the unpredictable course of scientific discovery.

The professional historian takes a more anti-realist view. He wants to concentrate on how science is done, and this has led to the operationalist or instrumentalist (or even structuralist) view becoming very popular. A seductive view, usually associated with the name of Thomas Kuhn, lays stress upon the fact that science is a *human* activity, and hence tries to develop a sort of sociology of science based on the activity of scientists. Kuhn is interested in the human bias towards particular scientific ideas, not on the scale of a particular concrete example as in our pictures above, but within an entire area of science. He believes that for most of the time scientists beaver away in a routine manner measuring and calibrating things, calculating details, filling in the gaps of knowledge. Gradually a critical state is set up as fundamental problems are identified which cannot be resolved within the existing picture. But then something revolutionary comes about. When the climate is right, a new idea or 'paradigm' emerges, perhaps because an old idea is indubitably disproved, or because someone comes up with a brand-new idea with far-reaching consequences. The direction of that subject area then undergoes a period of dramatic reorientation in which the new paradigm becomes the focal point for speculation, but this period is then superseded by a return to normal activity. Normal activity is distinguished by what Kuhn calls 'puzzle solving'. This term is judiciously chosen, for puzzles are problems that have assured solutions. Likewise, the activities of normal science are imagined to concentrate upon issues which must admit of solutions within the framework of ideas and puzzle-solving methods which define the current paradigm. Occasionally, difficulties will be encountered, and the paradigm will only be able to encompass certain 'anomalous' results if inconsistencies are ignored by being labelled 'problems' or 'paradoxes', or if *ad hoc* methods of analysis are employed. Eventually, the number of such unnatural acts will become more than the paradigm can bear, and, so Kuhn claims, a revolution will ensue in which a new paradigm will arise that is able to accommodate both the successes and the difficulties of the former. The paradigm is dead; long live the paradigm.

Kuhn's vocabulary, with its paradigms and puzzles, has been absorbed into the lingua franca of the scientist. Yet most scientists would instinctively repudiate the assumptions that lie at the heart of Kuhn's analysis. For Kuhn, science is neither right nor wrong. Problems are not resolved: merely dissolved. The laws of Nature as perceived by scientists are neither true nor false. Paradigmata are transient fashions, rather like artistic styles. According to Kuhn, every scientist views the Universe through the rose-tinted spectacles of some paradigm or other. Needless to say this view is anathema to the disciple of Karl Popper, whose realist base leads him to maintain that there are definite statements that can be universally agreed upon by different scientists, because they have recourse to the experimental test of falsification against the facts of reality.

There are fashions in the philosophy of science as well, and Kuhn's picture is just one of them. Eventually, if Kuhn himself is to be believed, his picture of the progress of science will become encumbered by contradictions and ambiguities, and it will be replaced by a new and better one. Indeed, this process has already begun. The popular view of the Kuhnian theory is invariably that painted in the first edition of his famous work *The Structure of Scientific Revolutions*, published in 1962. But this work was subjected to considerable criticism because of the vagueness and inconsistency of Kuhn's terminology (one critic, who Kuhn himself cited as being particularly cogent, identified twenty-one distinct usages of the term 'paradigm'!), and subsequent editions of the book were considerably revised. Finally, in 1974 Kuhn retreated from his former view of paradigms as underlying world-views driving the course of science through revolutionary discontinuities. Rather than aim to describe the sociology of science in terms of some global dynamic, Kuhn turned to the micro-sociological view of paradigms dictating the direction of research inside small but influential research groups. This view is less striking and considerably less controversial than its predecessor. None the less, it is still a philosophy that is ultimately self-refuting.

The crux of the Kuhnian view is that the transient paradigms which are periodically adopted and discarded are neither right nor wrong; they are just tools which have a temporary expediency determined by the existing climate of opinion. It is also an opinion concerning what scientists do. It takes the view that scientists do not discover anything that is really true. All our existing scientific viewpoints will eventually spawn anomalies, and will therefore be replaced by a new viewpoint, or paradigm, until it too is found wanting. Furthermore, the decision as to which of two competing theories is to be adopted is not taken solely on objective grounds. Additional subjective criteria like scope, simplicity, or symmetry are taken into account. Supporters of different paradigms may have different ideas as to what would constitute a decisive test of which one is superior.

The Kuhnian view describes the situation in the optical illusion in Figure 7.3: one group of scientists see the glamorous young woman while the others see the old hag. Who is right? It depends on what you are looking for, answers Kuhn.

There has been much learned argument as to whether this picture of normal activity assuming a particular world model, followed by mounting crisis resulting in a 'scientific revolution' and the adoption of a new set of assumptions, and so on, *ad nauseam*, is adequate. It attributes to science a sort of mob psychology that is really extremely vague. The reason why it is possible to squeeze the practice of science into its confines is probably because the same can be done with just about any activity: organized crime in Chicago, styles in high-jumping, terrorism, football tactics, car design, Paris fashions: you name it and Kuhn's theory applies. By seeking to treat the history of science like the history of art—as the story of the coming and going of styles—it seems to ignore the existence of an underlying core of fact, and its role in dictating the attitude of scientists and the direction of their interests. It is most often a change in the storehouse of facts that leads to a fundamental change within one area of science. It is also debatable whether there have really ever been any scientific revolutions of the Kuhnian variety. New theories usually contain the old ones within them as particular examples of a phenomenon that is more general than previously suspected. The march of progress resembles the redrafting and modification of a story rather than its succession by some stylistic revolution, like the making of it into a movie.

If Kuhn is correct, then it is largely irrelevant whether or not there exist laws of Nature, and what forms they take if they do, for science is an entirely human activity that cannot find them out. Kuhnian science is the scientist looking in a partially reflecting mirror. Whereas Popper would be willing to concede that we will almost certainly never discover *the* laws of Nature, because they are buried so deep in reality, none the less, unique and universal laws do exist. Kuhn, by contrast, regards laws of Nature as an ever-changing creation of the scientist's mind, part of the symbiotic psychological relationship between the observed and the observer. This is the most radical and general view that one could take about the subjectivity that is introduced into our study of Nature by our human intellectual tendencies: having recognized that there is a sociology of science it concludes there there is nothing more to science than its sociology.

The phantoms of the laboratory

> Science cannot solve the ultimate mystery of Nature. And it is
> because in the last analysis we ourselves are part of the mystery
> we are trying to solve.
>   Max Planck

When Francis Bacon abandoned the deductive logical reasoning of Aristotle and his medieval disciples, he acted out of a conviction that our knowledge must be

Figure 7.3 An example devised by the American psychologist E. G. Boring which manifests a perception ambiguity which results in a sudden Gestalt shift. The information available to the eye need not alter but a new arrangement of its results in a dramatic reorganization of it when the image shifts between that of a young and an old woman. [From Gregory, R. L. (1970). *The Intelligent Eye*. Weidenfeld and Nicolson, London, by kind permission of the publishers.]

founded upon the things we learn from Nature rather than the phantoms and prejudices we find nurtured within our minds by the philosophical systems of the past. He warned against four specific 'phantoms' which he believed adversely bias our thinking about Nature, and divert our search for the true laws of her operation down culs-de-sac and blind alleys. They are influences that stand between us and the raw truth about the world we seek to observe and understand.

First, Bacon warns us that we are inhabited by the *Phantoms of the Tribe*: those tendencies which are by-products of our human nature. Over the centuries we like to think that we have minimized these influences which had once made it so natural to believe that Man was the centre of the physical Universe and the focal point of all Nature's workings, and that everything else was specially designed for our convenience and benefit. This prejudice is not easy to overcome, as evolutionary biologists in the United States have recently discovered. In the last section we saw how our perception of visual patterns has been over-sensitized by millenia of natural selection. Phantoms of the Tribe need not be confined to 'tribal' prejudices; they can also include the necessary physiological properties which allow the survival of psychological prejudices.

Next, Bacon points to those personal prejudices each of us possesses: he calls them the *Phantoms of the Cave*. These are still with us. Some scientists just do not like certain lines of development in their disciplines, and even cease to contribute to their subject in the hope that certain unsavoury new ideas will pass out of fashion. Sometimes this individual bias will have a sound basis—the new research direction may simply involve the use of advanced mathematical or experimental techniques that the scientist may not have the experience or expertise to cope with; but often it has a much more human origin—the new development may have been initiated by an individual who our scientist just cannot abide, and the thought of working on that person's ideas is more than he can stomach!

Bacon calls the third source of distortion *Phantoms of the Market-Place*, and they are rather more subtle than the others. Perhaps they should now be called the *Phantoms of the University*. They emerge from the association of scientists with each other, and we witness them as the collective result of using a particular language, sharing particular concepts, and the common use of mathematics. In particular, this Phantom might animate the contemporary consensus of philosophers concerning the meaning and method of science. Kuhn seems to regard it as an overwhelming and unavoidable bias; paradigms are rather close to being Phantoms. Moreover, the later Kuhnian association of paradigms as arising through the influence of microcosms of scientists working within influential research groups also fits into this category of Phantom.

Last, we encounter what Bacon perceived to be the most pernicious influence upon our ability to think objectively: the *Phantoms of the Theatre*. The grand philosophical systems of the past and present can be seen as plays, in the sense that they create a world within the world: a world with its own scenery and characters. These philosophical models of the real world are, Bacon warns, just models. If we forget their divorce from the reality they seek to explain then we commit the error of a theatre-goer who confuses the scene on the stage before him with the real life it seeks to represent. The play may well be based upon reality; it may even be richer and bolder than reality, but it is a man-made world none the less. Before twentieth-century science began its dramatic rise, philosophical views were more explicit and important in the work of scientists. Today they are no less evident to the careful observer, but they are implicit or even deliberately obscured. Thus, for example, the belief that a unified Theory of Everything will explain the structure of the Universe uniquely and completely will appear unashamedly in scientific papers, but it is essentially a religious or metaphysical view, in the sense that it rests only upon an unstated axiom of faith.

What Bacon first so acutely recognized about science is that, by virtue of it being a human activity, its results are necessarily subject to human biases. Nevertheless, many subsequent generations of experimenters do not seem to have taken his warnings to mean very much more than 'Beware of Aristotle and the Scholastics'.

Errors

> It ain't what you don't know, that counts, it's what you know
> that ain't so.   Will Rogers

Men like Copernicus and Newton did not have a clear concept of what we would call 'scientific error'. To the experimental scientist the term 'error' has a wider meaning than the word conveys to the ordinary person in the street, for whom it means simply a 'mistake'—a blunder like misreading a thermometer, mixing the wrong chemicals, or allowing a nuclear reactor to overheat. But this is not the primary meaning of 'scientific error'. To the scientist the term means two other things.

The first is straightforward: the limiting accuracy to which a quantity can be measured. A simple example of this *experimental error* is illustrated by a statement of somebody's age. If we know only the year in which they were born, then we can only state their present age to lie somewhere within a *range* of twelve months. If we are told that a piece of wood is 20 feet long to the nearest foot we know only that its length lies between 19 feet 6 inches and 20 feet 6 inches. This type of 'error' specifies the limitations in accuracy of our measuring devices or the data we are given: it tells us the biggest mistake we could make even if we have carried out our measurements competently, and not done something stupid like misread the ruler. This type of error is not terribly interesting, but it is obviously one of the goals of experimental scientists to make it as small as possible. No observation or experimental measurement is of any use unless the possible measurement error is also given. To be told that the Conservative Party is five percentage points ahead of Labour in an opinion poll is a meaningless statistic unless the uncertainty, or experimental error, in the poll is also quoted.

We have already seen that the Heisenberg Uncertainty Principle of quantum mechanics ensures that there are inevitable errors associated with the measurement of all quantities even if the measuring instruments are perfect (and in practice, of course, they never are). This strange limitation arises because the very process of observation is inseparable from the state being measured. Perfect knowledge of the Universe is impossible because the act of knowing influences the Universe in an unknowable way. It is as if, by the time we record its state, it has changed slightly. Therefore every possible observation of the physical world must possess some finite measurement error. However, in practice, the measurement errors that limit the accuracy of scientific measurements, even in elementary particle physics, are considerably larger than the irreducible minimum imposed by the quantum theory. Only in the study of quantum liquids at temperatures close to absolute zero does experimental accuracy approach Heisenberg's limit.

The existence of practical limits to our measurement accuracy was not readily appreciated in the past. Copernicus knew that if the planets traced elliptical paths as they orbited the sun, then the observation of two points of a planet's orbit is sufficient to fix uniquely the particular elliptical path that is taken. And so he could not understand why the pairs of points he observed in each planet's orbit did not fix the actual elliptical path taken by the planet in the future. The reason for the disagreement was that the measurements made of the positions were not precisely accurate. Several slightly different ellipses are compatible with the range within which the real positions could have lain. Likewise, even Newton was puzzled as to why the measured motions of the celestial bodies did not quite fit his mathematical predictions: the answer was, again, that the measurements contained small errors due to the limiting accuracy of the instruments used to record them, but this idea seems to have been quite foreign to these early scientists.

The second form of error that besets experimental science is more subtle in its origins, and more serious in its consequences because one can never be certain that it has even been identified, let alone minimized or eradicated. This species of error we call a *'selection effect'* or *'systematic error'*, and its identification requires careful thought and a wide-ranging appreciation of the phenomenon under study. Its existence refutes the naïve idea that all that is required to test theories of science is the experimental method. Experimental arrangements and observational procedures have in-built propensities to gather certain types of facts more readily than others. In order to be sure that you are observing what you think you are observing, it is always necessary to have some theoretical understanding of the wider spectrum of phenomena that could be biasing your observations.

Imagine that you are embarked upon a project to learn all you can about the sizes of rats. Accordingly, you design lots of little rat-traps to catch specimens. Each trap is a small box four inches long, and has a cross-section two inches square at the ends. At one end there is a door, hinged at the top, poised to drop down and shut tight as soon as any rat venturing inside touches a little pile of food at the far end of the box. This stratagem is very successful. After a few weeks of baiting dozens of these rat-traps you find you have collected hundreds of live rats. You weigh them; you measure them; you observe them closely. But other pressing matters intervene. You must leave for an important conference in Tahiti, and are unable to finish the analysis of your measurements. Not having enough time to explain how the experiment was carried out you simply leave the list of the animals' sizes on an assistant's desk and ask him to do a preliminary analysis for you. When you return he is keen to see you, and claims to have discovered something very important about rats. He reports there to be overwhelming evidence for a definite maximum size of rat: rats of up to four inches in length were found with equal frequencies, and not a single rat was found to be greater than two inches in height or width.

A little thought convinces us that such a claim tells us nothing about rats. Our experiment was subject to a trivial example of a 'selection effect': namely, that no rats greater than two inches in height or width could get into the traps. The non-observation of any three-inch-fat rats is telling you something about your experiment, but nothing about rats. The overall picture of the range of rat sizes has been biased by the inability of the experiment to collect large rats.

In practice, such experimental biases are usually a little more subtle than this, and may even be unavoidable. For instance, an astronomer might be interested in discovering how the population of stars or galaxies is distributed according to their brightness. Any results that are obtained will be biased towards finding a disproportionate fraction of the brighter objects because they are easier to see. The nasty thing about such biases is that, no matter how careful you are, you can never be sure that they have all been completely eliminated. Indeed, the art of being a good experimentalist is to a large extent the art of sensing and eliminating sources of systematic error. In terrestrial scientific experiments or data collection these biases are usually determined by the particular apparatus and method used to measure something, and so if the measurement can be made by another independent experiment, preferably using a different method, and the same result is obtained, then the presence of significant systematic error is very unlikely. This is the main reason why scientists tend to be sceptical of dramatic experimental discoveries until they have been confirmed by other different experiments: they have no independent fix on the magnitude of possible systematic biases.

On occasion, the bias of an experiment may be coupled to the inevitability of measurement error. A famous example, again from astronomy, arose during the eighteenth and nineteenth centuries. Friedrich Bessel, a notable mathematician and astronomer who introduced one of the most useful equations of applied mathematics, and still had time to discover the star Sirius B, was a colleague and associate of the great (some would say the greatest) mathematician, Johann Karl Friedrich Gauss. Gauss was one of the first to understand and work out a mathematical treatment of measurement errors. This awareness he conveyed to Bessel, and it enabled him to resolve an awkward astronomical dilemma.

Towards the end of the nineteenth century, astronomers at the Royal Greenwich Observatory had found small but irritating systematic differences in their observations of star motions. Observers differed concerning the times they recorded for stars to pass between two lines drawn on the telescope's field of view. The Astronomer Royal of the day, Neville Maskelyne, had assumed that the intolerable differences between the transit times he measured and those recorded by his assistant were simply the result of incompetence on the part of his assistant. Bessel realized what the source of the discrepancies in these measurements really was, and more important—how to rectify it. The reaction speed of different astronomers was different. Some would always signal the star crossing the first reference line a fraction of a second earlier than others. In the days before the use of electronic timing methods in sport a similar problem used to beset athletic events. Some human time-keepers have a tendency to record faster sprint running times than others, and most time-keepers award faster times than electronic devices when both record the same race. The margin of error can be the difference between breaking the world record or simply recording a good performance. For this reason only electronically-timed performances are now eligible for the record books. Returning to astronomy: Bessel calibrated each astronomer with a so-called 'Personal Equation' to compensate for their individual biases in starting and stopping the clock when a star entered or left the cross-wires spanning the telescope's field of view. This procedure compensates for the problem of 'selection effects' introduced by individual idiosyncracies, and was used until the introduction of modern techniques of automatic electronic measurement.

The 'Groucho Marx Effect'

> I wouldn't want to belong to any club that would accept me as a
> member.
> Groucho Marx

Selection effects do not only arise in the experimental sciences: they also influence our theoretical and mathematical investigations. They are especially rife in theoretical physics. The situation in the fundamental areas of physics which seek to further our understanding of gravitation, quantum theory, and the micro-world of elementary particles is typically like this: we have an elegant system of non-linear mathematical equations in many quantities that have been derived by the application of some powerful invariance principle, and which encapsulate the essence of what is already known about the aspect of Nature under study. However, the outstanding problem is how one *solves* these equations, and so be in a position to work out what our grandiose theory has to say about unknown situations so that these predictions can be checked against what is seen to occur in those circumstances. For example, Einstein's theory of general relativity is a theory of gravitation which is equivalent to ten complicated partial differential equations that must be solved simultaneously to find out how the geometry of space and time responds to the presence of mass and energy within it, and how that geometry dictates where masses will move. In order to test the correctness of the equations of the theory, their solutions need to be found. We can find a particular solution that describes the gravitational field exerted upon the planets by the Sun if we idealize the Sun to be a sphere. It predicts that the motion of the planets should proceed in slightly different orbits than are predicted by Newton's theory of gravity. The difference is most marked for the motion of the planet Mercury, the closest to the Sun, for which the influence of the Sun's gravity is greatest. Observations confirm the existence of the small new effect predicted by Einstein's theory, and show the Sun's shape to be close enough to a perfect sphere for this idealization not to affect the conclusion significantly.

This procedure is clearly not perfect. The equations of physical theories like general relativity are too complicated to be completely solved. Hence, one must resort to significant idealizations and approximations in order to learn something about the features that are latent within the theory, and which could be present in the complete solution. A typical strategy is to try to solve the equations in a special situation; for instance, as with the idealized spherical Sun, when there is a simplifying symmetry. We might, for instance, find a solution that gives the gravitational field exerted by a *spherical* object, or by one that is not changing its shape with time. Idealizations of this sort forbid certain types of variation to arise in the equations, and simplify them dramatically with the result that they are usually soluble. In the case of Einstein's theory of general relativity the first solutions of the equations to be found were for unchanging spherical configurations of material. Since Einstein's equations were first written down in 1915 there has been a continual search for solutions to them. A few years ago four mathematicians co-authored a four-hundred page book devoted to displaying, classifying, and organizing the most important solutions that have been found to date. All the solutions in this book possess special simplifying properties. This is the reason why it has been possible for us to find them.

This is where the 'Groucho Marx Effect' comes in: the only solutions of the equations that we are clever enough to find always describe special idealized situations that will not generally arise in practice. The limited computational competence of human beings, and the inability of computers to do more than humans do (except to do it faster) means that the conclusions we draw from our theories of physics are conditioned to a considerable extent by the content of a relatively small number of exactly soluble examples. We can find solutions of Einstein's equations that describe exactly how a universe that is perfectly uniform expands in time. The real Universe is *almost* uniform over its largest dimensions—but not quite. There exist stars and planets, galaxies and clusters; these non-uniformities are, of course, essential for our own existence (there cannot be any observers in a perfectly uniform universe!). The question of how they arose, and from what initial state, is what most of modern cosmology is about. Unfortunately we know of no exact solutions of Einstein's equations that describe an expanding universe filled with a higgledy-piggledy collection of stars and galaxies. There undoubtedly do exist such realistic solutions, but they are mathematically too complicated for us to find. We rely heavily upon the real world in all its complexity being close to simple idealized situations. Often this is not the case. The horribly complicated turbulence that results at the base of a waterfall is so dissimilar to any smooth and idealized fluid flow that we have very little detailed understanding of it at all.

One way in which we can evade the need for idealization is by the use of approximations. Besides finding the exact solutions of equations for idealized situations one can also discover the approximate solution of the equations for situations that are *almost* ideal. Although the Universe does not expand at exactly the same rate in every direction, it almost does (to within one part in ten thousand, in fact), and so we believe that it will be approximately described by the idealized solution to about the same accuracy. In this case our expectation is borne out by the good agreement between our observations of the Universe and the predictions of the idealized model. However, it is quite possible that there exist idealized solutions that do not lie close, in any sense, to non-idealized ones. Such isolated examples would be atypical and quite misleading with regard to what the full solution of the equations of the theory are like. A good example is the famous special solution to Einstein's equations found by the logician Kurt Gödel. This solution showed that there is a particular solution of Einstein's equations which allows time-travel to occur. You could kill your own grandmother and create a paradox of fact (or even solve the problem of induction!). Gödel's solution describes a weird rotating universe which looks nothing like the one we live in, but this does not mean that we can stop worrying about time-travel. We need to know whether time-travel is a property of the full and realistic solutions of Einstein's equations—that would describe our own world—or whether it is a pathology of a small number of physically irrelevant solutions with weird properties.

This highlights another important point. Einstein's equations allow innumerable different solutions each describing different expanding universes. They have different initial properties: some possess galaxies while others do not. But there is, by definition, actually only one Universe. What is the selection principle that pins down the precise solution that describes *our* observed Universe, and it alone? This principle must come from outside the theory of general relativity.* The need for this 'selection principle' shows that Einstein's theory is not the best possible description of the Universe. It permits too many things that are not realized in Nature in addition to the things that are.

Many physicists have expressed the belief that when all the disparate theories of the different elements of Nature are unified in an all-encompassing 'unified field theory' then the constraints upon the shapes of the individual pieces in order that they can be fitted together will be so stringent that there will be only one possible unified picture. There may be one and only one possible theory of all fundamental phenomena, in which everything that is not forbidden will be compulsory. Steven Weinberg leans towards this hope, and once speculated that 'when you put quantum mechanics together with relativity, you find that it is nearly impossible to conceive of any possible physical system at all. Nature somehow manages to be both relativistic and quantum-mechanical; but these two requirements restrict it so much that it has only a limited choice of how to be—hopefully a very limited choice.'

If we are idealists then perhaps the future 'discovery' of such a theory is inevitable, but if we are realists then surely we can hold no such hope. For even if

---

*In fact, *any* space–time geometry is a solution of Einstein's field equations for *some* distribution of matter, just as any gravitational potential will solve Poisson's field equation for some density distribution. These equations only place restrictions upon Nature because most geometries and potentials are ruled out by the requirement that their associated matter distribution be physically realistic in various ways—for example by having the density of matter positive everywhere.

such a description of the ultimate workings of Nature does exist, who is to say it is within the grasp of human minds to find it? Indeed, we saw in the last chapter that it is possible to entertain the view that there is no such ultimate theory of everything at all. There is no reason, save our grandest presumption, that Nature should be fashioned with our computational incompetence in mind. And the final irony is that, even if such a super-theory does exist and we are able to write it down, we can never know that it is correct. The scientific method does not enable us to demonstrate that our theories are true: only that they are false. While we can come up with candidates for the ultimate theory that would be falsifiable, *the* ultimate theory would not be falsifiable at all. In the meantime, any theory that does not provide a unified description of everything must eventually be proven to conflict with some aspect of experience.

We must also be aware that while there may exist a unique *Theory* of Everything it may have an infinite number of cosmological *solutions*, and the actual Universe is only described by one of them when particular starting conditions are chosen. More problematic still is the fact that broken symmetry is ubiquitous in Nature. The key features of the Universe may owe their origin to random breakings of the underlying symmetry of the Theory of Everything.

The search for a completely determined set of natural laws does not proceed by logic and observation alone. Metaphysical and aesthetic criteria are called upon to guide theoretical speculation. But what are the right aesthetic criteria: beauty, harmony, symmetry, lack of symmetry, simplicity, computability, finiteness, brevity, minimal assumptions: who can say?

Beauty

> No doubt aardvarks think that their offspring are beautiful too.
>
> John Ellis

What criteria do scientists use to steer them towards successful and powerful theories of Nature? We have already seen that in practice such theories are necessarily mathematical in form. There is good mathematics and bad mathematics, ugly mathematics and beautiful mathematics. Difficult as it is to draw the precise line between these opposites, the professional mathematician finds it no harder to draw than the difference between night and day. The theoretical physicist Paul Dirac was one of the most outspoken supporters of the idea that the deployment of 'beautiful' mathematics—that which possesses symmetry, economy of form, a depth of interconnection with other parts of mathematics, and the maximum of structure from the barest of inputs—should be the priority of the theoretical physicist seeking a description of some physical phenomenon. This one should seek in conjunction with a desire for 'simplicity' that avoids superfluous ideas and hypotheses. Of the search for mathematical laws of Nature Dirac had this to say:

The dominating idea in this application of mathematics to physics is that the equations representing the laws of motion *should be of a simple form*. The whole success of the scheme is due to the fact that equations of simple form do seem to work. . . . The method is much restricted, however, since the *principle of simplicity* applies only to fundamental laws of motion, not to natural phenomena in general . . . What makes the theory of relativity so acceptable to physicists in spite of its going against the principle of simplicity is its great *mathematical beauty*. This is a quality which cannot be defined, any more than beauty in art can be defined, but which people who study mathematics have no difficulty in appreciating. The theory of relativity introduced mathematical beauty to an unprecedented extent into the study of Nature.

We can now see that we have to change the principle of simplicity into a *principle of mathematical beauty* . . . It often happens that the requirements of simplicity and of beauty are the same, but where they clash the latter must take precedence.

Dirac's position is quite extreme, since in some circumstances he did not wish even negative experimental evidence to deflect him from a particular line of inquiry that he had pursued because of its mathematical beauty and elegance. He continues:

If there is not complete agreement between the results of one's work and experiment, one should not allow oneself to be too discouraged, because the discrepancy may well be due to minor features that are not properly taken into account and that will get cleared up with further developments of the theory.

These two statements are very striking. They are good examples of the influence of selection effects of a particular sort. We certainly have an aesthetic sense that we cannot easily explain. Some of its features seem to be universal. Many would seek to argue that it is entirely a consequence of natural selection, but it seems to possess elements that are unnecessarily sophisticated for this purpose. In *The Sense of Beauty* George Santayana suggested that we perceive 'beauty' to reside in patterns and appearances that offer sufficient novelty to arouse our curiosity, but not so much that their complexity is beyond our understanding. He points to the starry night sky as a manifestation of this tantalizing property. The eighteenth-century Dutch writer Hemsterhuis defined beauty as that which provokes the greatest number of ideas in the shortest time. Scientists have found Nature as a whole to be beautiful in these peculiar senses: it presents challenging problems that offer the possibility of solution. It invites and satisfies curiosity. When all is said and done, scientists want equations that are analysable rather than merely simple or beautiful. And we should remember that it is the purest speculation that Nature and her laws are 'beautiful' in our (or indeed any) sense. Scientific observation alone cannot eliminate the possibility that one of David Hume's 'superannuated deities' produced the Universe we see as a faulty precursor to the real thing implemented in another time and another place!

Another definition of mathematical beauty is that suggested by the Indian

Nobel laureate, Subrahmanyan Chandrasekhar. He has proposed that Einstein's theory of general relativity has an aesthetic basis primarily because of the miraculous way in which it proved compatible with other laws of Nature which played no role in its conception and formation. However, it would be more accurate to say merely that within the compass of the general relativity theory there exists a core with this unforeseen harmony. There are other parts of general relativity, like the presence of solutions allowing time-travel and naked singularities, which would create conflict with all the other laws of physics if they were realized in Nature.

Relatively simple mathematics is useful in describing Nature, not just the most abstract and difficult—although some of that is useful as well. Perhaps this feature lured gifted individuals like Paul Dirac into the subject of mathematical physics. But it is more likely that, having found mathematics and physics to be the activity which most nearly fitted the perspectives he possessed, Dirac then sought to stress those aspects of physics that were closest to his ideals. And if they are often successful in producing great discoveries then they come to dominate one's view.

In this regard it is instructive to consider the interest that mathematicians and physicists have displayed towards the spectacular pictures of fractal curves that computers have recently generated in profusion. Collections of these have been displayed in many famous art galleries. The pictures have an undeniable beauty, which is enhanced by the skilful choice of false colour-coding introduced by the computer scientists. But beyond that there is an aspect that connects our aesthetic appreciation to that of Nature itself. The intricate structures of fractal curves, like the Mandelbrot set (p. 220), are closely related to the self-similar structures we see around us in the natural world—the branching of a leafless tree, the pattern of frosted snowflakes, the crenellated patterns of a mountain landscape—all exhibit a non-linear invariance which we find deeply appealing. It has taken us a long time to find the type of mathematical algorithms that can generate such structures systematically, but their discovery is surely important in locating the centroid of our aesthetic appreciation of visual symmetry and mathematical harmony.

We know of no reason why the laws of Nature should be either 'simple', 'beautiful', or anything else that appeals to us. Indeed, the theory that Dirac thought the ugliest and most unsatisfactory part of physics—quantum electrodynamics—is the most accurate of all the fundamental scientific theories that we possess. It is the quantum theory that describes the interaction between light and electric and magnetic forces. Its theoretical predictions are confirmed by experiment good to ten decimal places. Some physicists find this theory 'beautiful' too!

All we can conclude from this equally subjective discussion is that scientists do possess conscious and unconscious biases towards developing certain types of descriptions and laws of Nature. The more mathematical the science the more powerful will be these influences. The search for symmetries and invariances is a goal of the mathematical physicist. There is no real evidence that Nature is in any well-defined sense 'simple' or 'beautiful'. Indeed, the hallmark of most natural phenomena is a deep complexity masquerading as simplicity. One is reminded of the story of the astronomer who began a public lecture on the nature of stars with the statement 'Stars are very simple objects', only to be met with a cry from the back of the hall that 'you'd look pretty simple too from a distance of two hundred light-years!' What is more remarkable is the extent to which theories and descriptions that we know now to be seriously incomplete or just plain wrong proved to be such reliable temporary guides to a substantial fraction of the truth for so long in the past.

Dirac's remarks about the seductive character of elegant mathematics are the natural views of a realist who believes that Nature *is* intrinsically mathematical. Thus, Nature and her mathematical representation are regarded as equivalent, and the unquestioned beauty of the former can be sought in our gropings towards the latter. Yet, scientists of a different metaphysical persuasion have argued in quite a different direction to Dirac. Some operationalists, like Bridgman, actually regard the search for aesthetic mathematical structures in physical descriptions as a dangerous metaphysical diversion. With regard to general relativity, which both Dirac, Chandrasekhar, and most other physicists regard as the most 'beautiful' of theories, Bridgman writes,

> The metaphysical element I feel to be active in the attitude of many cosmologists to mathematics. By metaphysical I mean the assumption of the 'existence' of validities for which there can be no operational control ... At any rate, I should call metaphysical the conviction that the universe is run on exact mathematical principles, and its corollary that it is possible for human beings by a fortunate *tour de force* to formulate these principles. I believe that this attitude is back of the sentiment of many cosmologists towards Einstein's differential equations of generalized relativity theory—when, for example, I ask an eminent cosmologist in conversation why he does not give up the Einstein equations if they make him so much trouble, and he replies that such a thing is unthinkable, that these are the only things that we are really sure of.

Of course, Bridgman's philosophy of science regarded mathematics as simply a tool for the construction of operational definitions of physical quantities (although he had to admit that the operationalist doctrine could not cope with a subject like cosmology, where one needs to talk about quantities like 'the mass of the Universe' which cannot be defined operationally). He adopted a constructivist interpretation of mathematics which bordered upon formalism, and regarded this as the natural complement to his operationalist philosophy in physics. Proceeding from his recognition of our 'metaphysical' tendency to assume for Nature a mathematical structure, he argues that this unquestioned assumption is a dangerous human bias:

> I believe that there are dangers in any subject in which there is such an unavoidable

mixture of purely 'scientific' and 'human' elements. It seems to me that there is a particular danger of introducing actual inconsistencies into the structure if the metaphysical attitude with regard to mathematics is so far adopted as to obscure the perfectly legitimate use of mathematics in attaining simplicity of formulation.

Thus, while Dirac has so much faith in the mathematical beauty and economy of Nature that he is willing to follow it as a guiding principle that might at times rule over experiment, Bridgman argues that it is our willingness to follow such a seductive Pied Piper that undermines his faith in the intrinsically mathematical character of Nature.

Einstein had a view that intertwined the two considerations of Bridgman and Dirac. Unlike Bridgman, he believed that 'the supreme task of the physicist is to arrive at those universal elementary laws from which the cosmos can be built up by pure deduction'. He realized that, even if one adopted an operationalist stance to the creation of physical laws, they would still require a mathematical representation that could not be gleaned uniquely from experience. At this point we see that the physicist has some artistic licence in the way that he pursues the presentation and development of his theory.

> If, then, it is true that the axiomatic basis of theoretical physics cannot be extracted from experience but must be freely invented, can we ever hope to find the right way? Nay, more, has the right way any existence outside our illusions? Can we hope to be guided safely by experience at all when there exist theories (such as classical mechanics) which to a large extent do justice to experience without getting to the root of the matter? I answer without hesitation that there is, in my opinion, a right way, and that we are capable of finding it. Our experience hitherto justifies us in believing that nature is the realization of the simplest conceivable mathematical ideas. I am convinced that we can discover by means of purely mathematical constructions the concepts and the laws connecting them with each other, which furnish the key to the understanding of natural phenomena. Experience may suggest the appropriate mathematical concepts, but they most certainly cannot be deduced from it. Experience remains, of course, the sole criterion of the physical utility of a mathematical construction. But the creative principle resides in mathematics. In a certain sense, therefore, I hold it true that pure thought can grasp reality, as the Ancients dreamed.

Dirac's remarks about the secondary role of experiment relative to theory on some occasions are also worth expanding upon. What he means by them is not quite what Eddington implied by his famous warning against embracing sensational experimental results too readily, 'do not believe any experimental result until it is predicted by theory'; but rather that one may conceive a wonderful new mathematical model for the working of, say, the atomic nucleus, but find it to possess one obstinate defect which puts it in disagreement with some experiment. In this circumstance you should not necessarily lose faith in your good idea. It may be that you have omitted some comparatively minor (or even major) ingredient in your new theory, but this omission can be straightforwardly repaired in the future. This approach is much in evidence in modern theoretical research into elementary particle physics. In this area of research, speculative theory races way ahead of experiment because of the vast cost and sophistication of the necessary experiments. They require huge energies to be attained for particle bombardments, and batteries of sophisticated computers to record and interpret the results. Unaided by experimental data, a theorist may pick upon a gauge theory (of the sort we discussed in Chapter 4) defined by some group of mathematical operations, and work out some of its experimental consequences. Occasionally it will be found that almost all the pieces fall into place. Neat, unified explanations appear for facts that were previously independent and *ad hoc*. But there now exists some horrible new consequence that is at variance with experiment. Faced with this state of affairs the theorists may well ignore that problem and pursue the positive aspects of the theory in the hope that the awkward discrepancy can be sorted out later by a generalization of the theory that he has written down. This is often a good strategy, because the gauge theories under study do not claim to be theories of everything in the microscopic world. They are incomplete descriptions of Nature, and the theorist always likes to believe that it is this contemporary incompleteness alone that is the source of its bad predictions. When the theory is made greater in scope its superficial pathologies may well disappear. And sometimes they do. Part of the art of being an innovative theorist is to discern what are temporary difficulties born of the simple prototype model one is creating and what are fatal diseases endemic to any theory of the sort being proposed.

Dirac cites an example drawn from Schrödinger's experience to support his argument that formal beauty should sometimes override experimental data in the evaluation of a theory's merits:

> I heard from Schrödinger of how, when he first got the idea for his equation, he immediately applied it to the behaviour of the electron in the hydrogen atom and then he got results that did not agree with the experiment. The disagreement arose because at that time it was not known that the electron has a spin. That, of course, was a great disappointment to Schrödinger, and it caused him to abandon the work for some months. Then he noticed that if he applied the theory in a more approximate way, not taking into account the refinements required by relativity, to this rough approximation his work was in agreement with observation.

It is this story that provoked Dirac to make his much-quoted declaration; he continues:

> I think there is a moral to this story, namely that it is more important to have beauty in one's equations than to have them fit experiment. If Schrödinger had been more confident in his work, he could have published it some months earlier, and he could have published a more accurate equation.

There is another reason for sometimes taking Dirac's secondary view of experiment: experiments are sometimes wrong! It is interesting to recall a couple of cases where the existing experimental evidence was contrary to the predictions of theorists, but so strong was the belief in the elegance of the theoretical models that the adverse experimental evidence was correctly disregarded as unreliable.

When told of experimental evidence contradicting his theory of relativity Einstein's immediate reaction was that the experiments must be wrong—and he was right. Later, when the estimates of the age of the Universe obtained from the evolution of stars were found to be in disagreement with those predicted by relativistic cosmology, Einstein again stood by the predictions of his theory on the grounds that it was based upon a firmer theoretical foundation than the theory of stellar evolution required to interpret the observational data—and he was right again.

As a last example of the victory of theory over experiment (they are only notable because there are innumerable defeats) one can cite the 1958 paper of Richard Feynman and Murray Gell-Mann on the structure of the weak interaction. The theory was developed using what the two authors described as the 'predilection' of one them for a particular type of equation. However, the results of this theory were in disagreement with the observed distribution of electron neutrinos in the decay of helium-6. But the authors were not dismayed. On the contrary, they wrote of the disagreement in their paper, that

> These theoretical arguments seem to the authors to be strong enough to suggest that the disagreement with the He-6 recoil experiments and with some other less accurate experiments indicates that these experiments are wrong.

And, indeed, they were correct in this surmise, as later experiments revealed.

There is a further interesting consequence of erroneous experimental results that has become apparent in recent years on the interface between particle physics and cosmology. The 'Big Bang' theory of the origin and evolution of the Universe indicates that the Universe must have been a hotter, denser, and more crowded place in the distant past. As we extrapolate backwards in time we encounter universal conditions of ever-increasing temperature in which particles collide with each other at higher and higher energies. In short, the entire Universe resembled a gigantic experiment in ultra high-energy particle physics during its early stages. Cosmologists and particle physicists have therefore joined forces in the study of the early history of the Universe. The particle physicist sees a new environment about which predictions can be made using the very latest theoretical ideas and speculations. If these ideas lead to predictions that the present-day Universe should possess certain bizarre properties that it evidently does not—like containing no galaxies for instance—then we can discount these new ideas as false. Likewise, the cosmologist can examine the latest brain-children of the particle physicists to see if they result in the Universe containing new species of elementary particle which might resolve some of the gaps in our picture of the Universe's make-up. This close collaboration between particle physicists and cosmologists began in earnest in 1978.

Between about 1979 and 1983 a number of dramatic experimental results were announced. It was claimed that the neutrino possessed a small mass, and that different types of neutrino could transmute back and forth into each other; it was claimed that an isolated magnetic charge (the so-called 'magnetic monopole' whose existence had been proposed by Paul Dirac long before) had been detected; it was claimed that the effects of an electric charge asymmetry within the neutron had been detected, and there were claims to have seen protons decay. Each of these experiments generated an enormous amount of theoretical interest in the areas of cosmology and astrophysics they touched upon. Innumerable popular articles and books were written about them for the general public, and the number of conferences bringing together particle physicists and astrophysicists to discuss them spiralled to such an extent that it became possible to spend one's entire life either at or in transit between such gatherings! Looking back on this explosion of interest it must now be said that all of these stimulating experimental results have now either been withdrawn, or such doubt has been cast upon them that no theoreticians lean upon them at all. This disappointing situation is rather unusual, and merits some explanation. Fundamental experiments are generally performed carefully and correctly. All of the above-mentioned experiments were peculiar either in being chance observations or detections of events right at the limit of an experiment's sensitivity to resolve real measurements from the noise produced by the measuring instruments themselves. What is most notable about this collection of unfounded experimental claims is that they led to far more theoretical progress in exploring new ideas in both particle physics and cosmology than have correct experiments!

The Anthropic Principle

> There is no such person as a philosopher; no one is detached; the observer, like the observed, is in chains.   E. M. Forster

The Big Bang picture of the history of the Universe is the central paradigm within which cosmologists work to understand what we do and don't know about the Universe in which we find ourselves. But what has all this got to do with you and me? The problem of fitting human life into the impersonal tapestry of cosmic space and time has been pondered by mystics, philosophers, theologians, and scientists of all ages. The views they have come up with straddle the entire range of options. At one extreme is painted the depressing materialist picture of human life as a local accident, totally disconnected from and irrelevant to the inexorable march of the Universe into a future 'Big Crunch' of devastating heat or the eternal oblivion of the 'heat death', while at the other extreme is preached the anthropocentric teleological view that the Universe was tailor-made for human life by some form of providential design. The latter view was strongly held in many cultures, reaching its zenith amongst English scientists in the eighteenth century. It remained the view of many biologists until, in the mid-nineteenth century, Charles Darwin and Alfred Russel Wallace recognized the evolutionary adaptation of organisms to their environment by natural selection. Since that time biologists have rejected any notion of evolution as being goal-directed. There is no grand goal (Mankind?) to which the entire evolutionary process is directed. If the environment were to change in some unusual way so as to render intelligence a liability then we would cease to be well adapted to survive, and might well face the same sort of demise as did the dinosaurs.

The lesson we have drawn from the problem of 'selection effects' is that we must be aware of any in-built biases in our measuring instruments toward preferentially gathering evidence of a particular sort. If we were to observe the Universe only with the human eye then we would conclude that all the radiation in the Universe lies in that range of wavelengths which we call the 'visible waveband', spanning the spectrum from red to violet. But there exists radiation of other wavelengths that the human eye cannot detect. Light of longer wavelength has too little energy to record its reception on the rhodopsin molecules at the back of the retina, and light of much shorter wavelength than the visible is so energetic that its reception would destroy the eye. Our human physiology circumscribes the range of astronomical observations we can make unaided, and hence what the ancients could learn about astronomy.

Today, we compensate for the eye's limited observational range by building artificial 'eyes' of far greater power and scope. These take the form of traditional optical telescopes like those at Mt. Palomar, or radio telescopes at Jodrell Bank and Arecibo. These are now complemented by infra-red telescopes high in the mountains of Hawaii, and X-ray, infra-red, and ultraviolet detectors orbiting the Earth in satellites. But there is a more dramatic aspect of our human physiology that simply building better telescopes cannot overcome.

Human beings are complex biochemical computers. This is not all they are but it is their irreducible minimum specification. They are composed of self-reproducing helical molecules of DNA (= deoxyribonucleic acid) composed of atoms of carbon, nitrogen, phosphorus, and oxygen. How such intricate molecular structures arose on the Earth is not known for certain. It is possible they were generated initially by random interactions and mutations during the Earth's early history. If large numbers of different types of complex molecules were produced at different times in a primordial soup, it is quite plausible that those able to effect copies of themselves as a result of their interactions with other molecules would rapidly come to dominate the population at the expense of the non-replicators. But what is the origin of the carbon, nitrogen, oxygen, and phosphorus that compose the DNA molecules of life? Of this we are much more certain: it lies in the stars.

The atomic elements heavier than hydrogen and helium could not have been produced during the inferno of the Big Bang. The Universe expands and cools too rapidly for the heavier nuclei to be synthesized by nuclear reactions. The natural nuclear reactor we call the Big Bang shut itself down after the Universe had been expanding for about three minutes, but in that short time it transformed 25% of the mass of the Universe into helium, leaving almost 75% as hydrogen. I say 'almost' because the nuclear fusion of hydrogen into helium leaves tiny traces of deuterium, lithium, and the isotope helium−3 in relative abundances of 1/1000%, 1/100,000,000%, and 1/1000% by mass respectively. These predicted abundances correspond exactly to the fractions measured in the Universe today. This remarkable finding is one of the corner-stones of the Big Bang cosmological theory.

We have learned that the complex phenomenon we call 'life' is built upon elements that are heavier and more complex than the hydrogen and helium which emerges from the Big Bang. Most biochemists believe that the element carbon, on which our own organic chemistry is based, is the only viable foundation from which chemical life can arise *spontaneously*. Living systems on Earth are based upon the subtle chemical properties of carbon, and its relationships with hydrogen, nitrogen, oxygen, and phosphorus. Other elements play important roles but these five are the leading actors in the game of life. In order to create these five building blocks of life (and also silicon, if we foresee a future for the non-spontaneous evolution of 'life' evolved from current silicon technology), the simple nuclei made in the Big Bang must be cooked at high temperatures for billions of years. The furnaces that Nature has provided are the interiors of the stars. There, the hydrogen and helium surviving the Big Bang is slowly burnt into the heavier elements necessary for you and me. When stars have exhausted their nuclear fuel resources they implode at the centre, and expel their outer layers into space. These dramatic death throes, which we witness as supernovae, serve to disperse the biological elements through space where they become incorporated into planets, asteroids, and other forms of interstellar debris. Ultimately, they find their way into our bodies. We are the ashes of the stars.

The most important fact about this stellar alchemy upon which life hinges is the length of time it all takes. At least ten billion years of stellar burning are required to produce essential elements like carbon. It is this simple fact that renders our study of the Universe and its properties the victim of an all-embracing selection effect: our own existence. As an example, let us take the question of the *size* of the visible Universe.

The present speed of the Universe's expansion and its deceleration rate indicate that the expansion has been occurring for a time somewhere between 13 and 18 billion years. This state of expansion means that the size of the Universe is inextricably entwined with its age. The reason that the Visible Universe is more than 13 billion light-years in size today is that it is more than 13 billion years old. A Universe that contained just one galaxy like our own Milky Way, with its 100 billion stars, each perhaps surrounded by planetary systems, might seem a reasonable economy if one were in the universal construction business. But such a universe, with more than a 100 billion fewer galaxies than our own, could have expanded for little more than a few months. It could have produced neither stars nor biological elements. It could contain no astronomers. We should not be surprised to discover that the Universe is so vast in scale, because we could not exist in one that was significantly smaller. This realization, that some of the key structural features of the Universe may be necessary prerequisites for the existence of observers, must influence our view of many issues. Many a philosopher has argued against the ultimate significance of human life on the grounds that it occupies such an minuscule fraction of the known universe. Some modern astronomers see the vastness of the Universe as persuasive testimony to the overwhelming probability that the Galaxy is teeming with other intelligent life-forms with whom we might communicate. But the Universe needs to be as big as it is to support just one solitary outpost of life. It is a sobering thought that the global and possibly infinite structure of the Universe is so linked to the conditions necessary for the evolution of life on a planet like Earth.

This recognition, that there are types of universe which we could not expect to observe, is often called the *Weak Anthropic Principle*. It is at root an extension of our caution in requiring a full understanding of the in-built biases present in our measuring apparati when doing experimental science. It tells us that our astonishment at many properties of the Universe which appear unusual a priori must be tempered by the recognition that many of them simply must be present if a universe is to be studied by intelligent observers.

Cosmologists view the Weak Anthropic Principle as a qualification of the famous stricture of Copernicus, who by announcing that the Sun, and not the Earth, was at the centre of the solar system (which constituted the entire Universe as far as pre-nineteenth-century astronomy was concerned), removed the prejudice of centuries that humanity lay at the centre of the physical Universe. We should be careful not to confuse Copernicus's important lesson that we must not regard our position in the Universe as special in *every* way with the spurious belief that our position in the Universe cannot therefore be special in *any* way. We could not exist within a star; we could not exist when the Universe was less than a million years old. If the Universe did happen to possess a centre (there is no evidence that it does), and conditions were only conducive to the evolution and continued existence of life near that centre, then we should not be surprised to find ourselves living there.

It seems that this symbiotic relationship between the Universe and observers of it has other more mysterious features that are impressive but hard to evaluate objectively. The great success of Einstein's general relativity theory in describing the past and present of our Universe, with its highly regular expansion, low density, and patchwork of stars and galaxies, has provoked cosmologists to study the other types of universe that Einstein's equations can describe. The more we examine the other types of universe that the laws of physics appear to allow, so the more special and unusual do the properties of the actual Universe appear to be. Its uniqueness is impressed upon us most forcefully by the fact that we can seemingly conceive of so many alternatives. Whether the preponderance of theoretical possibilities in the face of the fact of the uniqueness of the Universe is telling us something about the initial conditions allowed for universes or the laws we think govern their evolution was something we discussed in Chapter 4.

For illustration we can pick on just a few unexplained large-scale properties of the Universe. We have seen why we must find it to be so large and so old. And we can see why it is unlikely that we would be around when it is more than ten times older and larger: by this epoch the stars will all have died; the resources of our planet will be unable to support us. It may well be that we will have become resourceful enough to continue living elsewhere in another way, but our continued existence will be more improbable than it is today when the Sun is shining in the prime of its nuclear life. But what of the galaxies? We don't know for certain how galaxies form, but we know a basic physical process that will lead to the development of structures like galaxies. If a collection of particles exert attractive forces upon one another then, unless they are distributed in a *perfectly* uniform manner, they will tend to become clumped and non-uniformly distributed as time passes. In the context of the Universe the particles are the atoms and grains of matter emerging from the Big Bang, and the attractive force is gravity. The result is a process of gravitational aggregation whereby the denser regions of space get denser at the expense of the sparser ones. By this inevitable process a Universe that is not perfectly smooth—and none could ever be so in view of the inevitable quantum ambiguity in the positions of elementary particles within it—will, over a period of billions of years, pass from being almost smooth and unstructured into a state in which matter is aggregated into dense islands scattered throughout a sea of lower density gas and dust. Only in these dense islands, which we identify with galaxies, can material attain the densities necessary to form stars. This much we understand, although not the later stages of how and why some dense islands of cosmic material wind up to resemble spiral galaxies whilst others end up as the giant ellipsoidal balls of orbiting stars we call elliptical galaxies. Yet although we know how early stages of the process of aggregation develop with the passage of time there is one important piece of the picture missing. How large was the irregularity at the beginning? Only when we know the answer to this question can we tell whether our explanation for the existence of galaxies is the correct one. One thing we can say about this starting value is that it must be very specially tuned if galaxies are to form in time. A small increase above the optimum value and the dense islands form too early, and collapse catastrophically under the pull of gravity to form giant black holes before stars can ever shine. A small decrease below the optimum value results in islands of denser material that are too feeble ever to condense into galaxies: stars never form. In either case there appears to be little chance of life evolving spontaneously. Likewise, the relative concentrations of matter and radiation emerging from the Big Bang are characterized by a particular value that inhabits a small niche that allows life as we know it to evolve. One of the goals of the inflationary universe theory is to provide an explanation of these small irregularities. The theory has the potential to predict the level of non-uniformity expected in the Universe. Unfortunately, so far the predictions obtained from the prototype theories give levels which are far too large, unless one hypothesizes the existence of matter fields with very special properties in the early universe, but since these properties have no other motivation than to get the 'right' answer to the cosmic inhomogeneity level, this is hardly very attractive.

Even if inflation can provide an answer to the problem of the inhomogeneity level of the Universe, it still requires an application of the Anthropic Principle. We recall that the idea of inflation is that each microscopic region of the Universe can, in its earliest stages, undergo an accelerated phase of expansion. Thus, the entire visible universe today may reflect the conditions that existed within a single prehistoric quantum fluctuation. However, the early universe possesses many (an infinite number if the universe is infinite) microscopic regions at the time when inflation can occur. Each will inflate by an amount determined by the local conditions within it. The result will be a universe resembling a foam of bubbles of all sizes. Some regions will inflate a lot, some only a little. There is no way in which the usual scientific method of prediction or falsification applies here. If we live in one of these bubbles it is a historical question, and history is the science of things that are not repeated. If inflation happened then we must find ourselves inhabiting one of the inflated bubbles which underwent at least ten or fifteen billion light-years of inflation. Beyond the horizon of our visible Universe there will exist other inflated bubbles with unpredictable sizes.

Coincidences

> Although we talk so much about coincidence we do not really
> believe in it. In our heart of hearts we think better of the universe,
> we are secretly convinced that it is not such a slipshod,
> haphazard affair, that everything in it has meaning.
>
> J. B. Priestley

The Weak Anthropic Principle should not be viewed as a falsifiable theory or theorem. It is a methodological principle which one ignores at one's peril. There have been examples of how a lack of recognition of it allows the cosmologist to speculate unnecessarily, and, in fact, to develop quite unnecessary new theories of gravity. Let us recall the most striking of these, because it was the stimulus for the first explicit cosmological statement of the Weak Anthropic Principle by Robert Dicke in 1957.

In the 1930s (while he was on his honeymoon, in fact) Paul Dirac drew attention to a peculiar coincidence of Nature: that the number of particles in the observable universe is roughly equal to 10^{78}, whereas the ratio of the strengths of electromagnetic to gravitational forces between two protons is close to 10^{39}. That these numbers are so huge is strange enough, but the fact that one is the square of the other suggests that they are not totally unrelated, perhaps by an equation like:

$$\text{(Ratio of strengths of electric and gravitational forces)}^2$$

$$= \hspace{5cm} (7.1)$$

$$\text{Number of atoms in the observable Universe}$$

However, the suggestion that the square of the first ratio is actually *equal* to the latter (as $10^{78} = 10^{39} \times 10^{39}$) creates a dilemma: the ratio of the intrinsic strengths of electric to gravitational forces is fixed by *constants of Nature* (the charge of the electron, the mass of the proton, and the Newtonian gravitational constant), and is believed to be the same everywhere and everywhen. By contrast, the number of particles in the *observable* Universe is continually increasing. At every moment, light rays that began their journey to us from huge distances away are reaching our telescopes for the first time. Objects that are further away from us than the speed of light multiplied by the time for which the Universe has been expanding have not yet been seen. We are surrounded by a spherical horizon about 15 billion light-years away which separates the observable part of the Universe (in its interior) from the, as yet unobserved, part beyond the horizon. But as time passes we can see farther and farther, and the number of atoms or particles in the visible part of the Universe will steadily increase—in direct proportion to the age of the Universe. The only way in which Dirac's equality (7.1) can hold is if either the strength of the electromagnetic force or of the gravitational force were to change with time.

Either suggestion is extremely radical. Dirac suggested that it was the intrinsic strength of gravity that weakened in inverse proportion to the age of the Universe. This idea subsequently generated a vast amount of theoretical and experimental physics as mathematicians showed how Einstein's theory of gravitation could be changed to include this feature, and various experiments were constructed to check whether the strength of gravity was changing as the Universe aged. To this day there is no evidence that gravity is weakening with time. As a result of the Viking space missions to Mars, we know that if gravity is weakening with time then it can have changed by no more than one per cent in the entire 15 billion year history of the Universe. This is a hundred times smaller than the change predicted by Dirac.

The interesting point about this story is that, in 1964, Robert Dicke, an American physicist working at Princeton, pointed out that Dirac's coincidence between the square of the relative strength of gravity and electromagnetism and the number of particles in the visible universe today was, in fact, one upon which our own existence depends. It tells us that we live close to the time when stars have started to burn their hydrogen into helium. Observers can only arise when the Universe has aged sufficiently for Dirac's coincidence to hold. Universes which do not display Dirac's coincidence are unlikely to contain observers. It is an anthropic selection effect, and no varying gravitational force strengths need be invoked to explain our observation of it.

The speculative Anthropic Principle

> There are two times in a man's life when he should not speculate:
> when he can't afford it, and when he can.   Mark Twain

The style of argument we have just been discussing is rather striking, and it provoked a number of cosmologists to indulge in more speculative extensions of it. Just suppose, they suggested, that there is an infinity of all possible universes having all possible sizes, ages, temperatures, shapes, and contents. Even imagine that the strengths of gravity and electromagnetism take on different combinations of values in each one of them. Then how large is the collection of possible universes within which observers could arise? In some sense it seems to be very small. If the fundamental forces of Nature are imagined to possess slightly different strengths than they do, then chemistry becomes impossible: there are no stars, no carbon compounds, and apparently no observers. Thus, it appears that of all the universes *we* can conceive, very few are able to support life. Most are stillborn, unable to produce the basic building-blocks of life or provide an environment in which evolution by natural selection can produce non-trivial results.

It is hard to know what to make of this type of argument at present. It may be true that the Universe could have been different. It may not. If it is true that universes with all possible structures can exist, and the collection of possible universes in which life can arise is very small, then there need be no further explanation for many of the Universe's observed properties. Indeed, if the Universe is infinitely large in spatial extent (a view that current observations favour) and its initial conditions are random, then somewhere within that randomly infinite set of starting conditions there must exist an infinite collection of sub-regions that will expand into the uniform and isotropic expanding visible region we call the presently observable universe. In this situation there is no deeper explanation for its *observed* large-scale properties. This is a rather unsavoury state of affairs for the scientist whose goal is to explain the complexity of the Universe we witness by recourse to a small number of all-embracing laws of Nature. There may exist such a 'simple' explanation, but then again there may not. We may well be living in a habitable portion of an infinite and random universe whose initial state obeyed no laws of Nature at all.

Life and observership

> I am always surprised when a young man tells me he wants to
> work at cosmology; I think of cosmology as something that
> happens to one, not something one can choose.
>
> W. H. McCrea

When we list the medley of conditions that must be satisfied in order that any type of chemical life evolve in the Universe, we find that a large number of very finely balanced 'coincidences' must exist in order that the Universe give rise to observers. If we were to imagine a whole collection of hypothetical 'other universes' in which all the quantities that define the structure of our Universe take on all possible permutations of values, then we find that almost all of these other possible universes we have created on paper are stillborn, unable to give rise to that type of chemical complexity that we call 'life'. This discovery led Brandon Carter to suggest that there might exist some more speculative metaphysical aspect to the Universe which he termed the *Strong* Anthropic Principle, to distinguish it from the uncontroversial Weak Anthropic Principle discussed above. The Strong Anthropic Principle suggests that, because there appear to exist such a large number of remarkable and apparently disconnected 'coincidences' which conspire to allow life to be possible in the Universe, the Universe *must* give rise to observers at some stage in its history.

Now this sounds rather strange. Cosmologists are talking about 'other universes'. Where are they? How can one say that our Universe is better suited to the evolution of life than another? We also speak of 'life' and 'observers' as though they have some role to play in physics. How can this be? In the search for answers modern physics points us in some surprising directions.

With regard to the way the Universe began we have two options which are considered seriously by theoretical cosmologists. Over the last twenty years favour has continually ebbed and flowed between the two. First, it could be that there is only one type of Universe that is logically possible. All the presently unexplained values of the fundamental constants of Nature would, in such a unique scheme, be found to possess no possible arbitrariness. There will be found to exist a single 'Theory of Everything', and this branch of scientific inquiry will then be complete. The current great excitement amongst theoretical physicists for 'superstrings' has arisen because this idea provides the first good candidate for a 'Theory of Everything'. The other possibility is that there are elements of randomness in the make-up of both the structure of the Universe and the fundamental constants of Nature. This randomness can emerge in various ways. The Universe as a whole may vary greatly in composition from place to place. If the constants of Nature arise from the breaking of some symmetry then this could have happened in different ways in different places, and may even be different elsewhere in the Universe today. The phenomenon of inflation could ensure that the laws and constants of Nature are similar only within the co-ordinated primordial region that inflated to form our visible Universe.

The conclusion of the second scenario is that the Universe could have been different. It is in some sense a particular asymmetric manifestation of deeper, but now partially hidden, symmetry. The symmetry of the laws of Nature are hidden by the need for particular outworkings of them to occur.

If the Universe is uniquely prescribed by some higher internal logic then we must judge ourselves extremely fortunate that this unique self-consistent arrangement happened to allow the evolution of observers to witness it, and we are unable to conclude anything further about the connection between life and the Universe without appealing to metaphysical or religious beliefs. If the Universe possesses some random aspect in its make-up then the verdict is rather different. We must accept that, contrary to the prejudice of many scientists, there are aspects of the large-scale structure of the Universe which do not have any explanation in the conventional sense. They arise as random events in the first moments of the Universe's history. They could have been otherwise (and may even *be* otherwise elsewhere in the Universe). We could not exist in the majority of possible universes, where the outcomes of such accidents lead to universes that cannot support life.

This still does not really introduce observers in any way that makes them necessary for the existence of the Universe, rather than just for the observing of it. We could imagine a Universe empty of life. It seems a lonely place—meaningless perhaps—but it doesn't seem logically impossible or physically inconsistent in any way. Or does it?

The greatest achievement of physical science in modern times has been the development and use of quantum theory. It is this branch of physics which underpins our everyday existence. Our understanding of its workings is so good that we are able to use it to develop lasers, transistors, microchips, and computers; the whole of our technological society is built upon it in a thousand different ways. But this totally pragmatic science which allows us to understand the microscopic structure of matter in fantastic detail, and which governs the behaviour of every atom of Nature and every DNA helix within our bodies, contains a deep mystery of physics at its heart. In Chapter 3 we saw how the standard form in which it was developed by Niels Bohr during the pre-war era maintains that no phenomenon exists until it is observed. And when it is observed, the state in which it is seen is determined unpredictably by the act of observership. All that definitely can be predicted about it is the probability that a particular measurement will be recorded when the state is observed.

According to Bohr, the only real properties of natural phenomena are observed phenomena. We can no longer maintain the old Cartesian view that we can observe Nature like a bird-watcher with a perfect hide. There is an unbreakable connection between the observer and the observed. The eminent American physicist and long-time co-worker of Bohr's, John A. Wheeler, has proposed that taken at face value this interpretation of quantum mechanics requires 'observers' in order to bring the quantum world into being. Thus, according to Bohr's extreme interpretation of quantum theory, the quantum reality of the distant stars and galaxies cannot be granted until they are 'observed'. In Wheeler's words 'observers may be necessary to bring the Universe into being'.

We still do not fully understand what properties are necessary to constitute an 'observer' in quantum physics. Some argue that any device for storing information will suffice, but others, most notably the Nobel laureate Eugene Wigner, have argued that the self-reflective property of human consciousness is necessary.

Bohr's interpretation of quantum theory is strange, but it is the one that working physicists adopt pragmatically without worrying about how it works. Such subtleties do not impinge upon its practical use in the laboratory. However, in recent years cosmologists have begun to consider the implications of applying quantum theory to the Universe as a whole—quantum cosmology. Such a programme immediately faces an impasse that can only be overcome by coming to grips with the meaning of quantum observership. If we only ascribe reality to what is observed, who observes the Universe? If we can only make statements about the probability of the Universe being observed in a particular state what does this mean when there is only one Universe? The 'Many Worlds' interpretation of quantum reality maintains that every time an observation is made there is a splitting of the observer or the world into two states—one for each possible outcome of the observation. Thus, the Universe evolves by successively splitting into an ever-increasing collection of different worlds in which everything that can logically occur eventually will. The randomness of quantum measurement, which Bohr regarded as intrinsic to the inseparability of the observer and the observed, is an illusion created by the fact that we experience just one path through the network of world-splittings. This interpretation of quantum reality is adopted by quantum cosmologists because it does not require the Universe to be observed. It ascribes an equal reality to universes other than the one we experience and observe. It ascribes no special significance to life at the quantum level, but there must exist branches where life evolves, because all possible outworkings of the laws of Nature are explored in the different branches of the Everett worlds. This array of universes is equivalent to the picture in which there are random elements in the make-up of our single Universe, although here the probability distribution is realized rather than potential.

As yet we do not know whether Bohr or Everett was right about the meaning of quantum mechanics. Whatever the answer to the riddle of quantum reality, the correct assessment of the role and meaning of observers in the Universe must await the outcome of the confrontation of the Cosmos with the quantum. The marriage of these unlikely partners will bring cosmologists face to face with the question of the origin of the Universe: a conundrum in which we are found to play a mysterious and unexpected part.

Is the Anthropic Principle an argument for the existence of God?

> Give us the power, O Lord, for if thou doest not give us the power we shall not give thee the glory, and who will be the gainer by that, O Lord?
> Old prayer

History reveals that most past cultures, be they Eastern or Western, harboured a deep intuitive belief that Nature was providentially designed for them by a benevolent God or gods. The beauty of Nature, the availability of natural resources, the recurrence of night and day, summer and winter, seed-time and harvest, seemed to bear eloquent witness to such a state of affairs. The Old Testament view that fashions so much of our own cultural and scientific heritage is no exception. It structured and reflected the early Jewish view of Nature. It would, of course, never have occurred to a pious Jew that Nature should be used to prove the existence of God. There was no doubt of His existence. Nature was something to be celebrated and to be part of. It was also something secular. There were no nature gods. Although the underlying assumption and belief was grounded in teleology it was not always naïvely anthropocentric, as the book of Job bears eloquent witness. Later, the Greek ideas of Aristotle attained a pre-eminence in philosophical thinking in Europe. Aristotle laid great stress upon the purpose of things as revealing their true meaning and significance. He was interested in 'why' things happened as well as 'how'. The Aristotelian view became merged with the Judaeo-Christian tradition, and was used to frame many arguments for the existence of God from the existence of apparent 'design' in Nature. Later, the revolutionary change of method and emphasis in science brought about by figures like Copernicus, Galileo, and Newton tailored the objectives of science to answering the 'how' questions and not the 'why'. One might have thought that this would lead to a demise of arguments for the existence of God being formed from the apparent life-supporting purpose of the world. Nothing could be farther from the truth. The dramatic unfolding of the laws of Nature, culminating in Newton's great works, led only to a change of emphasis. The evidence for the existence of a Deity was taken to be the meticulous mathematical precision and regularity of the laws of Nature themselves rather than individual events. Alongside this remained a more naïve argument which cited the match between human and animal physiology, and their needs for survival in the environments in which they were found, as evidence for the providential design of Nature. It was the latter view that Darwin's theory of natural selection completely undermined. However, as was widely appreciated at the time, the Darwinian revolution had nothing to say about the other type of

Newtonian Design Argument based upon the mathematical harmony of the unchanging laws of Nature.

For some there still exists an irresistible temptation to draw a strong metaphysical conclusion from the fact that we have found a whole collection of coincidences in the make-up of the physical Universe, which contrive in concert to make our existence possible. Is this not evidence for the existence of a God who has created the Universe with mortal man in mind, they ask? This type of 'natural theology', as it became known, originated as a systematic study in medieval times, but reached its peak amongst English scientists in the seventeenth century. Flushed with the success of the Newtonian revolution, they sought to understand the order they had found in terms of their religious beliefs. The meticulous clockwork precision and regularity of the underlying laws of Nature was cited as primary evidence for the existence of a Grand Designer behind the Universe, who was identified with the God of the theologians. Such ideas were absorbed within the mainstream of Protestant theology of the day and eloquently expressed by the hymn-writer, whose famous lines, 'Laws which never shall be broken/For their guidance hath He made' sprang from the contemplation of the newly revealed laws of Nature. And indeed, Newton was pleased with this use of his ideas. In the introduction of the *Principia* he remarked that in its writing he had an 'eye upon arguments' for belief in a Deity. And most intriguingly, he wrote to Richard Bentley that 'There is yet another argument for a Deity wch I take to be a very strong one, but till ye principles on wch tis grounded be better received I think it more advisable to let it sleep.' Newton never revealed this new argument. It is possible, in view of the context of the remark, that Newton had deduced an age of a hundred million years or so for the past lifetime of the solar system, based upon his law of gravitation.

Today, this type of theological deduction is still attractive to many individuals, and the merest suspicion of it seems to spur its opponents to man the trenches just as the theological exploitation of Newton's work led to the critical reaction of David Hume and Immanuel Kant against any argument for God from the so-called design of Nature. Heinz Pagels believes that some scientists regard the Anthropic Principle as a form of substitute religion; he claims,

Of course, some scientists, believing science and religion mutually exclusive . . . [when] . . . faced with questions that do not fit into the framework of science . . . are loath to resort to religious explanation; yet their curiosity will not let them leave matters unaddressed.

Maybe this is not so unusual a charge as it first sounds. Others have remarked upon the curious resemblance between traditional religious ideas about 'salvation' and the motivations of some eminent searchers for extraterrestrial intelligence, who believe that contact with advanced civilizations will reveal to us the secret of successful world government, and 'save us from ourselves'.

Motivated by these controversies, John Updike has recently felt the need to write an entire novel in which the English natural theological tradition is pitted against the Barthian stance of the utter transcendance and inaccessibility of God to mundane scientific arguments. The cyclopean enthusiasm of a young computer student out to develop a computer code that will lead from Nature up to Nature's God, aided by the anthropic coincidences amongst the fundamental constants of Nature, is foiled by the arid scepticism of a liberal theologian convinced of Barth's words, that there is 'no way from us to God', and alarmed at the prospect because, 'The god who stood at the end of some human way . . . would not be God'. It is interesting that Updike has used many (although often garbled) cosmological coincidences to decorate the dialogue of the novel, and acknowledges a selection of popular scientific articles as their source.

There are two simple things one can say about well-meaning logical and scientific quests for the existence of God (or gods). The logical arguments are all of a piece. They begin with some assumptions ('axioms' as the logicians like to call them), and then proceed to deduce the existence of God by a series of inexorable logical steps. But in the last analysis we are left not with a conclusion, but a choice. Only if we believe the assumptions at the outset must we believe the conclusions. There cannot be an ineluctable logical proof of God's existence or non-existence. There will always be a choice about the credibility of assumptions. Furthermore, one suspects that even the great propounders of logical arguments for the existence of God, like Thomas Aquinas, had a personal faith that would not have been perturbed one iota by the undermining of their logical or scientific demonstrations, because it was grounded elsewhere. By the same token, they could not honestly have expected their arguments to sway anyone else to accept their conclusion.

It is for reasons of this sort that the Strong Anthropic coincidences cannot be the basis of a cogent argument for God's existence from apparent anthropocentric design in the Universe, although they are quite consistent with such a conclusion. The wide range of remarkable coincidences between values of constants of Nature which have allowed complex living things to evolve are only conditions *necessary* for the existence of life. They are not sufficient to guarantee it. Modern biologists reject the notion that the evolution of life in the Universe is in any sense inevitable. Such a teleological view—that there is some future goal to which Nature is directed or magnetically attracted—finds no support in known facts, although it recaptured popular attention in the 1950s and 1960s following its semi-poetic espousal by the Catholic scientist and mystic Teilhard de Chardin.

The Anthropic Principle merely identifies coincidences which are *necessary* for the evolution of complex chemistry of the sort that biochemists believe to be essential for the spontaneous evolution of life by natural selection. The fact that it finds these 'coincidences' to be numerous and surprising does not allow the conclusion to be drawn that they also guarantee the presence of conscious observers in the Universe.

The time of your life

> I have not found out why we humans think of time as a line going
> from backwards, forwards, whilst it may be in all directions like
> everything else in the system of the world.
> Ferruccio Busoni

We are all aware of the subjectivity of time. Although we sense the arrival of the
future, we have no sure sense of the rate of passage of time. Yet, despite this we
hold to the belief (because we have read it) that there is a definite time behind the
subjectivity of our experience, and that time distinguishes the future from the past
in an absolute way. These thoughts take us back to issues that we have raised in
earlier chapters. In Chapter 3 we encountered the dilemma of the Second Law of
thermodynamics, which picks out an 'arrow' of time by the direction of entropy
increase. In Chapter 4 we encountered the discovery of the expanding universe,
which provides us with another arrow of time in the sense of expansion. The
paradox of quantum measurement which arose in Chapter 3, wherein the time-
reversible evolution of the quantum wave function is supplanted by the
irreversible effect of quantum measurement, gives rise to another breed of
irreversibility. In principle, all these 'arrows' which distinguish the future from the
past might be distinct from the subjective consciousness we have of the direction
of future time. Some have speculated that they might all be linked in some deep
way. In his suggestion that there exists a fundamental 'entropy' which gauges the
evolution of the complexity of the gravitational field of the Universe, Roger
Penrose hypothesizes that such a quantity plays some role in uniting the
irreversible antics of thermodynamics and quantum measurement with the
overall evolution of the Universe.

Connections between the local thermodynamic arrows of time and the
expansion of the Universe have been suggested before. It was once a popular
speculation to link the two with the result that, if the Universe were one day to
reverse its expansion into contraction, then the local arrows of time would also
reverse, and we would see entropy decrease in the future. Our desks would grow
spontaneously tidy; perpetual-motion machines would abound. Such a conclusion does not really stand up to close analysis though. The reversal of the
expansion dynamics of the Universe is a global phenomenon, whereas the arrow
of time in some microscopic physical process giving rise to frictional resistance
here and now is a local one. How can the local process 'know' that the Universe
elsewhere has expanded to its maximum extent. If we attribute the local arrow of
thermodynamics to the *local* expansion dynamics then we have a chaotic
situation. In a realistic universe some places will reach their expansion maxima
before others. The universe will be composed of regions, some expanding, some
contracting, with different thermodynamic arrows of time.

The idea that there might be different arrows of thermodynamic time is an old one which pre-dates the idea that time's arrow might be connected with the expansion of the Universe. In Chapter 3 we discussed how the thermodynamic arrow of entropy increase is a reflection of the relative probabilities of various states. Ordered states are far more improbable than disordered ones, and so it is far more likely that a system will evolve from a state of order into one of chaos. Boltzmann saw that this view precipitated a subjective view of the Second Law of thermodynamics and the direction of time which it defines. If the Universe varies from place to place in its initial state of disorder, then there will be some places which begin in an improbable state, and from which entropy and disorder tend to increase, but there will exist other regions which begin in probable disordered states from which the evolution can proceed to states of lower entropy and disorder. In this way the local thermodynamic arrows of time in the Universe would be different. Boltzmann explains:

> We have the choice of two kinds of picture. Either we assume that the whole universe is at present in a very improbable state. Or else we assume that the aeons during which this improbable state lasts, and the distance from here to Sirius, are *minute* if compared with the age and size of the whole universe. In such a universe, which is in thermal equilibrium as a whole and therefore dead, relatively small regions of the size of our galaxy will be found here and there; regions (which we may call 'worlds') which deviate significantly from thermal equilibrium for relatively short stretches of these 'aeons' of time. Among these worlds the probability of their state will increase as often as they decrease. In the universe as a whole the two directions of time are indistinguishable, just as in space there is no up or down. However, just as at a certain place on the earth's surface we can call 'down' the direction towards the centre of the earth, so a living organism that finds itself in such a world at a certain period of time can define the 'direction' of time as going from the less probable state to the more probable one (the former will be the 'past' and the latter the 'future'), and . . . he will find that his own small region, isolated from the rest of the universe is 'initially' always in an improbable state. It seems to me that this way of looking at things is the only one which allows us to understand the validity of the second law, and the heat death of each individual world, without invoking a unidirectional change of the entire universe from a definite initial state to the final state.

Events in the parts of Boltzmann's world where entropy decreases would be very strange. Poincaré argued that familiar concepts like 'prediction' would be hopeless. Friction would cease to be a retarding force. Objects would be spontaneously accelerated. In the future of our subjective time the oceans would not tend to equilibrate their temperatures. Inequalities would grow in an unstable fashion. This is not the type of world where life can either evolve or survive. Thus Boltzmann's world need not conflict with the world we see. The Weak Anthropic Principle persuades us that we could only exist in one of the entropy-increasing islands.

As a final speculation upon this world of many times we might consider what would happen at the interfaces between regions where the thermodynamic arrows of time differ. Norbert Wiener cites the following conundrum:

> It is a very interesting intellectual experiment to make the fantasy of an intelligent being whose time should run the other way to our own. To such a being all communication with us would be impossible. Any signal he might send would reach us with a logical stream of consequents from his point of view, antecedent from ours. These antecedents would already be in our experience, and would have served to us as the natural explanation of his signal, without presupposing an intelligent being to have sent it.

Reichenbach proposed one idea for communication which might be explored in order to ascertain that the other beings did have a counter-oriented thermodynamic arrow:

> That such a system is developing in the opposite time direction might be discovered by us from some radiation travelling from the system to us and perhaps exhibiting a shift in spectral lines upon arrival . . . the radiation travelling from the system to us would . . . not leave that system but arrive at it. Perhaps the signal could be interpreted by inhabitants of that system as a message from our system telling them that our system develops in the reverse time direction. We have here a connecting light ray which, for each system, is an arriving light ray annihilated in some absorption process.

It is not difficult to conceive of better tests than this which exploit the properties expected in sources of radio waves and noise in transmission signals, but we shall resist the urge to speculate further. It is left, in the immortal words of the textbook-writer, as an exercise for reader.

The misanthropists

> There is no possibility of reducing all laws to one law . . . no a priori means of excluding from the world the unique.
>
> Josiah Royce

The Weak Anthropic Principle recognizes the constraints that are placed upon what we can expect to observe in Nature by the selection effect of our own existence as observers made of carbon living billions of years after the Big Bang. This is an uncontroversial statement of truth, but is it a useful addition to our knowledge? Not everybody seems to think so. Heinz Pagels claims that

> As I thought more about the anthropic principle, however, it seemed less like a grand Darwinian selective principle and more like a far-fetched explanation for those features of the universe which physicists cannot yet explain. Physicists and cosmologists who appeal to anthropic reasoning seemed to me to be gratuitously abandoning the successful program of conventional physical science of understand-

ing the quantitative properties of our universe on the basis of universal physical laws ... We could debate its merits and demerits a long time. But such interminable debate is a symptom of what is wrong with the anthropic principle: unlike the principles of physics, it affords no way to determine whether it is right or wrong; there is no way to test it. Unlike conventional physical principles, the anthropic principle is not subject to experimental falsification ... the influence of the cosmological principle on the development of contemporary cosmological models has been sterile: it has explained nothing ... no knowledge has been gained by the adoption of anthropic reasoning. I would opt for rejecting the anthropic principle as needless clutter in the conceptual repertoire of science ... My own view is that although we do not yet know the fundamental laws, when and if we find them the possibility of life in a universe governed by those laws will be written into them. The existence of life in the universe is not a selective principle acting upon the laws of nature; rather it is a consequence of time.

This enthusiastic condemnation includes most of the standard objections to the use of the Anthropic Principle. We can abstract them explicity as follows:

1. Scientists spent centuries separating philosophy from science. The Anthropic Principle is undoing this by mixing them up again.

2. The Anthropic Principle is a form of teleological reasoning that Darwin overthrew.

3. The Anthropic Principle is not testable, therefore it is not scientific. It is a quasi-religious principle.

4. The Anthropic Principle is like the 'God of the Gaps'. With every new discovery that explains a previously unexplained large-scale property of the Universe, the need for the Anthropic Principle shrinks. Inflation explains most of the cosmological properties in a more attractive way.

5. The Anthropic Principle is an inappropriate methodology. Particle physics offers the prospect of a theory of everything in which the structure of the Universe, including all the values of its physical constants, will be determined uniquely and completely. The possibility of life evolving in the Universe will be built into these laws from the beginning. Life is only a consequence of the laws of Nature.

6. The Anthropic Principle makes statements of comparative reference to other hypothetical universes. We know of, and can know of, only one Universe.

7. The Anthropic Principle appeals to the Many Worlds interpretation of quantum mechanics, but we can never test that the other quantum worlds exist.

8. The Anthropic Principle takes a parochial view of life, and assumes that all life-forms in the Universe resemble ourselves.

The most common misconception regarding the Anthropic Principle, which features in the quotation given above is that it is in some sense a rival cosmological or particle physics theory which one is being offered as an alternative to the standard picture. This is rather misleading. All that is being claimed is that the Anthropic Principle must be used as a *complement* to the standard deductive theories, otherwise there is a real danger of drawing erroneous conclusions or, more commonly, providing elaborate 'explanations' for non-existent problems. A classic example is Dicke's demonstration that Dirac's Large Number Coincidences do not require any extreme hypothesis, like the time-variation of Newton's gravitation constant, to explain them. The Weak Anthropic Principle does not explain them, but it shows that a posteriori they are not surprising. A discovery that the gravitation constant was decreasing with time in the way predicted by Dirac would falsify the anthropic explanation for these coincidences. An anthropic explanation can be ruled out by observation.

The first three objections are all of a piece. We have become so indoctrinated by the philosophers of science, with their paradigms and exemplars, their emphases upon falsification and verification, that we can easily lose sight of the fact that they are methodological principles for the expedient *practice* of science. They need have nothing whatsoever to do with whether particular theories are actually true or false. If someone writes down the correct 'theory of everything', then that will not be falsifiable either. To believe that we will be able to test and falsify all theories is just the sort of anthropocentric view of the Universe that critics of the Anthropic Principle so roundly decry elsewhere. Why should Nature be constructed upon a scale that is spanned by human intelligence? Why should what is true also be humanly falsifiable or verifiable?

The fundamental questions of cosmology and particle physics are of a very special type. Any explanation for the origin and structure of the Universe is likely to be of a very unusual sort. We would be foolish to discard certain approaches to these problems simply because they do not have analogues in more mundane scientific investigation. It is surely right to study Nature with the confident assumption that it can be fully understood, but it is not correct to reject ideas because they do not fit in with the religious view that Pagels puts forward. Needless to say, the Weak Anthropic Principle does not claim that the Universe was constructed especially for life, human or otherwise, only that the existence of life may need to be included in a right evaluation of its global properties.

Objections 4 and 5 are the most interesting. Let us consider first the question of a 'theory of everything'. This is topical at present because of the impetus provided by superstring theories. Of course, the existence of a theory of everything is just an assumption. There is no evidence for it: it is a philosophical view that is mixed into science (compare Objection 1). Nevertheless, it is a reasonable one to entertain. But what is not reasonable to entertain is that a knowledge of laws of Nature in their unified entirety will suffice to provide a complete explanation for the structure of the Universe and our own evolution. Even if the laws of Nature are found to be uniquely determined, the solutions of those laws may not be. We know from our experience with particle physics that solutions of equations need not possess the same symmetries as the equations themselves. Even in the presence of a theory of everything it is quite reasonable to entertain the view that there will exist quasi-random elements in the Universe's structure, and even to some extent in its laws and 'almost' symmetries. This means that, even when in possession of the complete set of equations governing the evolution of the visible

Universe and the logically determined values of its fundamental constants, we will not be able to predict the structure of the actual Universe uniquely, any more than we can predict the direction in which the Earth is spinning from the law of angular momentum conservation.

One must also regard as speculation the assumption that a theory of everything will provide a set of unique initial conditions for the evolution of the Universe. Such a provision seems more unlikely still if the Universe had no beginning (so 'initial' conditions are set at past temporal infinity), or has tunnelled from some earlier quantum state, in which case we could have no more than a *probability* that any particular final state arises. In all these pictures where there is an element of chance in the gross structure of the Universe, it is quite possible a priori for the Universe to expand into a state that cannot evolve and support carbon-based life-forms. A correct explanation for its structure and evolution could not neglect the a posteriori fact of our own evolution.

As we look out into the Universe there are some things that we do not look to a fundamental law of Nature to explain—why it is raining today, why the Earth has a moon, the number of planets in the solar system, the number of galaxies in the Local Group of galaxies. These are chance events, in that they could have been different without doing violence to the laws of Nature. In the local cosmic environment from which these examples are drawn it is relatively easy to pick out such events, but when we consider the large-scale structure of the Universe it is not clear which aspects require a fundamental explanation in terms of laws of Nature and which do not. Pagels assumes everything about the large-scale Universe requires, and has, a fundamental and unique explanation, whereas the Anthropic Principle recognizes that there may be elements of the observed structure of the Universe that are chance outcomes of particular symmetry breakings. We are able to observe the outcomes that we do only because they fell out in a fashion that allows observers to arise subsequently. Both positions are assumptions, both employ unverified philosophical ideas, both could be wrong, or one might be right; it is too early to pass judgement with certainty.

With regard to the inflationary universe picture (see Objection 4) one can say more about the relationship with anthropic explanations. Far from being an alternative to the Anthropic Principle, in its purest form the inflationary universe actually has to employ the Anthropic Principle. Inflation assumes chaotic initial conditions in the Universe, and each local microscopic region then inflates by an amount determined by the degree of microscopic smoothness within it. Some regions inflate a lot, some only a little. The result is that the Universe ends up divided into domains which have very different conditions. We have to live in a domain that has inflated to at least thirteen billion light-years in extent in order that life can have formed. The inflationary hypothesis could be made to fail— simply assume that no domain inflates enough to explain the large-scale uniformity of the observable universe—but of course nobody countenances such an assumption. They use the Anthropic Principle *implicitly* to deduce that at least one region must expand to large size, and that we inhabit one of those that do. Since we cannot tell whether inflation did actually occur and create the particular properties of the observed universe, or whether they were fixed like that by initial conditions or some, as yet unknown, law of quantum gravitation, we cannot test the inflationary explanation decisively.

The issue of 'other worlds' also arises in inflationary explanations. Andrei Linde, one of the inventors of the current inflationary universe model, suggests that in an infinite universe we should regard the inflationary development of a large universe like our own as inevitable, because in a randomly infinite universe there will initially exist microscopic regions in all possible states of smoothness, which will therefore result in all possible inflated states.* We will inhabit one of the large ones for no other reason than the fact that a large universe is necessary for life to evolve. There is no reason at present to believe that a theory of everything will change this argument significantly.

We see from this idea that the infinite Universe can be recast into an infinite number of causally disjoint regions where different things happen. In this case the 'other worlds' are neither speculative nor mysterious. At present, in the absence of any definitive law of initial conditions, we regard the possibility of changing the possible initial conditions of the Universe as equivalent to changing the starting conditions in solutions of Einstein's equations. Each of the universes that results is regarded as a possible 'other universe'.

The Many Worlds interpretation of quantum mechanics has grown in popularity with the study of quantum cosmology. Again, it is an example of a type of theory that opponents reject basically because they do not like it, or because we might be unable to test the existence of other worlds. Clearly, if the Universe is perpetually branching every time a quantum interaction occurs we are again in the situation where a fundamental theory of everything is insufficient to explain the observed structure of the Universe. All possible degrees of inflation, all possible symmetry-breakings and the values of the fundamental constants they create actually occur in reality. The branch of it that we inhabit is chosen from the entire ensemble by the fact that the necessary conditions for the evolution of life are met within it. Whereas the standard picture described above,

---

*Linde's 'chaotic' version of inflation has a clear logical defect. It wishes to appeal to an infinite early universe in order to provide *necessarily* some region in which the state of smoothness allows inflation, because in an exhaustively random infinity everything that can occur with finite probability *will* occur (infinitely often in fact). The Anthropic Principle is then used to 'explain' why it is inevitable that we evolve in one of these particular regions that inflated to large size. However, in a randomly infinite initial state to the Universe there would exist, by the same token, an infinite number of regions with starting conditions which would evolve into a large, quiescent, isotropic universe populated by galaxies *even if inflation did not occur*. If one is going to resort to the statistics of infinite random sets, why bother with the extra 'epicycle' of inflation? The Anthropic Principle provides as good a justification for obtaining the present state from a randomly infinite beginning in the non-inflationary model as it does in the chaotic inflationary one. Only if the matter fields which give rise to inflation have to exist in Nature can the chaotic inflationary appeal to infinite random variations to evade Occam's razor.

in which there are quasi-random elements in the evolution of the Universe, has only one of the set of possibilities extant, the Many Worlds scenario has them all occurring. In defence of the Many Worlds interpretation, it is the simplest of the interpretations on offer because it uses the minimum of additional assumptions in order to explain the things that are seen.

A common reaction to the problem of the interpretation of quantum mechanics is that of the physicist who says that quantum mechanics works, and that is all that matters. The question of the *meaning* of quantum mechanics is not one that physicists should worry about. However, this is not an attitude that we are happy to adopt elsewhere. If a student comes and asks how to solve a quadratic equation, and says he just wants to know the formula that extracts the solution but he does not want to know why it works or where it comes from, we would take a very dim view of that student. The whole scientific enterprise is based upon rejection of the view that if it 'works' then that is good enough.

The question of whether all life-forms in the Universe have to resemble ourselves in respect of being carbon-based is an interesting one. The view of the biochemists is that only life that makes use of carbon can come into existence *spontaneously*. Thus, while in the future we may create forms of artificial silicon-based intelligence meriting the title 'life', this life is secondary: it could not evolve spontaneously. In fact this issue is a red herring as far as applications of the Weak Anthropic Principle are concerned. The arguments regarding the length of time required for the stellar synthesis of carbon apply equally to the origin of silicon, nitrogen, phosphorus, oxygen, and all the heavier elements as well. One can be very sure that there are no forms of atomic life that avoid the use of elements heavier than lithium, and large, Big Bang universes are necessary for the production of all these heavier elements.

In conclusion, it is important to stress once again that the fundamental problems at the frontiers of modern cosmology and particle physics are of a unique type. They are not like the problems of laboratory physics. They are not problems which always respect the traditional dogmas about the philosophy and practice of science. They are extraordinary problems, and they possess extraordinary solutions which it will require extraordinary methods to coax from the Universe. If our methods ultimately fail, then any boundary between fundamental science and metaphysical theology will become increasingly difficult to draw. Sight must give way to faith. Confronted with an emotionally satisfying mathematical scheme which is 'simple' enough to command universal assent, but esoteric enough to admit no means of experimental test and grandiose enough to provoke no new questions then, closeted within our world within the world, we might simply have to believe it. Whereof we cannot speak thereof we must be silent: this is the final sentence of the laws of Nature.

Select bibliography

This bibliography is designed to provide the reader with an introductory guide to interesting books and articles which relate to the material covered in the earlier chapters. The works are allocated to a single chapter, although some texts will be relevant to more than one. This list is intended to be neither a complete survey of the relevant literature nor a collection of references to material quoted in the text. However, if used as a basic introduction to the literature it will provide, through the secondary bibliographies contained within the works cited, a good survey of the relevant aspects of natural philosophy. The editions listed are not necessarily the first; only the most readily available. In practice, this meant simply that they were those editions which resided on the author's own bookshelves.

Chapter 1

Ayer, A. (ed.), *Logical Positivism* (Free Press, New York, 1959).
——, *Language, Truth and Logic*, rev. edn. (Dover, New York, 1946).
——, 'Laws of Nature', *Revue Internationale de Philosophie* (1956).
Barbour, I. G., *Issues in Science and Religion* (Prentice-Hall, New York, 1966).
Berofsky, B. (ed.), *Free-will and Determinism* (Harper & Row, New York, 1966).
Braithwaite, R. B., *Scientific Explanation* (CUP, Cambridge, 1956).
Bradley, F. H., *Appearance and Reality* (2nd edn., OUP, New York, 1969).
Bridgman, P., *The Logic of Modern Physics* (Macmillan, New York, 1927).
——, 'Operational Analysis', *Philosophy of Science* **5,** 114 (1938).
Butterfield, H., *The Origins of Modern Science* (Bell, London, 1957).
Campbell, N., *Physics: The Elements* (CUP, Cambridge, 1920).
——, *What is Science?* (Dover, New York, 1952).
Chalmers, A., *What is this Thing Called Science?* (Open Univ., Walton Hall, 1982).
Collingwood, R. G., *The Idea of Nature* (OUP, London, 1945).
Danto, A. and Morgenbesser, S. (eds.), *Philosophy of Science* (Meridian, New York, 1960).
Dretske, F., Laws of Nature, *Philosophy of Science* **44,** 248 (1977).
Ginsberg, M., 'The Concepts of Juridicial and Scientific Law', *Politica* **4,** 1 (1939).
Goodman, N., *Fact, Fiction and Forecast* (Harvard UP, Cambridge, Mass., 1955).
Hacking, I., *Representing and Intervening* (CUP, Cambridge, 1983).
Harré, R., *The Philosophies of Science* (OUP, New York, 1972).
Heisenberg, W., *Physics and Philosophy* (Harper & Row, New York, 1959).
Körner, S., 'On Laws of Nature', *Mind* **62,** 218 (1953).
Lewis, C. S., *The Abolition of Man* (Macmillan, New York, 1947).
Margenau, H., *The Nature of Physical Reality* (McGraw-Hill, New York, 1950).
MacKay, D., *The Clockwork Image* (IVP, London, 1974).

Meyerson, E., *Identity and Reality*, trans. Loewenberg, K. (Allen & Unwin, London, 1930).

Nagel, E., *The Structure of Science* (Harcourt Brace, New York, 1961).

Oldroyd, D., *The Arch of Knowledge* (Methuen, New York, 1986).

Pap, A., *An Introduction to Philosophy of Science* (Free Press, New York, 1962).

Pearson, K., *The Grammar of Science*, 3rd edn. (Macmillan, London, 1911).

Planck, M., *Philosophy of Physics*, trans. W. Johnston (Allen & Unwin, London, 1936).

Polanyi, M., *Personal Knowledge* (Univ. Chicago, Chicago, 1960).

Popper, K., *The Logic of Scientific Discovery* (Hutchinson, London, 1959).

——, *Objective Knowledge: An Evolutionary Approach*, 2nd edn. (OUP, Oxford, 1973).

Reichenbach, H., *Experience and Prediction* (Univ. Chicago, Chicago, 1938).

Rescher, N., *The Limits of Lawfulness* (Univ. Pittsburg, Pittsburg, 1983).

Ritchie, A. D., *Scientific Method: An Inquiry into the Character and Validity of Natural Laws* (Routledge, New York, 1923).

Rogers, E., *Physics For the Inquiring Mind* (Princeton UP, Princeton, 1960).

Russell, B., *Our Knowledge of the External World* (Allen & Unwin, London, 1914).

——, *Human Knowledge: Its Scope and Limits* (Allen & Unwin, London, 1948).

——, *An Outline of Philosophy* (Meridian, Ohio, 1960).

Schwartz, N., *The Concept of Physical Law* (CUP, Cambridge, 1985).

Scott, D., *Everyman Revisited: The Common Sense of Michael Polanyi* (Book Guild, Lewes, 1985).

Shapere, D., *Philosophical Problems of Natural Science* (Macmillan, New York, 1965).

Sullivan, J. W. N., *The Limitation of Science* (Penguin, London, 1938).

Toulmin, S., *Foresight and Understanding—An Enquiry into the Aims of Science* (Indiana UP, Bloomington, 1961).

Van Frassen, B. C., *The Scientific Image* (OUP, Oxford, 1980).

Wartofsky, M. W., *Conceptual Foundations of Scientific Thought* (Macmillan, New York, 1968).

Whittaker, E., *From Euclid to Eddington: A Study of the Conceptions of the External World* (CUP, Cambridge, 1949).

Chapter 2

Barfield, O., *Saving the Appearances: A Study in Idolatry* (Harcourt Brace, New York, 1965).

Barrett, W., *The Illusion of Technique* (Doubleday, New York, 1979).

Becker, C., *The Heavenly City of the Eighteenth-Century Philosophers* (Yale UP, New Haven, 1932).

Boutroux, E., *Natural Law in Science and Philosophy*, trans. Rothwell, F. (D. Nutt, London, 1914).

Burtt, E., *The Metaphysical Foundations of Modern Science*, rev. edn. (Routledge, London, 1932).

Clagett, M., *Greek Science in Antiquity* (Abelard-Schumann, New York, 1955).

Clarke, D. M., *Descartes' Philosophy of Science* (Manchester UP, Manchester, 1982).

Clavelin, A. C., *The Natural Philosophy of Galileo* (MIT, Cambridge, Mass., 1974).

Cohen, I. B., *Isaac Newton's Papers and Letters on Natural Philosophy and Related Topics* (Harvard UP, Cambridge, Mass., 1958).

Cohen, M. R. and Drabkin, I., *Source Book in Greek Science* (McGraw-Hill, New York, 1948).

Cornforth, F. M., *The Unwritten Philosophy and Other Essays* (CUP, Cambridge, 1950).

——, *Principium Sapientiae: The Origins of Greek Philosophic Thought* (CUP, Cambridge, 1952).

Dampier-Whetham, W. C., *A History of Science and Its Relations with Philosophy and Religion* (CUP, Cambridge, 1929).

Dawkins, R., *The Selfish Gene* (OUP, Oxford, 1976).

——, *The Blind Watchmaker* (Longmans, London, 1986).

De Santillana, G. and von Dechend, H., *Hamlet's Mill: An Essay on Myth and the Frame of Time* (Macmillan, London, 1969).

Drabkin, I. E. and Drake, S., *Galileo Galilei: On Motion and Mechanics* (Doubleday, New York, 1957).

Duhem, P., *The Aim and Structure of Physical Theory*, trans. P. Wiener (Princeton UP, Princeton, 1954).

——, *Medieval Cosmology*, trans. and ed. R. Ariew (Univ. Chicago, Chicago, 1985).

Farrington, B., *Greek Science* vols. 1 & 2 (Penguin, Baltimore, 1949).

Grant, R. M., *Miracle and Natural Law in Graeco-Roman and Early Christian Thought* (N. Holland, Amsterdam, 1952).

Frankfort, H., Frankfort, H. A., Wilson, J. A. and Jacobsen, T., *Before Philosophy*; originally entitled *The Intellectual Adventure of Ancient Man* (Penguin, Baltimore, 1949).

Frazer, J. *The Golden Bough* (Macmillan, New York, 1922).

Grant, E., *Physical Science in the Middle Ages* (Wiley, New York, 1971).

Henderson, J. B., *The Development and Decline of Chinese Cosmology* (Univ. Columbia, New York, 1984).

Herschel, J., *A Preliminary Discourse on the Study of Natural Philosophy* (London, 1830; repr. Johnson, New York, 1966).

Hertz, H., *The Principles of Mechanics* (Macmillan, New York, 1900).

Hooykaas, R., *Religion and the Rise of Modern Science* (Scottish Academic, Edinburgh, 1972).

Hume, D., *Dialogues Concerning Natural Religion*, ed. N. Kemp-Smith (Bobbs-Merrill, Indianapolis, 1947).

Jaki, S., *Science and Creation* (Univ. Edinburgh, Edinburgh, 1974).

——, *The Origin of Science and the Science of its Origin* (Scottish Academic, Edinburgh, 1978).

Jammer, M., *Concepts of Space* (Harvard UP, Cambridge Mass., 1954).

——, *Concepts of Force* (Harvard UP, Cambridge Mass., 1957).

Jevons, W., *The Principle of Science: A Treatise on Logic and Scientific Method* (London, 1874).

Kant, I., *The Critique of Pure Reason*, trans. N. Kemp-Smith (St Martin's, New York, 1961).

Kemble, E. C., *Physical Science: Its Structure and Development* (MIT, Cambridge, Mass., 1966).

Koyre, A., *From the Closed World to the Infinite Universe* (Johns Hopkins, Baltimore, 1957).

Kuhn, T. S., *The Copernican Revolution* (Harvard UP, Cambridge, Mass., 1957).

Lewis, C. S., *The Discarded Image* (CUP, Cambridge, 1964).

Losse, J., *A Historical Introduction to the Philosophy of Science* (OUP, Oxford, 1980).

Lovejoy, A. O., *The Great Chain of Being* (Harvard UP, Cambridge, Mass., 1936).

Needham, J., *Science and Civilization in China*, vols 1–7 (CUP, Cambridge, 1954–).

——, *Human Law and the Laws of Nature in China and the West* (OUP, London, 1951).

——, *The Grand Titration: Science and Society in East and West* (Allen & Unwin, London, 1969).

Newton, I., *Mathematical Principles of Natural Philosophy and His System of the World*, 2 vols., trans. and ed. F. Cajori (Univ. California, Berkeley, 1934).

Oakley, F., 'Christian Theology and the Newtonian Science: Rise of the Concept of Laws of Nature', *Church History* **30**, 433 (1961).

O'Connor, D. and Oakley, F., *Creation: The Impact of an Idea* (Scribner's, New York, 1969).

Peierls, R., *The Laws of Nature* (Scribner's, New York, 1956).

Peirce, C. S., *The Laws of Nature and Hume's Argument Against Miracles*, in *Selected Writings of Charles S. Peirce*, ed. P. P. Wiener p. 289 (Stanford, 1958).

Poincaré, H., *The Value of Science* (Dover, New York, 1958).

Ruby, J. E., 'The Origins of Scientific "Law"', *J. Hist. Ideas* **47**, 341 (1986).

Sambursky, S., *The Physical World of the Greeks* (Routledge, London, 1956).

——, *The Physical World of Late Antiquity* (Basic Books, New York, 1962).

Sarton, G., *A History of Science*, vols. 1 & 2 (Harvard UP, Cambridge, Mass., 1952).

Schlagel, R., *From Myth to the Modern Mind* (Lang, New York, 1985).

Schrödinger, E., *Nature and the Greeks* (CUP, New York, 1954).

Simpson, G. G., *The Meaning of Evolution*, rev. edn. (Yale UP, New Haven, 1969).

Taube, M., 'Dr Zilsel on the Concept of Physical Law', *Philosophical Review* **52**, 304 (1942).

Thorndike, L., *A History of Magic and Experimental Science* (Macmillan, New York, 1923).

Urmson, J. O., *Berkeley* (OUP, Oxford, 1982).

Westfall, R. S., *Never at Rest: A Biography of Isaac Newton* (CUP, Cambridge, 1980).

Wallace, W. A., *From a Realist Point of View* (Univ. Press of America, Lanham, Md, 1983).

Whitehead, A. N., *Adventures of Ideas* (CUP, New York, 1933).

——, *Science and the Modern World* (CUP, London, 1953).

Zilsel, E., 'The Genesis of the Concept of Scientific Law', *Philosophical Review* **51**, 245 (1942).

Chapter 3

Bohm, D., *The Special Theory of Relativity* (Benjamin, New York, 1965).

Bohr, N., *Atomic Physics and Human Knowledge* (Wiley, New York, 1958).

——, *Atomic Theory and the Description of Nature* (CUP, Cambridge, 1934).

Bergmann, P. G., *The Riddle of Gravitation* (Scribner's, New York, 1968).

Brillouin, L., *Science and Information Theory*, 2nd. edn. (Academic, New York, 1962).

Brush, S. G., *The Kind of Motion We Call Heat* (N. Holland, Amsterdam, 1976).

Campbell, L. and Garnett, W., *The Life of James Clerk Maxwell* (London, 1882).

Capek, M., *The Philosophical Impact of Contemporary Physics* (Van Nostrand, New York, 1961).

Cartwright, N., *How the Laws of Physics Lie* (OUP, Oxford, 1983).

——, 'Do the Laws of Physics State the Facts?', *Pacific Phil. Quart.* **61,** 75 (1980).

Davies, P. C. W. and Brown, J. R. (eds.), *The Ghost In the Atom* (CUP, Cambridge, 1986).

D'Abro, A., *The Decline of Mechanism in Modern Physics* (Van Nostrand, New York, 1939).

DeWitt, B. and Graham, N., *The Many-Worlds Interpretation of Quantum Mechanics* (Princeton UP, Princeton, 1973).

Dijksterhuis, E. J., *The Mechanization of the World Picture* (OUP, New York, 1961).

Eddington, A. S., *The Nature of the Physical World* (CUP, London, 1932).

D'Espagnet, B., *The Quantum Theory and Reality* (*Scientific American*, p. 158 (Nov. 1979)).

D'Espagnet, B., *In Search of Reality* (Springer, New York, 1983).

Doran, B. G., 'Origins of Field Theory', *Hist. Stud. Phys, Sci.* **6,** 133 (1975).

Einstein, A. and Infeld, L., *The Evolution of Physics* (Simon & Schuster, New York, 1938).

Feynman, R., *The Character of Physical Law* (MIT, Cambridge, Mass., 1965).

Folse, H., *The Philosophy of Niels Bohr* (N. Holland, Amsterdam, 1985).

Gooding, D. and James, F. A., *Faraday Rediscovered—Essays on the Life and Work of Michael Faraday, 1791–1867* (Macmillan, London, 1985).

Harman, P., *Energy, Force and Matter: the Conceptual Development of Nineteenth-Century Physics* (CUP, London, 1982).

Hawkins, D., *The Language of Nature* (Freeman, San Francisco, 1964).

Heimann, P. M., 'The Unseen Universe: Physics and the Philosophy of Nature in Victorian Britain', *Brit. J. Hist. Sci.* **6,** 73 (1972).

Herbert, N., *Quantum Reality* (Rider, London, 1985).

Hesse, M., *Forces and Fields: the concept of action of a distance in the history of physics* (Nelson, London, 1961).

Hesse, M. B., *Forces and Fields* (Philosophical Library, New York, 1962).

Hiebert, E., 'The Uses and Abuses of Thermodynamics in Religion', *Daedalus* **95,** 1046 (1966).

Holton, G., 'Mach, Einstein and the Search for Reality', *Daedalus* **97,** 636 (1968).

Jammer, M., *The Philosophy of Quantum Mechanics* (Wiley, New York, 1974).

Kilmister, C., *The General Theory of Relativity* (Pergamon, Oxford, 1973).

Maxwell, J. C., *Atom,* in *Encyclopedia Brittanica,* 1875.

McCormmach, R., *Night Thoughts of a Classical Physicist* (Harvard UP, Cambridge, Mass., 1982).

Mermin, D. N., 'Is the Moon there when nobody looks? Reality and the Quantum Theory', *Physics Today,* p. 38 (Apr. 1985).

Merz, J. T., *A History of European Thought in the Nineteenth Century* (London, 1907).

Niven, W. D. (ed.), *The Scientific Papers of James Clerk Maxwell,* 2 vols. (Dover, New York, 1966).

Pais, A., *Subtle is the Lord: The Science and Life of Albert Einstein* (OUP, Oxford, 1982).

Peynson, L., 'Relativity in Late Wilhelmian Germany: The Appeal to a Pre-established Harmony between Mathematics and Physics', *Archive for History of Exact Sciences* **27,** 137 (1982).

Rae, A., *Quantum Physics—Illusion or Reality?* (CUP, Cambridge, 1986).

Raine, D. and Heller, M., *The Science of Space-Time* (Pachart, Tucson, 1982).

Rosenthal-Schneider, I., *Reality and Scientific Truth: Discussions with Einstein* (Von Laue and Planck, Wayne State, Penn., 1980).

Russell, B., *The ABC of Relativity* (Allen & Unwin: London, 1926).

Sciama, D. W., *The Physical Foundations of General Relativity* (Heinemann, London, 1972).

Szilard, L., 'On the reduction of entropy of a thermodynamic system caused by intelligent beings', *Zeitschrift für Physik* **53,** 840 (1929).

Taylor, E. F. and Wheeler, J. A., *Spacetime Physics* (Freeman, San Francisco, 1966).

Turner, J., 'Maxwell on the Method of Physical Analogy', *Brit. J. Phil. Sci.* **6,** 226 (1955).

Wheeler, J. A., 'Niels Bohr, the man', *Physics Today*, p. 66 (Oct. 1985).

Wheeler, J. A. and Zurek, W. H., Quantum Theory and Measurement (Princeton UP, Princeton, 1983).

Wigner, E., *Symmetries and Reflections* (Indiana UP, Bloomington, 1967).

——, *Remarks on the Mind–Body Question*, in *The Scientist Speculates—An Anthology of Partly-baked Ideas*, ed. I. J. Good, p. 284 (Basic Books, New York, 1962).

Williams, L. P., *The Origins of Field Theory* (Univ. Press of America, New York, 1980).

Chapter 4

Aitchison, I. and Hey, A., *Gauge Theories in Particle Physics* (Hilger, Bristol, 1982).

Barrow, J. D., 'Cosmology and Elementary Particles', *Fundamentals of Cosmic Physics* **8,** 83 (1983).

—— and Silk, J., *The Left Hand of Creation: The Origin and Evolution of the Expanding Universe* (Basic Books, New York, 1983).

—— and Tipler, F. J., 'Eternity is Unstable', *Nature* **276,** 453 (1978).

—— and Turner, M. S., 'The Inflationary Universe: Birth, Death and Transfiguration', *Nature* **298,** 801 (1982).

Bondi, H., *Cosmology*, 2nd edn. (CUP, Cambridge, 1961).

Burrill, D. R., *The Cosmological Arguments: a Spectrum of Opinion* (Doubleday, New York, 1967).

Capra, F., *The Tao of Physics* (Wildwood House, Bungay, 1975).

Davidson, H. A., *Proofs for Eternity, Creation and the Existence of God in Medieval Islamic and Jewish Philosophy* (OUP, New York, 1987).

Davies, P. C. W., *Other Worlds: Space, Superspace and the Quantum Universe* (Dent, London, 1980).

——, *God and the New Physics* (Dent, London, 1983).

——, *Superforce* (Heinemann, London, 1984).

Dyson, F., 'Time Without End: Physics and Biology in an Open Universe', *Reviews of Modern Physics* **51,** 447 (1979).

Feynman, R., *QED: The Strange Story of Light and Matter* (Princeton UP, Princeton, 1985).

Frank, P., 'The Place of Logic and Metaphysics in the Advancement of Modern Science', *Philosophy of Science* **5,** 275 (1948).

Fritzsch, H., *Quarks* (Simon & Schuster, New York, 1983).

Green, M., Schwartz, J. and Witten, E., *Superstring Theory*, 2 vols. (CUP, Cambridge, 1987).

Gribbin, J., *In Search of the Big Bang* (Heinemann, London, 1986).

Grünbaum, A., *Philosophical Problems of Space and Time* (Knopf, New York, 1963).

Guth, A., 'The Inflationary Universe', *Physical Review D* **23,** 347 (1981).

Hawking, S. W. and Hartle, J., 'The Wave Function of the Universe', ibid. **28,** 2960 (1983).

—— and Israel, W. (eds.), *300 Years of Gravity* (CUP, Cambridge, 1987).

Heisenberg, W., 'The Representation of nature in Contemporary Physics', in Rollo May (ed.), *Symbolism in Religion and Literature* (Braziller, New York, 1960).

Isham, C., Penrose, R. and Sciama, D. W. S., *Quantum Gravity II* (OUP, Oxford, 1981).

Jaki, S., *Cosmos and Creator* (Scottish Academic, Edinburgh, 1980).

Mach, E., *Popular Scientific Lectures* (Open Court, New York, 1943).

——, *The Science of Mechanics, a critical and historical account of its development* (Open Court, New York, 1960).

——, *The Analysis of Sensation, and the Relation of the Physical to the Psychical* (Dover, New York, 1959).

MacKay, D., *Brains, Machines and Persons* (Collins, London, 1980).

——, *Human Science and Human Destiny* (IVP, New York, 1979).

McCrea, W. H., 'A Philosophy for Big-Bang Cosmology', *Nature* **228,** 21 (1970).

Milne, E., *Modern Cosmology and the Christian Idea of God* (OUP, Oxford, 1952).

Montagu, A. (ed.), *Science and Creationism* (OUP, New York, 1984).

Munitz, M. (ed.), *Theories of the Universe—From Babylonian Myth to Modern Science* (Free Press, New York, 1957).

North, J. D., *The Measure of the Universe: A History of Modern Cosmology* (OUP, Oxford, 1965).

Pagels, H., *The Cosmic Code: Quantum Physics as the Language of Nature* (Simon & Schuster, New York, 1982).

Pais, A., *Inward Bound* (OUP, Oxford, 1986).

Redfield, R., *The Primitive World and its Transformations* (Cornell UP, Ithaca, 1953).

Rowe, W., *The Cosmological Argument* (Princeton UP, Princeton, 1975).

Sorabji, R., *Time, Creation, and the Continuum* (Duckworth, London, 1983).

Stebbing, S., *Philosophy and the Physicists* (Methuen, London, 1937).

Tipler, F. J., 'Interpreting the Wave Function of the Universe', *Physics Reports* **137,** 231 (1986).

Toulmin, S., *The Return to Cosmology* (Univ. California, Berkeley, 1982).

Trefil, J., *The Moment of Creation: Big Bang Physics from Before the First Millisecond to the Present Universe* (Scribner's, New York, 1983).

Tryon, E., 'Is the Universe a Vacuum Fluctuation?', *Nature* **246,** 396 (1973).

Vilenkin, A., 'Creation of the Universe from Nothing', *Physics Letters B* **117,** 25 (1983).

Von Melsen, A. G., *From Atomos to Atom* (Harper, New York, 1960).

Weinberg, S. W., *The First Three Minutes* (A. Deutsch, London, 1977).

——, *The Discovery of Subatomic Particles* (*Sci. American Library*, New York, 1983).

Zukav, G., *The Dancing Wu Li Masters* (Morrow, New York, 1979).

Chapter 5

Andreski, S., *Social Sciences as Sorcery* (A. Deutsch, London, 1972).

Arnold, V. I., *Catastrophe Theory*, 2nd edn. (Springer, New York, 1986).

Aubert, K. E., 'Spurious Mathematical Modelling', *Math. Intelligencer* **6**, 54 (1984).

Bennett, C. H., 'The Thermodynamics of Computation—A Review', *International Journal of Theoretical Physics* **21**, 905 (1982).

Bennett, C. H. and Landauer, R., 'The Fundamental Physical Limits of Computation', *Scientific Amercian* **253**, (No. 1) 48 and (No. 4) 6 (1985).

Bernardete, J. A., *Infinity: An Essay in Metaphysics* (OUP, Oxford, 1964).

Birkhoff, G., 'A Mathematical Approach to Aesthetics', *Scientia* (Sept. 1931), p. 133.

——, 'The Mathematical Nature of Physical Theories', *American Scientist* **31**, 281 (1943).

Bishop, E., *The Foundations of Constructive Mathematics* (McGraw-Hill, New York, 1967).

——, 'The Crises in Contemporary Mathematics', *Historia Mathematica* **2**, 507, (1975).

Black, M., *The Nature of Mathematics* (Paterson, Littlefield, Adams & Co., New York, 1959).

Bochner, S., *The Role of Mathematics in the Rise of Science* (Princeton UP, Princeton, 1966).

Brillouin, L., *Scientific Uncertainty and Information* (Academic, New York, 1964).

Browder, F. E., 'Does Pure Mathematics Have a Relation to the Sciences?', *American Scientist* **64**, 542 (1976).

Carnap, R., *Foundations of Logic and Mathematics*, vol. I, no. 3 (*International Encyclopedia of Unified Science*, Univ. Chicago, Chicago, 1939).

Cassirer, E., *Determinism and Indeterminism in Modern Physics* (Yale UP, New Haven, 1956).

Chaitin, G., 'Information-Theoretic Limitations of Formal Systems', *J. of Assoc. for Computing Machinery* **21**, 403 (1974).

——, *Algorithmic Information Theory* (*Encyclopedia of Statistical Sciences* vol. 1, 38; 1982).

Davis, M., *The Undecidable* (Raven, New York, 1965).

—— and Hersh, R., *The Mathematical Experience* (Harvester, Brighton, 1981).

Deutsch, D., 'Quantum Theory, the Church–Turing principle, and the universal quantum computer', *Proceedings of the Royal Society London A* **400**, 97 (1985).

Dummett, M., *Elements of Intuitionism* (OUP, Oxford, 1977).

Dyson, F., 'Mathematics in the Physical Sciences', *Scientific American*, p. 129 (Sept. 1964).

Escher, M., *The Graphic Work of M. C. Escher* (Pan, London, 1961).

Feynman, R., 'Simulating Physics with Computers', *International Journal of Theoretical Physics* **21**, 467 (1982).

Fredkin, E. and Toffoli, T., 'Conservative Logic', *International Journal of Theoretical Physics* **21**, 219 (1982).

Geroch, R. and Hartle, J., 'Computability and Physical Theories', *Between Quantum and Cosmos*, eds. W. Zurek, A. van der Merwe, and W. A. Miller (Princeton UP, Princeton, 1988) pp. 549–67.

Hadamard, J., *The Psychology of Invention in the Mathematical Field* (Princeton UP, Princeton, 1945).

Hardy, G. H., 'Mathematical Proof', *Mind* **38**, 1 (1928).

Hempel, C. G., 'On the Nature of Mathematical Truth', *Amer. Math Monthly* **52**, 543 (1945).

Hofstadter, D., *Gödel, Escher, Bach: An Eternal Golden Braid* (Basic Books, New York, 1979).

Kitcher, P., *The Nature of Mathematical Knowledge* (OUP, Oxford, 1983).

Kline, M., *Mathematics in Western Culture* (OUP, New York, 1953).

——, *Mathematics and the Physical World* (Dover, New York, 1981).

Landauer, R., *Reversible Computation*, in *Der Informationsbegriff in Technik und Wissenschaft*, eds. Folberth, O. G. and Hackl, C., p. 139 (Oldenbourg, München, 1986).

Le Lionnais, F., *Great Currents of Mathematical Thought*, vols 1 & 2 (Dover, New York, 1971).

Mandelbrot, B., *The Fractal Geometry of Nature* (Freeman, San Francisco, 1982).

Merlan, P., *From Platonism to Neoplatonism* (Martinus Nijhoff, The Hague, 1960).

Myhill, J., 'What is a Real Number?', *Amer. Math. Monthly* **79**, 748 (1972).

Rucker, R., *Infinity and the Mind* (Harvester, Brighton, 1982).

Tarski, A., *Introduction to Logic and to the Methodology of the Deductive Sciences* (OUP, London, 1941).

Thom, R., *Structural Stability and Morphogenesis* (Benjamin, New York, 1975).

Traub, J. F. (ed.), *Algorithms and Complexity: New Directions and Recent Results* (Academic, New York, 1976).

Turing, A., 'On Computable Numbers, with an application to the Entscheidungproblem', *Proc. London Math. Soc (Ser. 2)* **42**, 230 (1936); erratum **43**, 546 (1936).

Wedberg, A., *Plato's Philosophy of Mathematics* (Greenwood, Westport Conn., 1977).

Weyl, H., *Philosophy of Mathematics and Natural Science* (Princeton UP, Princeton, 1949).

Whitney, H., 'The Mathematics of Physical Quantities', *Amer. Math. Monthly* **75**, 115 & 227 (1968).

Wigner, E., 'The Unreasonable Effectiveness of Mathematics in the Natural Sciences', *Communications on Pure and Applied Mathematics* **13**, 1 (1960).

Wolfram, S., 'Statistical Mechanics of Cellular Automata', *Rev. Mod. Phys.* **55**, 601 (1983).

——, 'Undecidability and Intractability in Theoretical Physics', *Phys. Rev. Letts.* **54**, 735 (1985).

Yanin, Y., *Mathematics and Physics* (Birkhauser, Boston, 1983).

Chapter 6

Barrow, J. D., 'The Lore of Large Numbers', *Quarterly Journal Royal Astron. Soc.* **22**, 388 (1981).

——, 'Natural Units Before Planck', *Quarterly Journal Royal Astron. Soc.* **24**, 24 (1983).

Bohm, D., *Causality and Chance in Modern Physics* (Routledge, London, 1957).

Born, M., *The Natural Philosophy of Cause and Chance* (OUP, London, 1949).

Bridgman, P., *Dimensional Analysis*, rev. edn. (Yale UP, New Haven, 1931).

Carnap, R., 'What is Probability?', *Scientific American* **189**, 128 (1953).

Davies, P. C. W., *The Edge of Infinity* (Dent, London, 1981).

Eigen, M. and Schuster, P., 'The Hypercycle, A Principle of Natural Self-Organization', *Naturewissenschaften* **64**, 541 (1977).

—— and Winkler, R., *The Laws of the Game* (Penguin, London, 1981).

Froggatt, C. D. and Nielsen, H. B., *Origin of Symmetries* (World, Singapore, 1990).

Hawking, S. W. and Ellis, G. F. R., *The Large-scale Structure of Space-time* (CUP, Cambridge, 1973).

—— and Israel, W. (eds.), *General Relativity: An Einstein Centenary Volume* (CUP, Cambridge, 1979).

Kaufmann, W., *The Cosmic Frontiers of General Relativity* (Little, Brown & Co., Boston, 1977).

Kippenhahn, R., *100 Billion Suns* (Basic Books, New York, 1983).

Iliopoulos, J., Nanopoulos, D. V. and Tamaros, T. N., 'Infrared Stability or Anti-Grand Unification', *Physics Letters B* **94**, 141 (1983).

Levy-Leblond, J. M., 'Constants of Physics', *Rivista Nuovo Cimento* **7**, 187 (1977).

McCrea, W. H. and Rees, M. J. (eds.), *The Constants of Physics* (The Royal Society, London, 1983).

Nicolis, G. and Prigogine, I., *Self-organization in Non-equilibrium Systems* (Wiley, New York, 1977).

Penrose, R., in, *General Relativity: An Einstein Centenary*, eds. S. W. Hawking and W. Israel (CUP, Cambridge, 1979).

Prigogine, I. and Stengers, I., *Order Out of Chaos* (Heinemann, London, 1984).

Schrödinger, E. *Science and the Human Temperament* (Norton, New York, 1935).

Tipler, F. J., Clarke, C. and Ellis, G. F. R., 'Singularities and Horizons—A Review Article', in *General Relativity and Gravitation: and Einstein Centenary Volume*, ed. A. Held, p. 97 (Plenum, New York, 1980).

Chapter 7

Bacon, F., *Novum Organum* (London, 1620).

Barrow, J. D., 'Life, the Universe and the Anthropic Principle', *The World And I* **2**, 179 (Aug. 1987).

—— and Bhavsar, S. P., 'What the Astronomers' Eye Tells the Astronomers' Brain', *Quarterly Journal Royal Astron. Soc.* **28**, 109 (1987).

——, 'Anthropic Definitions', *Quarterly Journal Royal Astron. Soc.* **23**, 146 (1983).

—— and Tipler, F. J., *The Anthropic Cosmological Principle* (OUP, Oxford, 1986).

—— and Tipler, F. J., *L'Homme et le Cosmos* (Imago, Paris, 1984).

Bridgman, P., *Reflections of a Physicist* (Philosophical Library Inc., New York, 1950).

Bronowski, J., *The Identity of Man*, rev. edn. (Natural History Press, New York, 1971).

Bunge, M., *The Myth of Simplicity* (Prentice-Hall, Englewood Cliffs, 1963).

Carr, B. J. and Rees, M. J., 'The Anthropic Principle and the Structure of the Physical World', *Nature* **278**, 605 (1978).

Carter, B., 'Large Number Coincidences and the Anthropic Principle in Cosmology', in *Confrontation of Cosmological Theories with Observation*, ed. M Longair, p. 291 (Reidel, Dordrecht, 1974).

Dyson, F., 'Energy in the Universe', *Scientific American* **224**, 50 (Sept. 1971).

Eccles, J., *The Human Mystery* (Springer, New York, 1979).

Gombrich, E. H., *Art and Illusion*, 2nd. edn. (Pantheon, New York, 1961).

Gregory, R. L., *The Intelligent Eye* (McGraw-Hill, New York, 1970).

——, *Mind in Science* (Weidenfeld & Nicolson, London, 1981).

Heisenberg, W., *Physics and Beyond* (Harper & Row, New York, 1962).

Henderson, L. J., *The Fitness of the Environment* (Macmillan, New York, 1913. repr. with introduction by G. Wald, Harvard UP, Cambridge, Mass., 1970).

Howson, C. (ed.), *Method and Appraisal in the Physical Sciences* (Cambridge UP, Cambridge, 1976).

Kuhn, T. S., *The Structure of Scientific Revolutions* (2nd, enlarged, edn., Univ. Chicago, Chicago, 1970).

Leslie, J., 'Observership in Cosmology: the Anthropic Principle', *Mind* **92,** 573 (1983).

Lovell, B., *In the Centre of Immensities* (Harper and Row, New York, 1983).

Luckiesh, M., *Visual Illusions* (Dover, New York, 1965).

Marr, D., *Visions* (W. H. Freeman, San Francisco, 1982).

Maurois, A., *Illusions* (Columbia UP, New York, 1968).

Mehra, J. (ed.), *The Physicist's Conception of Nature* (Reidel, Dordrecht, 1973).

Page, D., 'The Importance of the Anthropic Principle', *The World and I* **2,** 392 (Aug. 1987).

Pagels, H., 'A Cozy Cosmology', *The Sciences*, p. 34 (Mar./Apr. 1985).

Pagels, H., *Perfect Symmetry: The Search For the Beginning of Time* (Joseph, London, 1985).

Reichenbach, H., *The Direction of Time* (Univ. California, Berkeley, 1956).

Santayana, G., *The Sense of Beauty* (Dover, New York, 1955).

Shklovskii, I. S. and Sagan, C., *Intelligent Life in the Universe* (Dell, New York, 1966).

Thorpe, W., *Purpose in a World of Chance: A Biologist's View* (OUP, Oxford, 1978).

Updike, J., *Roger's Version* (André Deutsch, London, 1986).

Wallace, A. R., *Man's Place in the Universe* (Chapman and Hall, London, 1912).

Weyl, H., *God and The Universe: The Open World* (Yale UP, New Haven, 1932).

Wheeler, J. A., 'From Relativity to Mutability', in *The Physicist's Conception of Nature*, ed. J. Mehra, p. 202 (Reidel, Dordrecht, 1973).

Index

Note: footnotes are indicated by suffix 'n', Figures and Tables by *italic page numbers*.